The Essentials of
Technical Communication

THIRD EDITION

ELIZABETH TEBEAUX
TEXAS A&M UNIVERSITY

SAM DRAGGA
TEXAS TECH UNIVERSITY

New York   Oxford
Oxford University Press

Oxford University Press is a department of the University of Oxford.
It furthers the University's objective of excellence in research,
scholarship, and education by publishing worldwide.

Oxford   New York
Auckland   Cape Town   Dar es Salaam   Hong Kong   Karachi
Kuala Lumpur   Madrid   Melbourne   Mexico City   Nairobi
New Delhi   Shanghai   Taipei   Toronto

With offices in
Argentina   Austria   Brazil   Chile   Czech Republic   France   Greece
Guatemala   Hungary   Italy   Japan   Poland   Portugal   Singapore
South Korea   Switzerland   Thailand   Turkey   Ukraine   Vietnam

For titles covered by Section 112 of the US Higher Education
Opportunity Act, please visit www.oup.com/us/he for the latest
information about pricing and alternate formats.

Published by Oxford University Press
198 Madison Avenue, New York, New York 10016
http://www.oup.com

Oxford is a registered trademark of Oxford University Press.

Library of Congress Cataloging-in-Publication Data
Tebeaux, Elizabeth.
The essentials of technical communication / Elizabeth Tebeaux, Texas A&M
University, Sam Dragga, Texas Tech University. — Third edition.
    pages cm
Includes index.
ISBN 978-0-19-937999-6
1. Technical writing. 2. Communication of technical information.
I. Dragga, Sam.   II. Title.
T11.T295 2014
808.06′66—dc23

                                                        2014010176

Printing number: 9 8 7 6 5 4 3 2 1

Printed in the United States of America
on acid-free paper

Dedicated to
David H. Stewart
1926–2009

Our first English Department head, a scholar fully committed to the teaching of both writing and great literature. Without his support at the beginning of our careers, neither of us would have enjoyed as productive a life in technical communication as we have had. *The Essentials of Technical Communication* emerges from that support.

Brief Contents

Detailed Contents

Checklists

Preface

As we have emphasized throughout the first two editions of *Essentials of Technical Communication*, in the workplace, no one wants to read what you write—seriously. In addition, they will read as little of what you write as they possibly can. Even if your document will be of value to your readers, unless you make that point clear at the beginning of your document, it may be tossed.

Your boss or coworkers may not have the necessary background, the time, or the inclination to wade through your e-mail, memo, or report. We have developed *The Essentials of Technical Communication* with this in mind, as a practical introduction to all aspects of effective professional communication—a handbook to help you get your message across on the job, where time equals money and poorly crafted documents can have a host of unwelcome consequences.

As teachers of technical writing with more than 50 years of experience between us, we know that following a few simple guidelines leads to more efficient and effective communications. In this book we provide the guidelines you need as you plan, draft, and revise documents. Understanding these guidelines will help you avoid blank-page terror and enable you to write effectively and quickly—both requirements of employees who write for their jobs.

APPROACH

Our rationale is simple: we believe that the effective writer in a work situation must learn and internalize basic concepts of rhetoric and then apply these in developing documents. We've filled this brief book with memorable, concise guidelines. Each chapter in Part One focuses on basic rhetorical principles, and Part Two applies those principles to the planning and writing of particular types of documents.

A brief book enables instructors to adapt the book to their own uses. Many teachers want to build on principles by adding their unique approaches. This book provides the flexibility to allow for that possibility. In addition, many employees who did not study technical or business writing in college will find this book useful in learning how to write in the workplace.

ORGANIZATION

The book is organized into two parts. Part One (Chapters 1 through 6) lays out essential communication principles:

✦ Chapter 1, "Characteristics of Writing at Work," describes technical writing, or writing in the workplace, to show how it differs from academic writing.

✦ Chapter 2, "Writing for Your Readers," presents the essential elements of analyzing readers and then choosing content, format, and style as these meet the needs of the intended readers. We embed a discussion of the composing process in this chapter.

✦ Chapter 3, "Writing Ethically," discusses the ethics of technical documents. While most professionals have standards of good practice, writers should also follow principles of communication ethics.

✦ Chapter 4, "Achieving a Readable Style," explains how to write concise, pristine sentences and paragraphs.

✦ Chapter 5, "Designing Documents," illustrates basic principles for creating accessible and inviting documents. In a world of too much information, readers often miss or ignore important messages not presented in an easy-to-read format.

✦ Chapter 6, "Designing Illustrations," provides guidelines for developing effective visuals. Graphics software creates practically infinite possibilities for visuals, but effective use requires an understanding of fundamental graphic design principles.

Part Two (Chapters 7 through 12) then applies the principles from Part One to the types of documents most commonly prepared in the workplace:

✦ Chapter 7, "E-mails, Texts, Memos, and Letters," presents the basics of correspondence and demonstrates how to ensure that these routine messages are clear, readable, and effective.

✦ Chapter 8, "Technical Reports," presents the elements of report development along with examples, including an annotated abstract. We provide a formal report in Appendix C and on the book's companion website, **www.oup.com/ us/tebeaux**. (We also include links to documentation resources on the website. With the emergence of bibliography and citation software we believe extensive instruction in documentation is no longer needed in the text, though we do include a brief guide to the most common documentation systems in Appendix B.)

✦ Chapter 9, "Proposals and Progress Reports," provides guidelines for developing business proposals and status reports. In this chapter, we use several student examples as these respond to real situations in a university setting.

✦ Chapter 10, "Instructions, Procedures, and Policies," describes how to develop clear instructions for a variety of situations.

✦ Chapter 11, "Oral Reports," provides a short guide to developing and then presenting a concise, effective PowerPoint presentation.

✦ Chapter 12, "Résumés and Job Applications," describes how to prepare job application documents.

FEATURES

- **Sample Documents:** This text, although concise, includes a range of sample documents covering the essential types and styles you're likely to encounter in the workplace. Many of these documents are available for download on the book's companion website, **www.oup.com/us/tebeaux,** along with links to documentation resources.
- **Case Studies:** In Chapters 1, 2, 3, 7, 8, 9, and 10, case studies show how different types of documents function in different situations. These cases contextualize the documents to give you a sense of how and when the techniques we outline can and should be applied.
- **Checklists:** At chapter ends, we have included checklists—lists of questions you can use to ensure that your professional documents achieve your purpose. We hope you find that these are a handy reference tool. They are indexed in the front of the book.
- **Exercises:** Exercises at the end of each chapter guide practice in the techniques outlined in the text. Some of the exercises are designed to be done in class and could be done or discussed in small groups, and others are take-home assignments.
- **Appendices:** Three appendices contain a brief guide to grammar, punctuation, and usage (A); a synopsis of information literacy and three widely used documentation systems (B); and a sample report (C).
- **Companion Website:** The book's companion website at **www.oup.com/us/ tebeaux** offers additional resources for students, including chapter overviews, self-quizzes, downloadable versions of the checklists from the book, helpful links, annotated document pages, and downloadable sample documents, including those from the exercises at chapter ends. The site includes an instructor's manual, featuring downloadable PowerPoint files for use as lecture aids, chapter objectives, teaching strategies, workshop activities, writing projects, worksheets, and discussion questions. It also includes revision assignments, multimodal writing assignments, and multilingual writing assignments.
- **Instructor's Manual:** The instructor's manual is available in a CD version that includes a test bank.

NEW TO THIS EDITION

While improving upon our first two editions, we did not change those aspects of the book that made it popular with professors and students of technical communication. This new edition maintains the concise and practical nature of the first two.

But we have made several important changes based on the excellent suggestions from our expert panel of reviewers. We made each change to prepare students (1) to write in an increasingly dynamic, digital age and (2) to write for an increasingly diverse audience—both in the classroom and in the workplace.

Changes we've made for the third edition include the following:

- Chapter 1: We've included new material on the need for information security as the most important difference between writing at school and writing at work. We caution students about the risks of social media, texting, and e-mail, all of which in personal and business use carry legal liability.
- Chapter 2: We've added a new case document that incorporates both issues in e-mail design and audience perspective.
- Chapter 3: We've added case documents to this chapter to encourage analysis of ethical dilemmas and the professional obligations of technical communicators. We've also included the Code of Ethics of the National Society of Professional Engineers for comparison with the Society for Technical Communication's Ethical Guidelines.
- Chapter 4: We've added additional examples of common style problems: excessive "be" verbs and use of "there is, there are" constructions that reduce the directness of sentences and often increase sentence length. We have included a short report that includes excessive use of "be" constructions and other sentence problems. We've also added sample sentences and possible revisions based on the principles presented in this chapter as well as additional sentences that exhibit major errors for students to discuss and correct.
- Chapter 5: New here is greater emphasis on the complementary relationship of words and illustrations.
- Chapter 6: We've added project schedule charts and infographics to the wide array of illustrations that we examine with annotated examples.
- Chapter 7: We've added a claim letter to the case documents we examine in this chapter to reinforce the principle of clear and concise communication tailored to the audience.
- Chapter 8: We've included three case documents for analysis and discussion and new examples of informal reports. We've also added a new example of a student formal report in Appendix C.
- Chapter 9: We've included discussion of online multimedia progress reports. Appendix C also includes the proposal and progress report for the new final report on Alzheimer's disease.
- Chapter 10: We've added a new example of lab procedures prepared by a graduate student as many college lab instructions are hastily prepared by graduate students who do not think about the needs of undergraduate students in science lab settings.
- Chapter 11: We have new examples of PowerPoint slides, including a full slide presentation that exhibits qualities of good PowerPoint presentations and ineffective slides that demonstrate common errors. We've also provided a speech prepared to be read and invited students to improve it.
- Chapter 12: We've updated the examples of résumés and letters of application while reinforcing key rhetorical principles and the necessity of managing your professional identity on social networking sites.

In all chapters, we offer new and revised exercises. These require students to think critically about the topics discussed in the chapter.

Finally, the companion website and instructor's resources have been updated with new examples, exercises, and materials. Of particular note is the revised instructor's manual, which contains sections in each chapter on multimodal and multilingual writing, as well as new links, writing projects, and teaching strategies. The test bank has been updated as well and offers a revised and expanded selection of test questions.

Acknowledgments

We are grateful to the dedicated book publishers of Oxford University Press for their conscientious efforts to make this book eloquent, elegant, concise, and cogent. We continue to thank the reviewers commissioned by Oxford for the first and second editions of this text: Susan Aylworth, California State University, Chico; Latonia Bailey, Crowder College; Elizabeth Childs, Auburn University; Cathy Corr, University of Montana Missoula; Ed Cottrill, University of Massachusetts Amherst; Richie Crider, University of Maryland; Melody DeMeritt, California Polytechnic State University; Scott Downing, Kenai Peninsula College, University of Alaska Anchorage; Leslie Fife, Oklahoma State University; Maureen Fitzsimmons, Syracuse University; Elizabeth Holtzinger-Jennings, Pennsylvania State University; Danica Hubbard, College of DuPage; Kendall Kelly, Southwest Texas State University; Kevin LaGrandeur, New York Institute of Technology; Elizabeth Lopez, Georgia State University; Lisa McClure, Southern Illinois University, Carbondale; Raynette Meyer, Aiken Technical College; Elizabeth Monske, Northern Michigan University; Brenda Moore, New Jersey Institute of Technology; Marguerite Newcomb, University of Texas at San Antonio; Mark Noe, University of Texas-Pan American; Roxanna Pisiak, Morrisville State College; Liza Potts, Old Dominion University; Ritu Raju, Houston Community College; Denise Stodola, Kettering University; Leslie St. Martin, College of the Canyons; Dawn Taylor, South Texas College; Aaron Toscano, University of North Carolina at Charlotte; Michelle Weisman, College of the Ozarks; and Linda Young, Oregon Institute of Technology. And we extend our thanks to those who reviewed for this new edition: Robert Bleil, College of Coastal Georgia; An Cheng, Oklahoma State University Stillwater; Paul Dombrowski, University of Central Florida; Peter Dorman, Central Virginia Community College; Doris Fleischer, New Jersey Institute of Technology; Jennifer Haber, St. Petersburg College; Michael Klein, James Madison University; Kimberly Miller, Case Western Reserve University; Sally Palmer, South Dakota School of Mines and Technology; Joshua Prenosil, Creighton University; Kristen Proehl, Clemson University; Jeff Pruchnic, Wayne State University; Dirk Remley, Kent State University; Bonnie Startt, Tidewater Community College; Eleanor Sumpter-Latham, Central Oregon Community College; Tammy Winner, University of North Alabama; Susan Youngblood, Auburn University; and Pinfan Zhu, Texas State University.

We also thank the innumerable colleagues and students who have challenged and inspired us in the teaching of technical communication. And, as always, special thanks to Jene and Linda for their love and support.

PART ONE

Principles

Characteristics of Writing at Work

Technical, or *business*, *writing* describes writing that occurs in a business or work setting. University offices, corporations, research centers, hospitals, businesses of all sizes, even nonprofit organizations produce large quantities of technical writing, which differs from academic writing in a number of important ways. These differences mean that you cannot write on the job the way you have written in school. Writing in school and writing at work differ because the purposes and the context of each differ. Thus, the products of each contrast sharply.

Technical writing has similarities to any sport you want to pursue: you must first learn the foundational concepts, then understand how these concepts affect the sport before you apply the principles as a participant.

Quick Tips

On the job, keep in mind that **no one wants to read anything you write**. **Most of the time they will not read all of what you write**. They will read because they need to, not because they want to. They will read because you have information they need to take actions or make decisions. They don't get paid to read: they get paid to take actions and make decisions. The more time they need to read your document, the less productive time they have. Make sure everything you write is clear, correct, necessary, and polite. And never assume that anything you write is confidential.

Modern organizations have to keep their technical and business writing secure, whether it exists in paper or virtual form. Organizations that lose information to cyber thieves often face severe consequences.

Writing at Work versus Writing at School

Workplace writing requires that you continue to apply what you have learned about effective paragraph development, correct sentence structure, punctuation, and usage. As an educated adult, your writing should exemplify correctness. Beyond these fundamental principles, business or technical writing will differ from writing you have done as a student in five important ways.

Writing at work

1. Requires acute awareness of security and legal liability
2. Requires awareness that documents may be read by unknown readers, inside and outside the organization, for an infinite time
3. Achieves job goals
4. Addresses a variety of readers who have different perspectives from those of the writer
5. Requires a variety of written documents

Requires acute awareness of security and legal liability. The most fundamental characteristic of technical writing rests in the legal liability associated with workplace information. This liability extends from federal privacy acts, such as FERPA (Family Educational Rights and Privacy Act), that protect you as a student in college to protection of the research and intellectual property of the university from cyber thieves.

Chief information officers in educational, business, government, and research organizations work diligently to protect the privacy of information about their employees and the knowledge generated by these employees by following both federal and state privacy laws. Identity and information theft can occur at any time, despite the best efforts of any chief information officer's staff and security team. People throughout the world continue to attack computing systems to gain access to credit card numbers, personal and medical information, and transcripts of academic work, creative work, and research data—essentially whatever hackers can access, either for their own use or to sell to crime cartels.

Electronic communication has become a blessing and a curse. Today's workplace requires extensive technology. Research organizations, hospitals, banks, financial organizations, law firms, physicians, even small, locally owned businesses have to pursue strict security on all information they have about customers, clients, and patients. Organizations, like architectural firms, computer companies, engineering companies, and manufacturers, must protect their intellectual property from theft. The knowledge they produce for clients becomes the value of the organization. When you begin a job, you need to learn the security rules and follow them. For example, you will likely not be allowed to use your company e-mail for any purpose other than company business. Your company telephones will likely require the same restrictions. You should never access your personal blogs or social networking sites from your employer's computer.

To avoid potential security breaches,

- Remember that any text message you send will not be secure and may be legally accessed, whether the cell phone you use belongs to you or your employer. Company e-mail can be viewed by the company webmaster. Once you begin working for an organization, use caution in what you discuss via text messages and e-mail.
- Avoid blogs, unless your company uses secure blogs for creating collaborative reports, for example. Remember that others can see what you have written. Be sure that your comments exemplify tasteful, helpful, and accurate tone and content.
- Any electronic communication—texts, e-mails, and social media messages— can be subject to subpoena. Your Internet provider has to comply with "good cause" subpoenas. Again, what you say in cyberspace never goes away.
- Avoid using browsers available on company computers to locate information on any topic not related to your work.
- Use social media carefully. Your company may have a page on one of the social media sites, but do not use it or respond to it. First, ask the purpose of the site and the rules for its use by employees. Note: Many students have been expelled from their universities for inappropriate use of social media. A business organization, because of concerns for information security, will watch how employees use social media. You can lose your job if your comments on blogs, wikis, and other forms of social media disparage the organization and perhaps divulge proprietary information.
- Many organizations, before they hire new employees, will check social media to see what potential employees have said about themselves. Again, criminals across the world also check. Divulging confidential information, personal or professional, can have major consequences for you and organizations for which you work, have worked, or will work. Because virtual messages never go away, ask yourself, "If I decided to run for public office in 20 years, would I want people I don't even know to see what I said about myself today?"
- If you have a personal web page, be sure that what you place on the page makes a positive statement about you and does not discredit your employer in any way.
- Guard your external storage drives carefully. Never leave one in your computer when you work in a public place, even for a few minutes. When you purchase a flash drive, be sure it has been manufactured by a reputable company. Never buy nonpackaged flash drives. Never use a flash drive given to you as a gift from an advertiser. You do not know what material, malware, or viruses have been placed on the drive. Never forget that everything you write can be accessed by others. Tip: Always write as if someone you do not know might be reading over your shoulder. And follow all rules your employer stipulates. Accepting and agreeing to follow rules of confidentiality of company information may be a condition of employment with that organization. When you interview for a job, ask about the company's website, all social media sites, and management of those sites.

In school, your primary obligation is to avoid plagiarism. But what you write at work can be used against you in lawsuits. Once you sign your name to a report or letter, your signature makes you responsible for the content. Hostile readers can use what you say to support claims against you and the organization you represent. Because we live in an increasingly litigious society, designing documents that will prevent their misuse should be one of your primary goals.

Requires awareness that documents may be read by unknown readers. Always anticipate unknown readers who may receive copies of your reports or e-mail. Ask yourself this important question: "Does my report or e-mail contain any information that could be misconstrued and affect me or the organization adversely if unknown readers see my communication?" While academic writing responds to assignments, applicable only for a specific semester, course, and professor, workplace communications have no specific life span. They can be read and then used in ways you never intended or envisioned.

You cannot underestimate the problem that unknown readers present. Nearly everything you write for an organization will remain in the organization's archive indefinitely. Copies of your reports and letters will be placed in files accessible by readers who may not know anything about you or the situation you discuss in your document. These documents will often be used in assessing your performance and in determining your promotion potential. What you say suggests how well you have done your job. Unknown readers may also use your reports to gain understanding of a work situation they have inherited with a new job assignment. On the job, what you write becomes much more than a knowledge indicator for a grade.

Achieves job goals. In school, you write to show your professor that you know the subject matter and to make a good grade. But in the workplace, writing is the major way that people achieve their job goals and document their work. Writing becomes documentation that you have done your work and how you have done it. How well you write will suggest how well you have done your work. It will become part of the organization's permanent archives.

Addresses a variety of readers who have different perspectives. In college, you write your assignments for a single reader, a professor, a specialist in a subject area. But in a work setting, you can expect to write to readers who have varied educational and technical backgrounds, readers who have different roles inside and outside the organization, and readers who may know less about a topic than you do. Your supervisor, for example, may have majored in a field of study very different from yours, or your supervisor's responsibilities may have channeled his or her technical knowledge into other areas. For example, you may report directly to a person whose educational background has been in physical chemistry or electrical engineering but whose responsibilities may now be in personnel management, database administration, quality control, or financial analysis. Many technical people earn a master's degree in business administration to assist them in management roles.

In a work context, these readers will feel no commitment to read what you write unless your messages help them as they do their own work. They will generally not read all documents completely. Each will be interested in how your message affects his or her job goals. What seems clear and important to you may lack clarity and importance to others. Because e-mail has become a common way of communicating within organizations, you really have no idea who will read what you write as any message and its attachments may be forwarded. Documents posted online on an organization's website have no security from prying eyes and hackers.

We live and work in an information age where the quantity of information grows rapidly, where people have more to read than they can ever hope to read. As they see your report or your e-mail subject line, they will immediately ask themselves questions such as "What is this?" "Why should I read it?" "How does it affect me?" "What am I going to have to do?" Without a carefully stated subject line, your readers may delete your message before opening it. If they do open your e-mail, they will want to find the main points and ideas quickly; and they will become impatient if they are unable to find them by glancing at the page. They will not usually read any document completely or bother to respond to it unless, at the beginning, the message indicates that reading your document serves their best interests. How they respond to the first few sentences of your writing will often determine how much more of it they read.

On the job, your readers are not a captive audience, as your teachers have been. They do not have to read what you write. If you want your writing read, make your message clear and easy to read; make your message as interesting, relevant, and concise as possible. Because your readers often read selectively, conciseness and clarity are basic ingredients of effective business communication. Mechanical correctness remains a desirable quality, but correct writing that cannot be read easily and quickly will not be read.

Requires a variety of written documents.   Most academic writing includes essays, essay examinations, research papers, and laboratory reports. You direct your writing to your teachers. At work, however, employees can expect to write a variety of documents not relevant to academic writing assignments: letters, e-mails, information and procedure memos, proposals, progress reports, project reports, feasibility studies, economic justification reports, policy statements, travel reports, news releases, speeches, training procedures, budget forecasts, employee evaluations, user documentation, and perhaps articles for publication in trade journals. What you write will change with your responsibilities, the kind of job you have, and your position in the organization. How you write each document will depend on the topic discussed, the situation leading to the document, your readers' needs and perspectives, and your purpose in writing.

THE FOUNDATIONS OF EFFECTIVE WRITING AT WORK

Developing effective documents requires a process involving at least six stages:

1. Planning the document
2. Determining content

3. Arranging ideas
4. Drafting
5. Revising
6. Editing

While you may do each of these steps as a separate activity, when you write you will more than likely move back and forth from one activity to the other as you develop your document. Following this process will help ensure that content is appropriate as well as correctly and effectively presented.

THE QUALITIES OF GOOD TECHNICAL WRITING

Surveys show that organizations rank writing skills in this order of importance:

1. Accuracy
2. Clarity
3. Conciseness
4. Readability
5. Usability
6. Correctness

These qualities mean that a document

- makes a good impression when readers first interact with it: document is neat, readable, well organized, and inviting
- can be read selectively—for instance, by some users, only the summary; by other users, only the introduction and conclusions; by still other users, the entire report
- shows a plan that reveals the purpose and value of the document
- makes sense: ideas appear in a logical sequence immediately evident from the document design
- uses visuals, if necessary, to help readers understand ideas or data
- conveys an overall impression of authority, thoroughness, soundness, and honest work
- makes sense to people who were not part of the initial readership
- makes a positive statement about the writer and the organization
- enables people who need to use your writing to perform a task to do so

Beyond all these basic characteristics, good technical writing has no typographical errors, grammatical slips, and misspelled words. Little flaws distract attention from the writer's main points and call into question the writer's literacy.

As you study, understand, and practice writing for a workplace setting, keep in mind these qualities as well as the differences between the writing you do as an employee and the writing you have done as a student.

Case 1-1:

Bradshaw Engineering, LLP, understand the changes and challenges occurring in the field of chemical engineering. Matt Lunsford, one of the research engineers, tells Jerry Bradshaw, the senior principal, about a recent article he has read about carbon capture. Jerry has an established practice of posting summaries and articles on the company website to help employees remain informed. Matt provides a memo of transmittal to Jerry (Case Document 1-1A) and two short summaries of the article he has read and shared with Jerry (Case Documents 1-1B and 1-1C). Note how Matt develops the summaries with the knowledge level of both engineers and nonengineers in mind. He also uses basic principles of document design to create readable documents. How do his three documents reflect the qualities of good technical writing (see under "The Qualities of Good Technical Writing")?

Matt "transmits" his summaries with a memo of transmittal to Jerry. At work, you will need transmittal documents, discussed in Chapter 7 for sending documents, as they introduce the documents to readers.

CASE DOCUMENT 1-1A

DATE:	September 3, 2013
TO:	Jerry Bradshaw
FROM:	Matthew Lunsford
SUBJECT:	**Technical and General Summaries of Article over Carbon Capture Technology**

I have completed and attached two summaries of an article describing current technologies under development for carbon capture. The first, a technical summary, I have prepared for engineers here at BE. The second, a general summary, targets other employees. As you requested, all BE employees need to understand current research that relates to our work here at BE.

Subject and Purpose of Summaries

The summaries inform our employees about technological developments involving carbon capture. With the growing awareness of carbon dioxide emissions and their effect on the environment, reducing carbon dioxide emissions has been the subject of several important projects that BE has worked on in the last few years. Technology discussed in this article could become standard in industries with which we work, such as power generation; and all of us need to remain aware of advances in carbon capture technology.

Reason for Summaries

Your request for summaries of state-of-the art research helps all of us. Thus, BE encourages employees to stay up to date on current engineering developments relevant to our projects and planning.

CASE DOCUMENT 1-1B

C. Jones, "CO_2 capture from dilute gases as a component of modern global carbon management," *Annual Review of Chemical and Biomolecular Engineering*, Vol. 2, 31–52, July 2012.

Technical Summary

Article Purpose
This article describes methods currently under development for carbon capture. It also describes the need for more detailed research and analysis before these methods can be implemented on an industrial scale.

Rationale for the Topic
With the increasing concentration of carbon dioxide in the atmosphere, the well-being of society in the future largely depends on carbon management.

Major Points Discussed
- Carbon management includes two different processes: (1) reducing carbon dioxide emissions from large sources such as coal-fired power plants (carbon capture and sequestration) and (2) capturing carbon dioxide directly from the atmosphere (air capture).
- Researchers face a major challenge in developing methods of carbon capture that are both effective and energy-efficient. Currently developed methods have large energy penalties of ~30 percent, meaning 30 percent of power produced by a plant would go toward powering carbon capture.
- A relatively mature technology for carbon capture and sequestration (CCS) uses absorption of carbon dioxide by aqueous amine solution. This method will most likely dominate first-generation CCS technologies. However, other absorption methods under development involve use of room-temperature ionic liquids (RTILs) to dissolve carbon dioxide.

- Carbon dioxide–selective membrane technology represents another strategy that, when developed, will enhance CCS. Many other types of membranes have been studied. These include membranes composed of polymers only and those that also have amine solutions, enzymes, or RTILs as active media. Research has achieved sufficient selectivities for carbon dioxide versus nitrogen, but practitioners need increased throughputs for viable large-scale use. Also, further research is needed on the stability of these membranes in the presence of contaminants contained in plant flue gas.
- A final method being developed for CCS is adsorption of carbon dioxide using either supported amine adsorbents or metal–organic frameworks. Metal–organic frameworks are attractive because they can be easily tuned to obtain the exact adsorption properties desired. However, more research must show the stability of these materials in the presence of contaminants contained in plant flue gas.
- Air capture technology, which would actually decrease the concentration of carbon dioxide in the atmosphere, is also under development. Current methods typically involve an aqueous basic absorption process using metal hydroxides. In the future this process could complement CCS, which only limits increases in atmospheric carbon dioxide concentration. However, costs associated with air capture are currently much higher than those with CCS. Research also lacks quantitative analyses of hypothetical air capture processes.

Conclusion

The technology most likely to be applied in the near future for CCS is aqueous amine absorption. Other methods that involve the use of RTILs, membranes, and adsorbents require further research and development before their use on a reasonable scale. Air capture also requires further research and detailed descriptions of process designs before its viability can be determined.

CASE DOCUMENT 1-1C

C. Jones, "CO_2 capture from dilute gases as a component of modern global carbon management," *Annual Review of Chemical and Biomolecular Engineering*, Vol. 2, 31–52, July 2012.

General Summary

Article Purpose
This article describes advances in an emerging technology called "carbon capture," which involves the removal of carbon dioxide from plant emission sources and from the atmosphere.

Rationale for the Topic
With the increasing concentration of carbon dioxide in the atmosphere, the well-being of society in the future largely depends on our ability to manage the amount of carbon in the atmosphere.

Major Points Discussed
- Carbon management includes two different processes: (1) reducing carbon dioxide emissions from large sources such as coal-fired power plants (carbon capture and sequestration) and (2) capturing carbon dioxide directly from the atmosphere (air capture).
- Researchers face a major challenge in developing methods of carbon capture that provide both effective carbon capture and energy efficiency.
- Some relatively mature technology exists for carbon capture and sequestration (CCS). This technology involves the use of solvents to absorb carbon dioxide escaping from emission sources such as power plants. Other methods still in the development stage include use of membranes and adsorbents. These substances also selectively capture carbon dioxide while allowing other gases to pass through to the atmosphere.
- Air capture technology, which would actually decrease the concentration of carbon dioxide in the atmosphere, is also in development. In the future this process could complement CCS, which now only limits increases in atmospheric carbon dioxide concentration. However, the costs associated with air capture are currently much higher than those with CCS, and further research must continue.

Conclusion

The technology most likely to be applied in the near future for CCS is absorption of carbon dioxide. Other methods that involve the use of membranes and adsorbents require further research and development before they can be scaled to industry size. Air capture also requires further research before researchers can determine its viability.

EXERCISES

1. Visit the website of one of the employers listed below. What kinds of writing do employees of these firms have to do?

    Virginia Department of Transportation (www.vdot.virginia.gov/projects)
    McKinsey (www.mckinsey.com)
    Deloitte (www.deloitte.com)
    Chevron (www.chevron.com)
    National Institutes of Health (www.nih.gov)

    Prepare a memo to your instructor explaining the kinds of writing you find on the website. For format, use the memo of transmittal in Figure 1–1 as an example.

2. Based on the description and qualities of good technical writing discussed in this chapter, write a memo to your instructor listing the problems you see in Figure 1–1.

CPSC Safety Alert

Avoiding Tap Water Scalds

The majority of injuries and deaths involving tap water scalds are to the elderly and children under the age of five. The U.S. Consumer Product Safety Commission (CPSC) urges all users to lower their water heaters to **120 degrees Fahrenheit.** In addition to preventing accidents, this decrease in temperature will conserve energy and save money.

Most adults will suffer third-degree burns if exposed to 150 degree water for two seconds. Burns will also occur with a six-second exposure to 140 degree water or with a thirty second exposure to 130 degree water. Even if the temperature is 120 degrees, a five minute exposure could result in third-degree burns.

Various procedures for lowering water temperature in the home exist, depending on the method of heating. Here are some suggestions:

Electric water heaters. Call your local electric company to adjust the thermostat. Some companies offer this service at no-charge. Hot water should not be used for at least two hours prior to setting. To make the adjustment yourself, start by shutting off current to the water heater, then turn off the circuit breaker to the heater or remove the fuse that serves the heater. Most electric water heaters have two thermostats, both of which must be set to a common temperature for proper operation. To reach these thermostats you must remove the upper and lower access panels. Adjust the thermostat following the instructions provided with the appliance. Hold a candy or meat thermometer under the faucet to check water temperature.

Gas water heaters. Because thermostats differ, call your local gas company for instructions. Where precise temperatures are not given, hold a candy or meat thermometer under faucet for most accurate reading first thing in the morning or at least two hours after water use. If reading is too high, adjust thermostat on heater, according to manufacturer's instructions, and check again with thermometer.

Furnace heater. If you do not have an electric, gas, or oil-fired water heater, you probably have an on-line hot water system. Contact your fuel supplier to have the temperature lowered. If you live in an apartment, contact the building manager to discuss possible options for lowering your tap water temperature. Reducing water temperature will not affect the heating capacity of the furnace.

CPSC notes that a thermostat setting of 120 degrees Fahrenheit (49 degrees Celsius) may be necessary for residential water heaters to reduce or eliminate the risk of most tap water scald injuries. Consumers should consider lowering the thermostat to the lowest settings that will satisfy hot water needs for all clothing and dish washing machines.

Never take hot water temperature for granted. Always hand-test before using, especially when bathing infants and young children. Leaving a child unsupervised in the bathroom, even if only for a second, could cause serious injuries. Your presence at all times is the best defense against accidents and scalding to infants and young children.

Publication 5098   009611 032012

U.S. CONSUMER PRODUCT SAFETY COMMISSION • (800) 638-2772 • www.cpsc.gov • www.SaferProducts.gov

FIGURE 1–1

Writing for Your Readers

Before you begin to plan your document, first think about your readers and then your purpose. Every decision you make in developing your document should reflect your audience, their needs, and your purpose.

Avoid becoming absorbed in ideas and information you plan to include. Never forget that the person or group who will read the document may have a very different perspective of the content. Your readers cannot climb into your mind and know your thoughts. When you carefully analyze your readers, you can often design and write your report in a way that helps them understand what's in your mind. Reexamine Case Documents 1-1A, 1-1B, and 1-1C at the end of Chapter 1. In what ways has Matt shown his awareness of the needs of readers?

Quick Tips

To develop any communication, you have three main goals that connect reader, purpose, and context:

1. You want your readers to understand your meaning exactly in the way you intend.
2. You want your writing to achieve its goal with the designated readers.
3. You want to keep the goodwill of those with whom you communicate.

UNDERSTAND YOUR READERS—THE HEART OF THE PLANNING PROCESS

To achieve the three goals just listed, you must pursue the following four tasks, both before you begin to write and while you are actually composing your document:

1. Determine as fully as possible who will read what you write.
2. Know the goals you want your writing to achieve and the business context in which you need to communicate.
3. Understand your role in the organization as a writer and how your role should be reflected in what you write.
4. Determine the content by considering your readers' frame of reference and your purpose in writing.

Keep in mind that business readers want answers now. Employees in most organizations, particularly large ones, have more to read than they can and will read. If you want your letter or report to be read, be sure that important information occurs on the first page of a report and in the first paragraph of an e-mail, memo, or letter and that you answer the following questions your readers will ask:

• What is this?
• Do I have to read it?
• How does it affect me?
• What will I have to do?
• What are the main ideas?

If you answer their questions at the beginning of your document, readers may be willing to read more of your report, e-mail, or letter.

To answer these questions in the minds of your readers, you will have to answer three questions yourself about the document you are writing:

• Who will read what I write?
• Who will act on what I write?
• Who else may read what I write?

In many cases, your primary reader will transmit your document to someone else for action. Perhaps this individual is one of your secondary readers or someone unknown to you.

Determine your readers and their perspectives. When you consider your readers, determine as much as you can about them.

• How much do your readers know about your topic?
• Do your readers have expertise in this area?

Readers with technical expertise in the area you discuss have different needs (and often different perspectives) from readers who lack technical expertise.

- How much do your readers know about your topic?
- Do you know or can you estimate your readers' educational levels?
- Do you know your readers' cultural backgrounds?

If you work in an organization that does business with readers from other cultures, plan to do background reading on these cultures.

- Will your readers have interest in what you write? If not, how could you present your message to make it appealing?
- What kind of relationship do you have with these readers? Do you know the readers' attitude toward you, the subject matter you need to communicate, the job you have, and your area within the organization? Do you have credibility with these readers?

A host of factors determine your readers' perception: education, family, geographical and cultural background, job responsibilities, rank in the organization, age, life experiences, and gender—just to name a few demographics that define how people see the world. How much your readers know about your topic determines what you say and the technical level of your presentation.

- How well do you know your readers?

You may not know your readers personally. However, if you know an individual's level in the organization, the responsibilities associated with that level, and the kind of technical expertise your reader has, this information will help you decide what you need to say and how to present your information. Knowing your readers' responsibilities in the organization can help you anticipate their attitude—if your subject will interest them. Because people tend to read only what will help them, try to relate your message to your readers' job. Knowing the readers' attitude toward the topic addressed in your message will help you determine how to present your information.

- Who else might read what you write?

Most reports and letters have distribution lists: the names of those who receive copies. A person on the distribution list may be the person who will ultimately act on what you write. Thus, the needs and perceptions of those who receive copies should be considered.

- Why is each person on the distribution list receiving a copy?
- How much does each person on this list know about your topic?

Sometimes your primary reader may know the situation you are discussing, and the purpose of the report may be to inform others within the organization by going through proper channels.

- What situation led to the need for this document?

Often, you can better understand your reader's perspective if you understand the situation that requires you to write this document. The need for written communications develops from interactions of people involved in a work environment. To be able to select content, level of language (technical or general), and amount of explanation needed in a business context, a writer must be careful to determine the needs of each reader. Closely examining a situation requiring a written response may even help you determine what you need to tell your readers and how to present your message.

Determine your purpose.

- Do you know why you need to write this document?
- What do you want to achieve with your document?

Determining why you are writing—what you want to achieve—equals the importance of determining your readers. *Purpose always relates to readers.* And you may have more than one purpose. For example, you may be writing to inform, to provide information, to recommend action. In addition, what you say may serve as documentation—proof of your efforts—to show that you have provided the information requested. Written messages that document employees' activities serve a major function in today's business organizations. Without documentation, you may have difficulty proving that you performed specific tasks.

Understand your role as a writer.

- What position do you have in the organization?

As an employee, you will be hired to perform the duties that define a particular job. As the one responsible for performing specific tasks, you will be communicating with employees above you, below you, and on your own level. In writing to individuals in any group, you will communicate not as you would with a friend or family member but as the person responsible for the work associated with that position. When you write, you create a personality that should fit the position you hold.

To have credibility as a writer in an organization, the image that you project should be appropriate to your position. What you say and how you say it should reflect your level of responsibility in the organization—the power relationship that exists between you and the reader. The image you project will change, depending on your readers. You will project the image of a subordinate when you write to those higher than you, but you will transmit the image of the supervisor to those who work directly under you. When you communicate with others on your own job level, you will convey the image of a colleague. Good writers have the ability to fit their message to each reader.

Plan the content.

- What ideas should you use to achieve the goals of the message?
- What ideas should you omit?
- How should you arrange your ideas?

Once you have analyzed your readers and your purpose, you can begin to decide what you want and need to say, then how you will phrase and arrange your ideas.

• How do you want your message to sound?

Knowing how your message should sound will always be critical. Always try to convey a respectful tone commensurate with your position in the organization. How a message is conveyed may often be as important as the content.

Case 2–1 shows how a writer's assessment of audience and purpose changes the content and presentation in the e-mails for each reader involved in the situation.

Anticipate the context in which your writing will be received.

• How will readers use your writing?

Once your document has reached its primary destination, it may be placed in a stack for later reading; it may be skimmed and then routed to the person who will be responsible for acting on it; it may be read, copied, and distributed to readers unknown to you; it may be read and used as an agenda item for discussing a particular point; or it may be read carefully and later used as a reference. Knowing how readers to whom you write will use the documents they receive can often guide you in deciding not only what to say but also how to organize the information and arrange it on the page.

THE BASIC PARTS OF THE COMPOSING PROCESS

The composing process, integral to your analysis of audience, has six main stages:

1. **Analyzing** the situation
2. **Choosing/discovering** content
3. **Arranging** content
4. **Drafting**
5. **Revising**
6. **Editing** the finished draft

A writer who tries to do all stages at once usually creates a document that will fail. Research has shown that good writers usually follow a standard process—one that will make your writing tasks easier and the results more effective.

Analyzing the writing situation—purpose, readers, and context. The first step in composing is the most critical. In this step, you need to know *why* you need to write: what you want to achieve with your document, what situation or problem has led to the necessity of your writing this document. Then, you need to consider your readers—those who will or may read your document.

Every technical or workplace document responds to a specific situation. Each document has a targeted audience. Writing responds to both—the situation and the readers in that situation. When you write, you do not simply compile information about a subject.

Choosing/discovering content. You select content for your document based on your purpose, *what* your reader needs and *how* you think your reader perceives the subject.

As you search for information, remember your purpose, what you want your reader to know and do with what you write.

Then, begin to list ideas you can use to develop your topic. Based on these ideas, ask yourself what additional information you will need to locate. Don't like what you wrote? Delete it. You may want to begin your document by writing your purpose at the beginning to help you stay on track.

Case 2-1

As Case Document 2-1A indicates, Running Brook Neighborhood needs money to maintain the entrance to the subdivision. Charles Fields, treasurer of the property owners' association, has decided to write a flyer to place in each mailbox of the Running Brook homes. He has received very little money for the required upkeep. He drafts his request, but before he makes copies for the project, he asks a neighbor for her opinion of his flyer. She offers to revise the flyer (Case Document 2-1B) because she sees the difference between his version and what she thinks should be written. He makes the required number of copies, places one in each mailbox, and receives more than the amount RB needs for its entrance maintenance.

CASE DOCUMENT 2-1A

Subject: Running Brook Subdivision Market (RBSM) Fund

I have been remiss in reporting the status of the RBSM fund. In the past, an oral report was delivered at the annual subdivision picnic. I did not make such a report at the last picnic since I felt the information should be provided to all RB residents and not just to picnic attendees. The report for 2011–2012 will be distributed with the flyers that announce the 2013 Running Brook picnic. Since the subdivision has several new families, a historical element was added to this report.

The POA supplies each subdivision with a marker, which generally is located in the middle of a small landscaped area. Maintenance, decoration, and any embellishment of the marker area is the responsibility of the residents of the subdivision. Resident involvement in the care of the RBSM began in December 2009 when a waterline and meter were installed at a cost of $450.00 and obligation to pay an annual water usage fee was assumed. The fee allows for consumption of 4,000 gallons of water each month. Water consumption above 4,000 gallons incurs an additional charge, an event that has occurred during the hot summer months of the past 3 years. The fee has increased from $106.98 to $123.30 over the past 3 years, and I pay it from my checking account every 2 months throughout the entire year.

As of October 2012, cost of care of the marker was $712.34 (waterline and meter, water usage, plants, and landscaping material). This cost was defrayed by donations amounting to $705.00 from 70 percent of the residents, and money ($123.78) from the 2010 picnic. The account had a positive balance of $116.44 ($828.78 minus $712.34).

Expenditures for the period October 2011 through October 2012 included water usage ($126.72) and plants ($88.76) for a total of $215.48. For a time, the account had a negative balance of $99.04. Donations amounting to $245.00 were received from only 16 of the 62 families (25 percent) in the subdivision. The account was healthy again with a positive balance of $145.96.

Expenditures for the period October 2012 through October 2013 were $230.12 (water usage $372) and plants, fertilizer, hat, pumpkins, etc. $93. Once again, the fund had a negative balance ($84.16). Recent donations from 9 residents totaled $135.00. The current fund balance is $50.84.

Projected costs from now until next October include water ($123.30) and plants in the spring and fall ($90.00–$100.00) or approximately $250.00. Clearly, the present fund balance will not support these costs. Accordingly, contributions are needed. The treasurer for the fund is Joann Fields. Hopefully, every RB family will participate. I sense we all believe that participation provides a measure of the RB community spirit.

Seasons Greetings,

Charles Fields

2 Roaring Brook Court, phone 933–3314, e-mail clfields@earthlink.com.

CASE DOCUMENT 2–1B

Running Brook Marker Fund Needs Your Donations

The Problem
- Our marker maintenance fund currently has a balance of **$50.84**. Our projected maintenance cost this year is $250.00.
- Since 2000, the number of residents contributing has dropped from 70 percent of the residents to only 9 residents. Without contributions, maintenance cannot occur.

How You Can Help
Please contact Joann Fields, treasurer, at 922–3314 or clfields@earthlink.com. Or you can mail her a check, made out to the Running Brook Marker Fund, for $10. This amount will cover the cost of water, plants, and maintenance. Any money left over will be applied to next year's maintenance.

Expenditures and Balances to Date			
	Costs	Donations	Account Balance
Oct. 2010	$712.34	$705	$116.44
Oct. 2011	$215.48	$245	$145.96
Oct. 2012	$230.12	$135	$50.84

Information for New Running Brook Residents

Each subdivision in Hot Springs Village has a marker that denotes the entrance to the subdivision. The Property Owners' Associations provide the markers, but residents must maintain and decorate the area surrounding the marker.

A few years ago, we installed a waterline and meter. Water costs have increased, particularly during the summer. Other expenses include plants, fertilizer, hay, and pumpkins.

Keeping our marker area well maintained provides a positive first impression of our neighborhood.

Contact Information—Give us a call if you have questions

Charles and Joann Fields
2 Roaring Brook Court
Hot Springs Village
phone 933–3314, e-mail clfields@earthlink.com

Why is Case Document 2–1B better than Case Document 2–1A? Why would you be more inclined to read and respond to Case Document 2–1B? Charles also decides to paper-clip an empty envelope with his name and address to the request. Why would this decision help Charles achieve his goal?

Case 2-2

The director of online sales at Pine Avenue Books would like to perform a customer survey regarding the store's website in order to improve the efficiency of the online shopping experience. She drafts the following e-mail that she will send to all customers who made online or on-site purchases at the store in the last 2 years (Case Document 2–2A). The store manager, however, tells her that her draft, if sent by e-mail, will likely not be read: (1) The subject line does not encourage busy readers to open the e-mail. (2) The purpose of the message does not appear until the second paragraph. (3) The e-mail as a whole appears dense and difficult for readers to skim. The online sales director decides to revise the original to respond to the three issues.

CASE DOCUMENT 2–2A

SUBJECT: Please take a 5-minute, multiple-choice survey; details follow.

I am the director of online sales at Pine Avenue Books and am conducting a study of the online shopping sections of the store's website. If you have purchased books online in the last 2 years, please consider participating and passing the attached invitation on to your friends for their potential participation.

This study wants to determine how customers view the store's website and, in particular, the sections related to online shopping. Your participation may provide useful information about online shopping and help us to improve your online shopping experience in the future. I am looking for customers who have bought books online at Pine Avenue Books in the last 2 years.

If you participate, your obligations will be low. You will complete a short, anonymous survey via the Internet that will require approximately 5 minutes of your time and will be returned to me via Survey Monkey, an online data collection service. If you complete a survey, your responses will be returned to me anonymously (I will not be able to identify your e-mail address, your IP address, or any other information that would inform me as to your identity or your location). If you agree to do so, you will also participate in a 15- to 20-minute follow-up session (this meeting will occur by telephone or e-mail as you choose). All data for surveys and follow-up interviews will be strictly confidential. Your identity will never be revealed in any results, discussions, or presentation of the research. In addition, all information will be destroyed after I have analyzed the data.

Completion of the survey and postsurvey interviews is voluntary; you may skip questions and can quit any portion of the study at any time. If you are willing to participate in this study, please click on the following link and complete the survey: http://www.surveymonkey.com/PineAvenueBooks. If you wish to participate in a follow-up interview, please include contact information at the end of the survey.

You may contact me via e-mail (DHS48@packard.com) or phone (512-823-9235)

CASE DOCUMENT 2–2B

SUBJECT: Please take a 5-minute survey about online shopping at Pine Avenue Books.

SURVEY PURPOSE
This study seeks to gauge your opinion of the online shopping sections of the Pine Avenue Books website (www.pineavenuebooks.com). Your participation will help us to improve your future online shopping experience.

If you have purchased books online in the last 2 years, please consider participating and passing the attached invitation on to your friends for their potential participation.

SURVEY REQUIREMENTS
If you participate, your obligations will be minimal.

1. You will complete a short, anonymous survey via the Internet that will require approximately 5 minutes of your time and will be returned to me via Survey Monkey, an online data collection service.
2. If you complete a survey, your responses will be returned to me anonymously. (I will not be able to identify your e-mail address, your IP address, or any other information that would inform me as to your identity or your location.)
3. If you agree to do so, you will also participate in a 15- to 20-minute follow-up session (this meeting will occur by telephone or e-mail as you choose).
4. All data for surveys and follow-up interviews will be strictly confidential. Your identity will never be revealed in any results, discussions, or presentation of the research. In addition, all information will be destroyed after I have analyzed the data.

5. Completion of the survey and follow-up interviews is voluntary; you may skip questions and can quit any portion of the study at any time.

6. If you are willing to participate in this study, please click on the link http://www.surveymonkey.com/PineAvenueBooks and complete the survey. If you wish to participate in a follow-up interview, please include contact information at the end of the survey.

You may contact me via e-mail (DHS48@packard.com) or phone (512-823-9235)

Arranging content. As you collect and begin summarizing information and data, consider how to arrange the material. In what order should you present your content? Memos, for example, need to begin with the news or essential information to ensure that readers at least read what's most important before they start skimming the document or stop reading it altogether. Most reports begin with an introduction, followed by a summary of the report. Alternately, you can combine the introduction with a summary of the report. The discussion section, in which you present the supporting information, follows. Most reports follow some version of this plan. Many business organizations have templates for reports.

You can arrange content by placing material in "stacks," which can be used as a resource when you begin writing. If you know what arrangement you want or have to use, sort material so that you can easily find it when you begin drafting specific segments of your document. You can also sort material electronically: create folders for each segment of your report, then arrange material within each folder before you begin drafting. This method allows you to track material you use and insert appropriate citations when you use material from a specific source.

If you use electronic articles from your library's database, insert these articles into files that can be accessed later when you begin to draft your document.

Drafting. Every individual drafts differently. Most writers work on a document in a start/stop fashion. When you begin your draft, open your file and save it with the name of your report. Then, begin typing ideas or sections. (You may wish to move/paste material you listed, arranged, and then developed in step 2.) You may wish to type the names of your main segments, boldface those, and insert information beneath the appropriate segment. This method helps you keep track of the information that you are using to develop your draft. Note that some of the ideas in your list become headings. Some may be combined with other ideas. You can arrange, delete, and add ideas as you need to.

As you continue to draft, you will revise. But during the drafting stage you should revise only to improve the meaning. Try to avoid worrying about sentences that don't sound "quite right." If the sentence you write captures what you want to say, even clumsily, don't stop to revise. You can "clean up" these sentences later. Don't attempt to correct mechanical problems unless you feel you can do so without slowing your ability to transfer your ideas from your mind to the screen. Focus on presenting your material to your readers; then you can begin a formal revision process once you believe you have your basic ideas on the screen or page.

Revising. During the formal revision process you need to revise several times and focus on different issues:

- **Logic.** Does your presentation make sense? Try reading paragraphs aloud that seem to be "scrambled." Hearing what you have written often tells you if/ where problems in logic occur. Does your material occur in the appropriate order for your purpose and for your readers?
- **Completeness.** Does your presentation seem complete, in terms of your purpose and your readers' needs and requirements? Have you checked all your information for correctness? Does your document contain all requested information?
- **Style.** Examine each paragraph and each sentence. Are your paragraphs really paragraphs? Do they have topic sentences? Do all the sentences in the paragraph pertain to the meaning you are building in the paragraph? Start each paragraph with a topic sentence. Eliminate or recast sentences that provide little support for the topic sentence. Today's readers usually dislike wordy, dense, complicated sentences. Make your sentences clear, concise, and precise to encourage your readers to follow your ideas. Also, watch the length of your paragraphs. Long paragraphs discourage readers and tend to become incoherent.
- **Visuals.** Do you need visuals—photos, graphs, drawings, pictorial illustrations— to help your reader "see" and remember key ideas? Visuals combined with text often provide the best means of communicating with your readers.
- **Document design.** When you began drafting, if you used headings or names of report segments to help you organize your draft, you began at that point to design your document. *Document design* refers to the way you arrange information and display it on the page. The importance of how information looks on the page cannot be stressed enough. If you want your writing to be read, design the page so that information is inviting and accessible.

Editing. Get into the habit of performing several "edits" for any document: one for mechanics—spelling, usage, punctuation, and sentence structure.

A second edit focuses on the document as a whole. How does it look? How does it sound? Is the important information easy to locate? Have you included all needed information in the document?

Another edit focuses on citing sources: check your documentation to be sure that you give credit or sources of all information you have used. Be sure that when you use illustrations and ideas from other sources you give credit to the source.

In short, don't try to check for every error at once, in one reading. Editing requires care, objective reading, and diligence.

For an example of how the composing process works when it is applied to a routine business document, read Case 2–3 and track the development of this memorandum. Note how the message looks after Bob's initial revision of his draft.

Case 2-3

Bob Johnson, an engineering project manager with a local civil engineering firm, has been named local arrangement chair of the forthcoming construction engineering conference. Two months before the conference, Bob needs to send a memo to everyone in his group to let them know what responsibilities they will have at the conference. While all employees know about the conference, Bob has informed his group via e-mail that each office group will have responsibilities throughout the conference. Bob wants the memo to inform his group specifically about what they will need to do.

Bob uses the planning stage of developing the memo to decide what topics he wants to present to his group. He will send this memo as an e-mail attachment. As he plans his memo, Bob types the following list of topics:

Location, date, time info of conference
Specific duties of the SE Group
General instructions
Conference schedule
Other information SE Group needs to know

As he develops the memo, he inserts information beneath each heading and highlights incomplete sections. He writes notes to himself to allow him to keep track of ideas he may need to add. His first draft looks like this (Case Document 2-3A):

CASE DOCUMENT 2-3A

Bob's First Draft

Conference Location, Date, Time

CE conference—October 28–29—Lancaster Center. Our group's responsibility—serve as greeters, help prevent glitches. Over 150 engineers have already registered, and the cutoff date is still 3 weeks away. We need to be sure we are organized. Conference may be larger than last year. We want to do our part to ensure the success of the meeting. Help all attendees have a good conference. Be proactive in anticipating problems with people getting where they need to be:

Please be at the Lancaster Center at the following times:

Oct. 28: noon–end of the day
Oct. 29: 7:00–end of the conference. Last session begins at 3:30
Refreshments will be available during break periods. WATER IN ALL ROOMS.

Our Responsibilities

Helping visitors locate the section meetings, answer any questions, deal with any hotel reservation glitches, transportation problems, questions about restaurants. Remain available until after the dinner on Oct. 28. Oct. 29: On site throughout the day.

General Instructions Information
- Number expected to register and attend: 200+.
- Visitors will arrive at the hotel by mid-morning on the 28th. Some will come the evening before. [Contact Ralph to see if we need to be at the hotel on Oct. 27 after 5:00? Check information folder for sponsor letters.]
- Be available no later than noon on 28th. If possible, arrive at the Lancaster Center earlier than that. Dress is business casual.
- If those flying in arrive late, contact Jim or Joanna via their cell phones to ensure that registrations are not canceled. Jim: 228–3459; Joanna: 322–1875.

Conference Schedule

Oct. 28
Light lunch: noon–1:15
Opening session: 1:30–3:00
Second session: 3:30–5:00 [check sponsors for all sessions. These have to be correct!!!! Check with central planning group.]
Dinner: 6:30—Holcomb Room, 2nd floor of the LC

Oct. 29
Breakfast: Room 104 of the LC, 7:00–8:00 [sponsor? Be sure we have a complete list with all names spelled correctly. Contact info for each.]
Third session: 8:30–10:00 [session sponsor?]
Closing session: 10:30–noon [sponsor?]
Lunch: noon

Displays

Nine vendors will display software all day the 29th in room 106. Consultants will be on hand to discuss compatibility issues. Be available to help vendors set up.

Conference Materials

Will be available at the check-in desk at the front door.
Each folder will contain brochures about new products and a schedule. Add a list of restaurants downtown?

Other

Breakout rooms will be available for the second part of the sessions.
Phone and faxes are available in Room 110, 8:00–5:00.

Bob's revision follows (Case Document 2–3B). How could he have improved it with the following goals in mind? He wants his employees to

- report on time
- know what they should do when they arrive
- have cell phone numbers for quick contact purposes
- know whom to contact throughout the conference

Does the memo enable Bob's employees to achieve those goals?

CASE DOCUMENT 2–3B

Bob's Revision

TO:	SE Group	**DATE:**	October 1, 2011
FROM:	Bob Johnson		
SUBJECT:	Preparations for the Construction Engineering Conference		

Conference Location, Date, Time

The construction engineering conference is scheduled October 28–29 at the Lancaster Center. Our group will serve as greeters. Over 150 engineers have already registered, and we still have 3 weeks before the cutoff date for registration. We need to be sure we are organized to help visitors as they arrive.

Please be at the Lancaster Center at the following times:

Oct. 28: noon–end of the day
Oct. 29: 7:00–end of the conference. Last session begins at 3:30

SE Group—Specific Duties

We will be responsible for helping visitors locate the section meetings, answering any questions, and dealing with any hotel reservation and transportation glitches and will remain available until after the dinner on Oct. 28. Oct. 29— We need to be on site throughout the day and help guests who need to leave promptly at the close of the morning session.

General Information

- Number expected to register and attend: 200+.
- Visitors will arrive at the hotel by mid-morning on the 28th. Some will come the evening before.
- Be available no later than noon. If possible, arrive at the Lancaster Center earlier than that.

If those flying in arrive late, contact Jim or Joanna via their cell phones to ensure that registrations are not canceled. Jim: 228–3459; Joanna: 322–1875.

Conference Schedule

Oct. 28
Noon–1:15	Light lunch—Mellon Room (Sponsor: KLM Ltd.)
1:30–3:30	Room 105, Opening session
3:30–5:00	Room 105, Second session (Sponsor: Bickle and Lauren)
6:30	Buffet in Holcomb Room, 2nd floor of the LC

Oct. 29
7:00–8:00	Breakfast: Room 104 of the LC
8:30–10:00	Room 105, Third session (Sponsor: MERK Inc.)
10:30–noon	Room 105 Closing session (Sponsor: Malcolm, Fisher, & Peabody)
Lunch: noon	Mellon Room

Software Displays
Nine vendors will display software all day the 29th in Room 106. Consultants will be on hand to discuss compatibility issues. Be available to help vendors with set-up.

Conference Materials
- Available at the check-in desk at the front door.
- Registration folder with name tags: will contain brochures about new products and a schedule of activities. List of restaurants in town for those who are staying for the weekend.

Other Information
- Phone and faxes are available in Room 110, 8:00–5:00.
- Refreshments will be available during break periods. Bottled water in all rooms.

Call me on my cell phone if anything comes up that isn't covered here.

PLANNING AND REVISION CHECKLIST ✔

Analyzing the Situation

☐ What is your subject or topic?
☐ What is the purpose of the document?
☐ Who are your readers, known and potential?
☐ Why are you writing? Why is this document required? What is the situation that led to the need for this document? Who cares?

Selecting Content

☐ What topics do you need to cover? What do your readers need to know? What do you want your readers to do?
☐ What structure do you plan to use? If you have required report sections, what are they?
☐ What information resources do you have available? What resources do you need to locate?
☐ What types of visuals (e.g., graphs, photos, diagrams) are you considering using? Will they help convey the message?

Arranging Content

☐ In what order should the information be placed? What does your reader need to know first?
☐ Have you sorted your material into specific groups?
☐ Can you see a plan for headings that announce the content to your reader?
☐ Does all content have relevance to your purpose?

Drafting

☐ Have you inserted information under each of your headings?
☐ Have you recorded the sources of all information you will use so that you can develop correct citations after you have completed your draft?
☐ Have you noted where you will use graphics? Have you noted the source of each graphic you use from another source?

Revising

☐ Have you stated clearly the purpose of your report?
☐ Does your content support your purpose?
☐ Is your tone appropriate?
☐ Will your readers be able to follow your logic?

continued

PLANNING AND REVISION CHECKLIST ✔   *continued*

☐ Have you included all required items—report sections and required information?

☐ Have you checked all facts and numbers?

☐ Could any material be deleted?

☐ Is your document easy to read? Are your paragraphs well organized and of a reasonable length?

☐ Have you had someone read your draft and suggest improvements?

Editing

☐ Have you checked for misspellings and for other mechanical errors, such as misplaced commas, semicolons, colons, and quotation marks?

☐ Have you checked all points of the completed draft where your word processing program suggests that you have errors in either sentence structure, mechanics, or spelling?

☐ Have you included all the formal elements that your report needs or is required to include?

☐ Is your system of documentation complete and accurate (if you are following a style sheet)?

☐ Are your pages numbered?

☐ Are all graphics placed in the appropriate locations within the text?

☐ Is the format consistent—font selected, size, placement of headings?

EXERCISES

1. Phoenix Publishers has always provided free child care to all employees with children ages 3 months to 5 years. For its 50 years of operation, the company has taken great pride in being a family-friendly employer. Tough economic times for the book publishing industry and rising costs of operation for the child-care center, however, now require that the company begin charging parents $100 per month per child for the services of the child-care center. According to the president of the company, it was either that or freeze wages for all employees or lower the already slim dividend paid to the company's stockholders and risk a loss of investors. The president of the company directs you to write three letters regarding this important change: one to parents using the child-care center, one to all employees, and one to the stockholders. Note that parents will also

receive the letter addressed to employees. Note also that some employees are also stockholders. The president recognizes the sensitivity of this policy change and thus will also expect from you a memo justifying the variations you made in the three letters.

2. Find a blog posting aimed at specialists in your field. Revise the posting for a specific group of readers outside your field who might need to know this information (e.g., business managers, city officials, school administrators). Attach the original to your revision, and submit both to your instructor.

3. Examine Figure 2–1 (addressed to parents) and Figure 2–2 (addressed to employers). What similarities and differences do you notice in the verbal and visual information included? How do the writers adapt information to each audience? What changes, if any, would make the two documents more suitable for their specific audiences?

4. Since Figure 2–2 addresses employers, how would you adapt this information for a message addressed to employees?

Handwashing: A Family Activity
Keeping Kids & Adults Healthy

Handwashing is an easy, inexpensive, and effective way to prevent the spread of germs and keep people healthy.

For kids, washing hands can be a fun and entertaining activity. It is simple enough for even very young children to understand. Handwashing gives children and adults a chance to take an active role in their own health. Once kids learn how to properly wash their hands, they can—and often do—show their parents and siblings and encourage them to wash hands, too.

Parents can help keep their families healthy by:

- Teaching them good handwashing technique
- Reminding their kids to wash their hands
- Washing their own hands with their kids

Improving Health

- Handwashing education in the community:
  » Reduces the number of people who get sick with diarrhea by 31%
  » Reduces diarrheal illness in people with weakened immune systems by 58%
  » Reduces respiratory illnesses, like colds, in the general population by 21%

Saving Time and Money

- Handwashing is one of the best ways to avoid getting sick and spreading illness to others.
- Reducing illness increases productivity due to:
  » Less time spent at the doctor's office
  » More time spent at work or school

Helping Families Thrive

Children who have been taught handwashing at school bring that knowledge home to parents and siblings. This can help family members get sick less often and miss less work and school.

Despite widespread knowledge of the importance of handwashing, there is still room for improvement. A recent study showed that only 31% of men and 65% of women washed their hands after using a public restroom.

For more details, visit _vwww.cdc.gov/handwashing_.

CDC Department of Health and Human Services
Centers for Disease Control and Prevention

CS234835-A

FIGURE 2-1 Document for Exercise 3

Source: US Department of Health and Human Services, Centers for Disease Control and Prevention. Handwashing: A Family Activity. Washington, DC: GPO, 2012. http://www.cdc.gov/healthywater/pdf/hygiene/hwfamily.pdf

Handwashing: A Corporate Activity
Improving Health & Increasing Productivity

Handwashing is an easy, inexpensive, and effective way to prevent the spread of germs and keep employees healthy.

Handwashing gives people the opportunity to take an active role in their own health. Most handwashing studies have focused on child care or health care settings. The few that have looked at corporate settings show that promoting clean hands results in fewer employee sick days.

Improving Health

Germs can spread quickly. A healthier community means healthier employees. Handwashing education in the community:

- Reduces the number of people who get sick with diarrhea by 31%
- Reduces diarrheal illness in people with weakened immune systems by 58%
- Reduces respiratory illnesses, like colds, in the general population by 21%

Saving Time and Money

Handwashing is one of the best ways to avoid getting sick and spreading illness to others.

Sick employees are less productive even when they come to work. They may also spread illness to others at work. One recent study promoting clean hands in corporate environments showed:

- Fewer employee illnesses
- Less use of sick days

Helping Families and Workforces Thrive

Employees with healthy children spend less time away from work taking care of sick children, are more productive at work when not dealing with family illness, and get sick less often themselves.

Employers should promote employee handwashing and encourage them to also:

- Teach their children good handwashing technique
- Remind children to wash their hands
- Wash hands with their children

Despite widespread knowledge of the importance of handwashing, there is still room for improvement. A recent study showed that only 31% of men and 65% of women washed their hands after using a public restroom.

For more details, visit www.cdc.gov/handwashing.

Department of Health and Human Services
Centers for Disease Control and Prevention

CS234835-B

FIGURE 2-2 Document for Exercise 3

Source: US Department of Health and Human Services, Centers for Disease Control and Prevention. Handwashing: A Corporate Activity. Washington, DC: GPO, 2012. http://www.cdc.gov/healthywater/pdf/hygiene/hwcorporate.pdf

Writing Ethically

Quick Tips

On the job, you typically won't have time to analyze all the issues or answer all the questions relating to a given ethical dilemma. You might have to make a decision quickly—in minutes or seconds.

In cases like this, think of a person in your company or in your profession you admire for his or her integrity and good judgment (e.g., a colleague, a supervisor, a mentor). Ask yourself, "How would he or she manage this dilemma?" Allow your answer to this question to guide your actions.

YOUR PROFESSIONAL OBLIGATIONS

None of us are isolated individuals, operating entirely separate from the traffic of human society. Your ethical obligations are several, often intersecting, and from time to time competing. Consider, for example, your duties to the following:

- **To yourself:** You will have to make decisions and take actions that allow you to support yourself financially while establishing (and maintaining) your reputation in your field. You can't quit (or lose) your job every time you object to a policy of your boss.
- **To your discipline and profession:** As a member of your profession, you have a responsibility to advance the knowledge and reputation of your field. You must share information with your colleagues that will improve the practices of your profession, clarify understanding, offer new insights, and promote better training of new students of your discipline. You must communicate in a manner that brings credit to your profession and inspires the next generation to want to study and join your profession.

- **To your academic institution:** You have a moral obligation to the institution that trained you for your profession. Your successes or failures will be indicative of the merits of that institution and its faculty. If you disgrace yourself by illegal or unethical actions, for example, investigating officials and the public might ask why you weren't taught better behavior.
- **To your employer:** Your responsibility as an employee is to serve the interests of your organization, to help it make money, to promote its products and services, and to shield confidential information and intellectual property, especially if doing so offers a competitive advantage.
- **To your colleagues:** You have a duty to your colleagues on the job to do your fair share of the work assigned and to do it with integrity, accuracy, and efficiency. You also have a responsibility to use no more than your fair share of the resources allotted and to take no more than your fair share of the credit (or blame) given.
- **To the public:** Your obligation to society is to promote the public good through greater safety, fuller liberty, and a better quality of life. Your decisions and actions on the job could allow communities to thrive in resilient and sustainable environments or to be poisoned by private greed and callous disregard for civic aspirations.

It is also important to keep in mind that something could be legal but still unethical: for example, zoning regulations in your city might allow your company to build a chemical storage facility adjacent to a hospital, but the danger to immobilized patients from a chemical explosion—and their inability to evacuate quickly—might make this choice of location unethical. You would have to weigh the risk to the public versus the jobs that would be created in your city and the potential profits for your company.

In communicating on the job, your choices from time to time might be obvious, but you will typically have to decide among several satisfactory choices or several unsatisfactory choices. You will have to juggle your various obligations and determine which has priority. You can't simply do whatever the boss tells you to because you also have important responsibilities to yourself, your profession, your schools and teachers, your colleagues, and the public itself. You will have to make every effort to avoid being either submissive or self-righteous.

CODES OF CONDUCT AND STANDARDS OF PRACTICE

Both your professional association and your employing organization likely have codes of conduct or ethical guidelines that specify their expectations regarding appropriate behavior on the job. Many disciplines, such as engineering, accounting, medicine, and financial management, have specific guidelines for professional conduct and for standards of practice. Continuing education and periodic testing on these rules may be a part of your certification. Violation of these rules can carry heavy penalties, particularly because these disciplines involve clients' health and financial resources.

Examine the code of ethics for the National Society of Professional Engineers (Figure 3–1). Even if you are not an engineering major, you can see how the National Society of Professional Engineers dictates how engineers conduct their work. In contrast, the Society for Technical Communication, a leading international association for technical writers and graphic artists, has composed its guidelines as a list of six principles (Figure 3–2). Other professional associations publish highly detailed regulations to specify ethical practice in their field, such as the 400-page code of conduct of the American Institute of Certified Public Accountants or the 500-page code of medical ethics of the American Medical Association.

Your employer may also have a code of conduct that identifies the ideals of the organization and the behaviors expected of every employee. This could be a brief list of dos and don'ts or a comprehensive discussion of policies and procedures. Both codes of conduct and standards of practice might also be the entry point to a wider array of available ethics resources, including frequently asked questions and help lines for reporting violations and soliciting advice.

Familiarize yourself with the codes of conduct or standards of practice that regulate ethical communication for your company and your profession because from time to time you might have to interpret and cite their guidelines to justify your decisions regarding ethical dilemmas on the job.

**National Society of
Professional Engineers®**

Code of Ethics for Engineers

Preamble

Engineering is an important and learned profession. As members of this profession, engineers are expected to exhibit the highest standards of honesty and integrity. Engineering has a direct and vital impact on the quality of life for all people. Accordingly, the services provided by engineers require honesty, impartiality, fairness, and equity, and must be dedicated to the protection of the public health, safety, and welfare. Engineers must perform under a standard of professional behavior that requires adherence to the highest principles of ethical conduct.

I. Fundamental Canons

Engineers, in the fulfillment of their professional duties, shall:

1. Hold paramount the safety, health, and welfare of the public.
2. Perform services only in areas of their competence.
3. Issue public statements only in an objective and truthful manner.
4. Act for each employer or client as faithful agents or trustees.
5. Avoid deceptive acts.
6. Conduct themselves honorably, responsibly, ethically, and lawfully so as to enhance the honor, reputation, and usefulness of the profession.

II. Rules of Practice

1. Engineers shall hold paramount the safety, health, and welfare of the public.
   a. If engineers' judgment is overruled under circumstances that endanger life or property, they shall notify their employer or client and such other authority as may be appropriate.
   b. Engineers shall approve only those engineering documents that are in conformity with applicable standards.
   c. Engineers shall not reveal facts, data, or information without the prior consent of the client or employer except as authorized or required by law or this Code.
   d. Engineers shall not permit the use of their name or associate in business ventures with any person or firm that they believe is engaged in fraudulent or dishonest enterprise.
   e. Engineers shall not aid or abet the unlawful practice of engineering by a person or firm.
   f. Engineers having knowledge of any alleged violation of this Code shall report thereon to appropriate professional bodies and, when relevant, also to public authorities, and cooperate with the proper authorities in furnishing such information or assistance as may be required.
2. Engineers shall perform services only in the areas of their competence.
   a. Engineers shall undertake assignments only when qualified by education or experience in the specific technical fields involved.
   b. Engineers shall not affix their signatures to any plans or documents dealing with subject matter in which they lack competence, nor to any plan or document not prepared under their direction and control.
   c. Engineers may accept assignments and assume responsibility for coordination of an entire project and sign and seal the engineering documents for the entire project, provided that each technical segment is signed and sealed only by the qualified engineers who prepared the segment.
3. Engineers shall issue public statements only in an objective and truthful manner.
   a. Engineers shall be objective and truthful in professional reports, statements, or testimony. They shall include all relevant and pertinent information in such reports, statements, or testimony, which should bear the date indicating when it was current.
   b. Engineers may express publicly technical opinions that are founded upon knowledge of the facts and competence in the subject matter.
   c. Engineers shall issue no statements, criticisms, or arguments on technical matters that are inspired or paid for by interested parties, unless they have prefaced their comments by explicitly identifying the interested parties on whose behalf they are speaking, and by revealing the existence of any interest the engineers may have in the matters.

4. Engineers shall act for each employer or client as faithful agents or trustees.
   a. Engineers shall disclose all known or potential conflicts of interest that could influence or appear to influence their judgment or the quality of their services.
   b. Engineers shall not accept compensation, financial or otherwise, from more than one party for services on the same project, or for services pertaining to the same project, unless the circumstances are fully disclosed and agreed to by all interested parties.
   c. Engineers shall not solicit or accept financial or other valuable consideration, directly or indirectly, from outside agents in connection with the work for which they are responsible.
   d. Engineers in public service as members, advisors, or employees of a governmental or quasi-governmental body or department shall not participate in decisions with respect to services solicited or provided by them or their organizations in private or public engineering practice.
   e. Engineers shall not solicit or accept a contract from a governmental body on which a principal or officer of their organization serves as a member.
5. Engineers shall avoid deceptive acts.
   a. Engineers shall not falsify their qualifications or permit misrepresentation of their or their associates' qualifications. They shall not misrepresent or exaggerate their responsibility in or for the subject matter of prior assignments. Brochures or other presentations incident to the solicitation of employment shall not misrepresent pertinent facts concerning employers, employees, associates, joint venturers, or past accomplishments.
   b. Engineers shall not offer, give, solicit, or receive, either directly or indirectly, any contribution to influence the award of a contract by public authority, or which may be reasonably construed by the public as having the effect or intent of influencing the awarding of a contract. They shall not offer any gift or other valuable consideration in order to secure work. They shall not pay a commission, percentage, or brokerage fee in order to secure work, except to a bona fide employee or bona fide established commercial or marketing agencies retained by them.

III. Professional Obligations

1. Engineers shall be guided in all their relations by the highest standards of honesty and integrity.
   a. Engineers shall acknowledge their errors and shall not distort or alter the facts.
   b. Engineers shall advise their clients or employers when they believe a project will not be successful.
   c. Engineers shall not accept outside employment to the detriment of their regular work or interest. Before accepting any outside engineering employment, they will notify their employers.
   d. Engineers shall not attempt to attract an engineer from another employer by false or misleading pretenses.
   e. Engineers shall not promote their own interest at the expense of the dignity and integrity of the profession.
2. Engineers shall at all times strive to serve the public interest.
   a. Engineers are encouraged to participate in civic affairs; career guidance for youths; and work for the advancement of the safety, health, and well-being of their community.
   b. Engineers shall not complete, sign, or seal plans and/or specifications that are not in conformity with applicable engineering standards. If the client or employer insists on such unprofessional conduct, they shall notify the proper authorities and withdraw from further service on the project.
   c. Engineers are encouraged to extend public knowledge and appreciation of engineering and its achievements.
   d. Engineers are encouraged to adhere to the principles of sustainable development[1] in order to protect the environment for future generations.

FIGURE 3-1 National Society of Professional Engineers' Code of Ethics

Source: http://www.nspe.org/Ethics/CodeofEthics/index.html

3. Engineers shall avoid all conduct or practice that deceives the public.
   a. Engineers shall avoid the use of statements containing a material misrepresentation of fact or omitting a material fact.
   b. Consistent with the foregoing, engineers may advertise for recruitment of personnel.
   c. Consistent with the foregoing, engineers may prepare articles for the lay or technical press, but such articles shall not imply credit to the author for work performed by others.
4. Engineers shall not disclose, without consent, confidential information concerning the business affairs or technical processes of any present or former client or employer, or public body on which they serve.
   a. Engineers shall not, without the consent of all interested parties, promote or arrange for new employment or practice in connection with a specific project for which the engineer has gained particular and specialized knowledge.
   b. Engineers shall not, without the consent of all interested parties, participate in or represent an adversary interest in connection with a specific project or proceeding in which the engineer has gained particular specialized knowledge on behalf of a former client or employer.
5. Engineers shall not be influenced in their professional duties by conflicting interests.
   a. Engineers shall not accept financial or other considerations, including free engineering designs, from material or equipment suppliers for specifying their product.
   b. Engineers shall not accept commissions or allowances, directly or indirectly, from contractors or other parties dealing with clients or employers of the engineer in connection with work for which the engineer is responsible.
6. Engineers shall not attempt to obtain employment or advancement or professional engagements by untruthfully criticizing other engineers, or by other improper or questionable methods.
   a. Engineers shall not request, propose, or accept a commission on a contingent basis under circumstances in which their judgment may be compromised.
   b. Engineers in salaried positions shall accept part-time engineering work only to the extent consistent with policies of the employer and in accordance with ethical considerations.
   c. Engineers shall not, without consent, use equipment, supplies, laboratory, or office facilities of an employer to carry on outside private practice.
7. Engineers shall not attempt to injure, maliciously or falsely, directly or indirectly, the professional reputation, prospects, practice, or employment of other engineers. Engineers who believe others are guilty of unethical or illegal practice shall present such information to the proper authority for action.
   a. Engineers in private practice shall not review the work of another engineer for the same client, except with the knowledge of such engineer, or unless the connection of such engineer with the work has been terminated.
   b. Engineers in governmental, industrial, or educational employ are entitled to review and evaluate the work of other engineers when so required by their employment duties.
   c. Engineers in sales or industrial employ are entitled to make engineering comparisons of represented products with products of other suppliers.
8. Engineers shall accept personal responsibility for their professional activities, provided, however, that engineers may seek indemnification for services arising out of their practice for other than gross negligence, where the engineer's interests cannot otherwise be protected.
   a. Engineers shall conform with state registration laws in the practice of engineering.
   b. Engineers shall not use association with a nonengineer, a corporation, or partnership as a "cloak" for unethical acts.

9. Engineers shall give credit for engineering work to those to whom credit is due, and will recognize the proprietary interests of others.
   a. Engineers shall, whenever possible, name the person or persons who may be individually responsible for designs, inventions, writings, or other accomplishments.
   b. Engineers using designs supplied by a client recognize that the designs remain the property of the client and may not be duplicated by the engineer for others without express permission.
   c. Engineers, before undertaking work for others in connection with which the engineer may make improvements, plans, designs, inventions, or other records that may justify copyrights or patents, should enter into a positive agreement regarding ownership.
   d. Engineers' designs, data, records, and notes referring exclusively to an employer's work are the employer's property. The employer should indemnify the engineer for use of the information for any purpose other than the original purpose.
   e. Engineers shall continue their professional development throughout their careers and should keep current in their specialty fields by engaging in professional practice, participating in continuing education courses, reading in the technical literature, and attending professional meetings and seminars.

Footnote 1 "Sustainable development" is the challenge of meeting human needs for natural resources, industrial products, energy, food, transportation, shelter, and effective waste management while conserving and protecting environmental quality and the natural resource base essential for future development.

As Revised July 2007

"By order of the United States District Court for the District of Columbia, former Section 11(c) of the NSPE Code of Ethics prohibiting competitive bidding, and all policy statements, opinions, rulings or other guidelines interpreting its scope, have been rescinded as unlawfully interfering with the legal right of engineers, protected under the antitrust laws, to provide price information to prospective clients; accordingly, nothing contained in the NSPE Code of Ethics, policy statements, opinions, rulings or other guidelines prohibits the submission of price quotations or competitive bids for engineering services at any time or in any amount."

Statement by NSPE Executive Committee

In order to correct misunderstandings which have been indicated in some instances since the issuance of the Supreme Court decision and the entry of the Final Judgment, it is noted that in its decision of April 25, 1978, the Supreme Court of the United States declared: "The Sherman Act does not require competitive bidding."

It is further noted that as made clear in the Supreme Court decision:
1. Engineers and firms may individually refuse to bid for engineering services.
2. Clients are not required to seek bids for engineering services.
3. Federal, state, and local laws governing procedures to procure engineering services are not affected, and remain in full force and effect.
4. State societies and local chapters are free to actively and aggressively seek legislation for professional selection and negotiation procedures by public agencies.
5. State registration board rules of professional conduct, including rules prohibiting competitive bidding for engineering services, are not affected and remain in full force and effect. State registration boards with authority to adopt rules of professional conduct may adopt rules governing procedures to obtain engineering services.
6. As noted by the Supreme Court, "nothing in the judgment prevents NSPE and its members from attempting to influence governmental action . . ."

Note: In regard to the question of application of the Code to corporations vis-a-vis real persons, business form or type should not negate nor influence conformance of individuals to the Code. The Code deals with professional services, which services must be performed by real persons. Real persons in turn establish and implement policies within business structures. The Code is clearly written to apply to the Engineer, and it is incumbent on members of NSPE to endeavor to live up to its provisions. This applies to all pertinent sections of the Code.

National Society of Professional Engineers®

1420 King Street
Alexandria, Virginia 22314-2794
703/684-2800 • Fax:703/836-4875
www.nspe.org
Publication date as revised: July 2007 • Publication #1102

FIGURE 3-1 Continued

ETHICAL PRINCIPLES

As technical communicators, we observe the following ethical principles in our professional activities.

LEGALITY

We observe the laws and regulations governing our profession. We meet the terms of contracts we undertake. We ensure that all terms are consistent with laws and regulations locally and globally, as applicable, and with STC ethical·principles.

HONESTY

We seek to promote the public good in our activities. To the best of our ability, we provide truthful and accurate communications. We also dedicate ourselves to conciseness, clarity, coherence, and creativity, striving to meet the needs of those who use our products and services. We alert our clients and employers when we believe that material is ambiguous. Before using another person's work, we obtain permission. We attribute authorship of material and ideas only to those who make an original and substantive contribution. We do not perform work outside our job scope during hours compensated by clients or employers, except with their permission; nor do we use their facilities, equipment, or supplies without their approval. When we advertise our services, we do so truthfully.

CONFIDENTIALITY

We respect the confidentiality of our clients, employers, and professional organizations. We disclose business-sensitive information only with their consent or when legally required to do so. We obtain releases from clients and employers before including any business-sensitive materials in our portfolios or commercial demonstrations or before using such materials for another client or employer.

QUALITY

We endeavor to produce excellence in our communication products. We negotiate realistic agreements with clients and employers on schedules, budgets, and deliverables during project planning. Then we strive to fulfill our obligations in a timely, responsible manner.

FAIRNESS

We respect cultural variety and other aspects of diversity in our clients, employers, development teams, and audiences. We serve the business interests of our clients and employers as long as they are consistent with the public good. Whenever possible, we avoid conflicts of interest in fulfilling our professional responsibilities and activities. If we discern a conflict of interest, we disclose it to those concerned and obtain their approval before proceeding.

PROFESSIONALISM

We evaluate communication products and services constructively and tactfully, and seek definitive assessments of our own professional performance. We advance technical communication through our integrity and excellence in performing each task we undertake. Additionally, we assist other persons in our profession through mentoring, networking, and instruction. We also pursue professional self-improvement, especially through courses and conferences.

Adopted by the STC Board of Directors
September 1998

FIGURE 3-2 Society for Technical Communication's Ethical Principles for Technical Communicators

Source: http://www.stc.org/about-stc/the-profession-all-about-technical-communication/ethical-principles

Recognizing Unethical Communication

Throughout this book, we emphasize that communication and work cannot be separated. Writing clearly, accurately, and effectively constitutes a crucial part of your job. And once a document leaves your control—you disseminate it by e-mail, fax, or regular mail—it takes on a life of its own. It provides documentation of your work and your ability to perform work. The quality of your work becomes inseparable from your communication of your work. To safeguard your professional reputation, you must recognize the ways in which individuals on the job might violate standard practices of ethical communication. Chief among the possibilities include plagiarizing, deliberately using imprecise or ambiguous language, manipulating statistics, using misleading visuals, promoting prejudice, and distributing misinformation.

Plagiarism and theft of intellectual property. On the job, you may be responsible for the security of five kinds of intellectual property:

1. Copyrightable Material: A composition of original material fixed in a tangible medium, such as books, journals, software applications, computer programs, video or audio recordings, and illustrations. This includes materials available in digital files, e-mail messages, and website pages.
2. Trademark: A display of words or symbols communicated in text, illustrations, or sounds that identifies and distinguishes the goods and services of a manufacturer or supplier, such as the name or logo of a company.
3. Trade Secret: A design, formula, list, method, pattern, or process that offers a competitive advantage over parties who don't have the same information, such as a special recipe.
4. Invention: A new and unique design, device, method, or process that is subject to patent protection.
5. Tangible Research Property: Tangible items created during research related to copyrightable materials, trademarks, trade secrets, and inventions, such as databases, diagrams, drawings, notes, prototypes, samples, and associated equipment and supplies.

Copyrightable material is unique in that for certain purposes (e.g., criticism, news reporting, research, teaching) you have the right to borrow limited portions for presentation or publication without the explicit permission of the owner. If the borrowing is extensive, however, permission is necessary.

On the job, writers will often recycle the words and images from various documents of their company without identifying the original source. They will readily lift paragraphs from the corporate website, for example, to use in a business letter to a potential client, or they will borrow a table from the annual report to use again in a proposal to a potential funding agency. Such recycling of material—or *boilerplate*—is efficient, and it is entirely legal and ethical as long as the participating writers recognize and allow this sharing of effort. In such cases, the company has ownership of the words and images that are being recycled. (If you have doubts about the propriety of such recycling within your company, ask the writer directly for his or her permission.)

If it isn't your company's materials that are being used, however, you must acknowledge the sources of borrowed words, images, and ideas. In the majority of documents you write (such as letters, e-mail messages, blog postings, and memos), the acknowledgment may be a brief and simple introduction to the borrowed material: for example, "As Dr. Shirley Olson discovered, it's possible to vaccinate mosquitoes to prevent their developing and passing the disease on to human beings."

In formal reports, however, some official system of documentation would be necessary to identify the source of the information and to give full credit to Olson. Your organization might develop a special style for such source citations or adopt a standard style guide such as *The Chicago Manual of Style* of the University of Chicago Press, the *Publication Manual of the American Psychological Association*, or the *MLA Style Manual* of the Modern Language Association.

To use the words, images, or ideas of others without attribution is *plagiarism*. It constitutes a theft of intellectual property and is highly unethical and potentially illegal. Your intentions are immaterial; that is, it would be plagiarism if it were deliberate or if it were entirely inadvertent. You must, therefore, be especially cautious to avoid plagiarism: your organization could find itself the subject of a criminal case or a civil suit, and you could lose your job and your reputation.

Note also that material is automatically copyrighted as soon as it is created: it need not carry a copyright notice, and the copyright need not be registered with the US Copyright Office (copyright.gov), though both actions will help to deter infringement.

If you quote a source (including material from e-mail messages, blog postings, or video clips), put the borrowed material inside quotation marks (or display it in a separate indented paragraph) and specify the source. If you paraphrase or summarize, you don't need quotation marks but you still must specify the source. Make sure that even your acknowledged paraphrases and summaries differ from the wording and phrasing of the original passage so that you can't be accused of plagiarism.

Here, for example, is the original passage from a source:

Ethylene oxide has a boiling point of 51°F. It is processed as a liquid through the application of pressure. As the temperature of the ethylene oxide increases, the pressure in the feed line will correspondingly increase. At the time of the explosion, the ambient temperature was around 93°F. The feed line was not insulated or cooled.

(From United States Environmental Protection Agency, Office of Solid Waste and Emergency Response. 2000. EPA chemical accident investigation report, Accra Pac Group, Inc. North Plant, Elkhart, Indiana. EPA 550-R-00-001. Washington, DC.)

The following summary would constitute plagiarism even if the source were cited. Note how this passage changes words here and there but essentially duplicates the original passage:

The boiling point of ethylene oxide is 51°F. Pressure is applied to process it as a liquid. As the temperature of the ethylene oxide rises, so does the pressure in the feed line. The temperature in the vicinity was roughly 93°F at the time of the explosion, and the feed line was neither insulated nor cooled.

The next summary would be considered ethically appropriate. Note how the order of the sentences as well as the words have been substantially changed. Nevertheless, the source of the information must still be identified:

Ethylene oxide was pressurized as a liquid in a feed line that was neither insulated from external temperatures nor subjected to any kind of special cooling. Ethylene oxide boils at 51°, but the ambient temperature rose to approximately 93°F. Increased pressure from the boiling chemical inside the feed line caused the line to rupture, resulting in the explosion.

If the borrowing is a substantial portion of the original source (e.g., several paragraphs or a single image), you will have two ethical and legal duties:

1. Acknowledge the source.
2. Request permission from the owner of the intellectual property.

The extensive borrowing of copyrighted material will ordinarily be permitted but often with restrictions and for a cost. You may contact the copyright owner directly or make your request through a service such as the Copyright Clearance Center (copyright.com). In this case, the borrowed material carries a note indicating both the source and the receipt of permission:

From *Understanding Intercultural Communication*, by Stella Ting-Toomey and Leeva C. Chung. Copyright 2007 by Oxford University Press. Reprinted by permission.

If you change the original material (e.g., summarizing passages, revising illustrations, pulling still slides from digital video), you would specify the adaptation of the borrowing:

From Lisk, John. (2011, March 10). Ways to Relieve Stress [Video]. CNN.com. http://www.cnn.com/video/#/video/health/2011/03/10/hm.stress.buster.cnn. Adapted by permission.

You do not need permission for material in the public domain (i.e., intellectual property for which copyright protection has already expired or material created by agencies of the government of the United States), but such sources must always be acknowledged.

Deliberately imprecise or ambiguous language.   Ordinarily, unclear and ambiguous language is a result of the writer's negligence, but it could also be a sign of a deliberate effort to mislead or manipulate the reader by hiding or disguising information.

Writers can imply that things are better or worse than they really are through the choice of words. For example, a writer answering inquiries about the company's electric vehicle could reply, "This automobile was designed to travel 1,000 miles on a single electric charge." It could be that the vehicle was designed with this objective in mind, but if it doesn't genuinely operate with this level of efficiency, the writer has made a false implication without telling a straight lie. And if this promised 1,000 miles doesn't include the weight of the average driver, the deception is all the greater.

Negative assertions can often appear deceptive. For example, consider a marketing claim such as "No antivirus software does more or costs less than Infolator." With the words "does more" and "costs less" in the claim, you might be inclined to think that Infolator is truly better and lower in price. Notice, however, that Infolator doesn't really have to be better or lower in price to make its claim: it could have the same functions and be the same price as other antivirus software. The real claim here isn't that it's a better and cheaper product but that it's just as good and just as cheap.

Manipulation of numerical information. A leading way for individuals to be deceived is through the manipulation of statistics. For example, imagine the writer of a recommendation report who would like to give the impression that a controversial change in security practices really has wide support. She surveys only the managers in the company and finds that 51 percent of the 19 percent who returned the survey favor the change. In the report she writes, "A majority of those who completed the survey favored the change." By not revealing that this "majority" is only 51 percent and that it represents only about 10 percent of the company's managers, she magnifies the thin support for the controversial change. She has not exactly lied, but she has likely deceived most readers of the report.

Use of misleading illustrations. Like words, illustrations have the capacity to misrepresent and mislead. For example, a company of one hundred employees has only two who are African Americans. In the recruiting materials it carries to college campuses, the company shows a dozen of its employees doing different jobs, including both of its African Americans. This portrayal constitutes unethical communication because it implies that African Americans constitute almost 17 percent of the employees—a gross distortion of the real situation. Prospective job candidates would be substantially deceived about the diversity of colleagues and the working environment this company offers.

Or consider a line graph, such as Figure 3–3, that might be given to prospective clients of ABC Manufacturing. Here, we see a picture of volatile change year to year.

If the increments on the line graph are changed from every year to every 4 years, however, ABC's disturbing record of erratic production almost vanishes entirely (see Figure 3–4). Viewers of this line graph would be given only the impression of relative stability. Prospective clients would be thoroughly deceived and would be quite surprised if ABC failed to meet its production goals in subsequent years.

FIGURE 3-3 ABC Production Reported in 1-Year Intervals

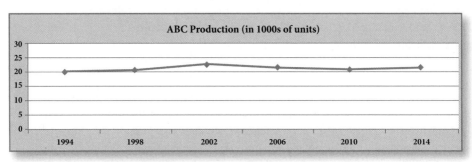

FIGURE 3-4 ABC Production Reported in 4-Year Intervals

Promotion of prejudice. Writers also communicate unethically by voicing prejudice through their choice of words and illustrations. If you use titles for men but not for women (e.g., Mr. William Jones, advertising manager, and Harriet Smith, operations manager), you make women seem less credible and authoritative. If photographs in the company's annual report always show women sitting at desks staring at computer screens while men sit at big conference tables listening to reports and making decisions, the pictures send the message that men occupy (or deserve) important positions as executives while women function best in clerical positions.

As a communicator, make sure that you don't reinforce or inspire prejudice and bigotry. Your ethical obligation requires you to offer only valid and reliable findings, fair and unbiased analyses, and logically justified conclusions.

Uncritical use of information. On the Internet especially, you can find all kinds of erroneous information, manipulated images, distorted depictions, and dubious claims. If you use the information you find without judging its accuracy, assessing the credibility of your sources, or verifying it across multiple sources, you could easily be distributing dangerous and damaging misinformation. Your negligence here makes you look both unethical and inept. To build a reputation for integrity, you must accept responsibility for the accuracy of the information you communicate.

Case 3-1: Accuracy in Reporting

You currently work as an environmental engineer for Ventus Power, which operates wind energy facilities in five states: Arizona, Nevada, New Mexico, Texas, and Wyoming. Included in your responsibilities is monitoring of wildlife impact at the Sweetwater facility and the periodic preparation of reports to be submitted to the US Department of Energy and the US Fish and Wildlife Service (USFWS).

You have just prepared a report indicating that fatal bird collisions with the wind turbines at Sweetwater are at a new high for the third consecutive quarter. You know this evidence of a persistent and growing hazard to migratory birds will trigger a visit from regulators and a potential interruption of operations. And mitigation efforts will likely cost the company at least $500,000.

Your supervisor, Garrett Driscoll, says the higher level could be regarded as a "temporary blip" and asks you to "adjust" the data so that the facility appears to be running safely. He explains, "We can't afford the expensive repair. It might also cost us jobs. And the publicity will be terrible—antagonizing the environmentalists and discouraging investors."

You decide to write a memo to Driscoll telling him that, as a professional engineer, the National Society of Professional Engineers (NSPE) code of ethics does not allow you to misrepresent information. You must report this information as you know it. You recognize that Driscoll will disagree with your decision and will likely share the letter with his bosses. In your message to Driscoll, you explain why you must report this information according to your profession's standards of practice. Before you compose your letter, you again review the NSPE ethics guidelines (Figure 3-1).

Question: Why should you write to Driscoll instead of just telling him?

CASE DOCUMENT 3-1

DATE:	October 24, 2012
TO:	Garrett Driscoll
SUBJECT:	**High Levels of Fatal Bird Collisions at Sweetwater**

Garrett:

I understand your concern regarding the cost of solving our facility's problem with fatal bird collisions. The $500,000 reparation expenditure will indeed affect our company financially. However, the high mortality rate cannot be viewed as a mere technicality. The facility's unintentional bird kills are again at a new high, and the number will persist or grow if we do not take measures to mitigate. Excessive fatal collisions put us in legal jeopardy as well as do callous injury to bird populations. Adjusting the data so that the facility appears to be operating without incident would be an explicit violation of an engineer's ethical standards.

As an environmental engineer, I adhere to the Code of Ethics instituted by the National Society of Professional Engineers. I will report this data to the USFWS and the Department of Energy as is. As soon as we have the okay from the corporate office for expenditures, I will assess the situation and determine a suitable fix.

I believe that accurately reporting the statistics will positively contribute to our reputation and the integrity of this company.

Questions: Do you believe this response will be satisfactory? How could you improve it?

Case 3-2: Teaching Ethics by Example

You complete your report and submit it to the vice president of operations for signatures and submission to the USFWS. You also keep a copy on your computer. You are satisfied that you have followed the actions required by the NSPE code of ethics. Driscoll is annoyed but quiet, and you assume that he has resigned himself to your decision. Soon thereafter you leave Sweetwater for a supervisory position at the Albuquerque facility.

A year later, you read in the newspaper that the US Department of Justice is investigating Ventus Power for violating the Migratory Bird Treaty Act. It appears that scores of fatal collisions of migratory birds, including eagles, at the Sweetwater Wind Facility were never reported to the Fish and Wildlife Service and no mitigation efforts were instituted.

Almost immediately you find yourself interrogated by investigators from the Department of Justice and the USFWS as well as corporate officials at Ventus Power. Luckily, you still have a copy of your report on your computer as well as a copy of your message to Driscoll.

After both legal entities clear you of any responsibility for the failure to report, you decide to write an internal memo to the engineers who answer to you in your new job about the importance of effective communication and of following engineering ethics in all situations.

CASE DOCUMENT 3-2

Date: Monday, April 12, 2014
To: Ventus Engineering Corps
Subject: **Proper Reporting Procedures for Ethics Violations and Other Issues**

Purpose

A recent incident with our corporation has prompted me to write this open memo to the engineering corps of our company to explain proper reporting procedures. In case you were unaware, Ventus Power has been cited by the US Department of Justice for violations of the Migratory Bird Treaty Act in failing to report fatal collisions of migratory birds (including eagles) with the turbines at the Sweetwater Wind Power Facility. An investigation by the company identified the persons at fault, and they have been appropriately reprimanded. As some of you may know, I served at one point as the engineer in charge of wildlife impact monitoring at the Sweetwater facility and, as such, found myself one of the first to be questioned about bird collisions. I had to report unacceptable mortality levels and advised mitigation efforts while in my position. Because I followed proper reporting procedure, the company will not bring any litigation or punishments against me. The following guidelines make sure that necessary information makes its way to the right people and that we deal with problems properly before they can harm the people or environments we operate around. These guidelines can also safeguard you.

Introduction

At some point in your career you may have to report an issue you are experiencing, especially issues of ethical violations. Remember that we are at all times bound to our code of ethics. These guidelines keep us, our customers, and our company safe. Operations or procedures done in an unethical manner may lead to people being hurt or killed, can bring litigation against our company, and will lead to the termination of any employees found in violation. For this reason, it is extremely important that everyone follow proper procedures for communications and reporting so that problems can be dealt with as quickly as possible and so that you can be sure you will not be considered a responsible party.

Everyday Communication

These communications include project reports and other regular incremental reports. They need to be written as simply as possible with an emphasis on any items or issues that need immediate attention from your supervisor. Supervisors use their direct subordinates as their eyes and ears on projects, looking for any problems or issues that will require their action. Remember, if something needs to be fixed, it should be mentioned. These simple guidelines should cover nearly 99% of reports you will find yourself writing, but there are special cases that require a more involved approach.

Special Cases

These special cases occur whenever a serious problem exists. This should be obvious, such as anything that could lead to lawsuits against our company, any time product quality is not at an acceptable level, and especially any time you feel you are being asked to violate the code of ethics. These situations require your full attention. For these reports, tell events exactly as they appear to you, including background information that clearly defines the problem you are facing, any efforts you have made to correct the problem, and any assistance you need to solve it. Send copies of your letter to whoever needs to know this information. Normally this will be your supervisor and anyone else in your department who works with the affected system. If your supervisor asks you to in anyway cover up or ignore this problem, it is your responsibility to report the problem up the chain of command to your supervisor's supervisor since that is what they were supposed to do. You can decide how to handle your supervisor's response, although if you feel that he or she has truly violated the code of ethics, it is your ethical responsibility to report it.

Conclusion

In conclusion, I appreciate the hard work every one of you puts into your job and I hope that this level of dedication continues. With proper reporting procedures we can make sure that all problems are dealt with properly and that everyone remains safe. And as a warning to those who think that sweeping an issue under the carpet or agreeing to a supervisor's request to do so is acceptable, it is possible that nothing will happen to you in the short run; but problems always find their way to the surface, and once they do it's too late to protect yourself. So be smart and keep our customers, our company, and your career safe by always reporting issues. Thank you for your time and continued efforts to make this company the best that it can be.

Questions: How effective is this memo at explaining the importance of following a code of conduct, of acting ethically, and of communicating clearly? What changes would make it more persuasive or inspiring for the intended readers?

Managing Unethical Situations

Always keep in mind that unethical behavior has potentially serious consequences. It could cause persistent injury to your reputation, loss of your job, forfeit of the right to practice your profession, and inability to find a new job. You could also be subject to civil suits for damages caused by your unethical behavior. If your actions were also illegal, you could be fined or jailed.

If you find yourself asked by a boss, colleague, or client to do something that you don't consider right, therefore, don't hesitate to ask polite questions. Don't accuse or immediately declare the request unethical. Explain your dilemma. Ask the individual to clarify. He or she might be unaware of the difficult moral position you're being put in by such a request or might immediately modify the request.

Note also that people of different cultures might have different ethical perspectives. For example, the dominant ethical perspective in the United States assumes that to be fair you must consider all people as equal and give privilege to none, but in China this idea could be considered unfair because family, friends, and neighbors are ordinarily thought to deserve special consideration. In this situation, you would be wise to acknowledge each side's intention to be ethical as you try to find a solution that is satisfactory to both. Here again it might be a case of competing right answers instead of a question of right or wrong.

If explanations and negotiations don't satisfy you and if time allows, visit with a supervisor or mentor for guidance. Review the code of conduct of your profession or company for passages that might support or challenge your position.

If all of your investigation and deliberation fail to quiet your moral doubts, explain (in writing, if possible) that you don't feel comfortable doing X but you could do Y. That is, identify both the thing you can't do (the unethical action solicited) and the thing that you could do (the ethical alternative).

If asked to justify your decision, cite the appropriate passages of your profession's or company's code of conduct. That is, make it clear that it isn't only you who is rejecting the request but that it is your profession or your company that proscribes this behavior. Again, don't be impolite and don't accuse.

If you see something occurring on the job that you don't think is right, discuss the situation with your supervisor or mentor (in writing, if possible). If your investigation fails to satisfy you that the activity is justified, consider your several ethical obligations, including to your profession and to the public. Always make the decision that you could live with if your decision were made public—the decision that would make your parents and teachers proud of you.

A final piece of advice: Keep in mind that you bring two important credentials to a job, a knowledge of your field and a reputation for integrity. If you don't have a reputation for integrity, your knowledge of your field can't be trusted.

ETHICS DECISION CHECKLIST ✔

☐ What is the nature of the ethical dilemma?

☐ What are the specific aspects of this dilemma that make you uncomfortable?

☐ What are your competing obligations in this dilemma?

☐ What advice does a trusted supervisor or mentor offer?

☐ Does your company's code of conduct address this issue?

☐ Does your professional association's code of conduct address this issue?

☐ What are you unwilling to do? What are you willing to do?

☐ How will you explain or justify your decision?

EXERCISES

1. Visit the website of two major employers in your field. Locate the code of conduct for each. What are the similarities and differences in the two codes of conduct? What are the values that each employer espouses? What does each employer expect from its employees? What does each employer expect from its executives? What additional ethics resources does each employer make available to employees? What can you tell about each employer from the code of conduct? Based on your review of the two codes of conduct, which employer would you prefer to work for? Why? Summarize your findings in a memo, and share this memo with your colleagues.

2. The Ethisphere Institute (ethisphere.com) every year publishes a list of the world's most ethical companies. Check this year's list, and choose one of the companies to investigate. What makes it one of the world's most ethical companies? Do you agree or disagree with this assessment? Report your findings in a slide presentation addressed to majors in your field.

3. You have been assigned as the project manager for a new bridge that your company hopes to build in a foreign city. You visit that city with your company's vice president of international operations, scouting the site for the bridge and meeting with city officials in preparation for making a bid on the project. In a private meeting with the minister of public projects, the vice president pulls out a platinum wristwatch in a box from Tiffany's and gives it to the minister. "This is for you from the president of my company!" announces the vice president. The minister seems quite pleased with the expensive gift and assures the vice president that the bid from your company would be "positively winning" or something to that effect. You can't swear to it because the minister's English is frail, but it does seem that your company was being promised the bridge project.

You know that your company has a policy barring gifts to foreign officials, but you don't know that it necessarily applies in this case. In the taxi on the way to the building site, you ask the vice president about the gift. He advises you that policies are "nice words" but that "this is how we have to do business here" and mentions that the minister and the president of the company are personal friends. You smile and change the subject of conversation, but you are still worried that the gift giving was unethical and might also be illegal.

Managing this project would give a real boost to your career. It could lead to more international opportunities and higher visibility in your profession as well as executive positions at your company and possibly job offers from prestigious competitors. It could also terminate your career if it were determined that you were involved in bribing a foreign official.

What should be your next steps in addressing this dilemma? What sources of information could you consult—about applicable laws, about ethics, about intercultural communication? Who could you talk to? What documents or messages, if any, should you write?

4

Achieving a Readable Style

Style refers to the overall way you express your ideas in a document, from the paragraph level to sentence structure and the words and phrases you choose.

THE PARAGRAPH

A reminder: We define a *paragraph* as a group of sentences that begins with a central, or topic, statement of their content. The supporting sentences build on the idea stated in the topic sentence and should occur in a logical order. In short,

- Begin each paragraph with a topic sentence that summarizes content to come.
- Include only information relevant to the topic sentence.
- Place sentences in a logical order.
- Avoid long paragraphs.

Examples for study. Effective report segments result from effective paragraphs. Examine the following introduction to a technical report (Figure 4–1). Because each paragraph begins with a topic sentence (easy to locate and retrieve),

Introduction

In 1996, the U.S. Department of Energy (DOE), Office of Fossil Energy, asked Argonne National Laboratory (ANL) to conduct a preliminary technical and legal evaluation of disposing of nonhazardous oil field wastes (NOW) into salt caverns. The conclusions of that study, based on preliminary research, were that disposal of oil field wastes into salt caverns is feasible and legal. If caverns are sited and designed well, operated carefully, closed properly, and monitored routinely, they can be a suitable means for disposing of oil field waste (Veil et al. 1996). Considering these findings and the increased U.S. interest in using salt caverns for nonhazardous oil field waste disposal, the Office of Fossil Energy asked ANL to conduct a preliminary identification and investigation of the risks associated with such disposal.

As Chapter 8 discusses, report introductions must have purpose statements to tell readers what the report will do. Having the purpose statement at the beginning of a paragraph helps readers find the purpose statement.

Report Purpose

The purpose of this report is to evaluate the possibility that adverse human health effects (carcinogenic and noncarcinogenic) could result from exposure to contaminants released from the caverns in domal salt formations used for nonhazardous oil field waste disposal. The evaluation assumes normal operations but considers the possibility of leaks in cavern seals and cavern walls during the post-closure phase of operation. It does not consider the risks associated with emissions from surface equipment operating at the site, nor does it consider the risks associated with surface oil leaks or other equipment-related spills or accidents.

The study focuses on possible long-term risks to human health. It does not address potential ecological effects, although such effects could result. Also, risks associated with naturally occurring radioactive materials (NORM) are not addressed. This preliminary assessment estimates risks associated with disposal in a single generic cavern only. No attempt has been made to address the possibility or likelihood that several caverns may be located in relatively close proximity and that more than one cavern could be a source of contamination to a given receptor. Also, no attempt has been made to evaluate the possible impacts of synergistic effects of multiple contaminants on a single receptor.

Because the history of salt cavern use for solid waste disposal is very limited, no readily available data could be accessed for this study. As a result, data from similar operations and professional judgment were used to develop the possible release mechanisms assumed in this hypothetical, generic analysis. The validity of the results would be enhanced if real data could be used. As data are generated on the use and post-closure operations of salt caverns used for solid waste disposal, they should be incorporated to update this study.

Report Development Process

Moves from broad to specific in supporting information

In this assessment, several steps were followed to identify possible human health risks. At the broadest level, these steps include identifying a reasonable set of contaminants of possible concern, identifying how humans could be exposed to these contaminants, assessing the toxicities of these contaminants, estimating their intakes, and characterizing their associated human health risks. The risk assessment methodology and

FIGURE 4–1 Effective Introduction to a Technical Report

techniques used in this report are based in large part on two documents. The first document is a training manual that was developed for a risk assessment workshop sponsored by DOE (DOE 1996). The second is the Risk Assessment Guidance for Superfund (U.S. Environmental Protection Agency [EPA] 1989).

Report Plan

The remainder of this report consists of nine sections. Section 2 provides background on the development, use, and closure of salt caverns that may be used for disposal of nonhazardous oil field wastes and possible cavern release scenarios. Section 3 identifies contaminants of potential concern that could cause harm to human health. Sections 4, 5, and 6 provide information for assessing potential exposure pathways that the contaminants of concern could take to reach human populations. Specifically, Section 4 describes fate and transport mechanisms of the contaminants of concern; Section 5 describes specific hydrogeologic conditions of locations where salt caverns are most likely to be used for oil field disposal (Gulf Coast, Texas, and New Mexico); and Section 6 describes potential release modes that could cause contaminants to leak from the cavern and be transported to areas where human populations may be exposed. Section 6 also estimates possible concentrations of the contaminants to which humans could be exposed under various release scenarios. Section 7 describes the toxicity of those contaminants that could come in contact with humans, given the fate and transport mechanisms identified in Section 5 combined with the potential exposure pathways described in Section 6. Section 8 estimates the potential intakes of those contaminants by humans and characterizes the risks to which those humans may be subjected on the basis of the intake of the contaminants (the potential for harm), their toxicities, and the release assumptions. Section 9 addresses the sensitivity of the estimated risks to operating procedures and potential regulatory structures, and Section 10 summarizes the results of the analyses.

> Topic sentence helps readers anticipate presentation of the nine sections. Naming each section allows the writer to show relationships among the sections.

FIGURE 4-1 *continued*

you can read just the topic sentences and have a clear sense of the content of the introduction. While the final paragraph is the longest of this introduction, its structure allows you to follow easily the development of the paragraph.

The next example paragraph (Figure 4–2) uses a list to draw the reader's eyes to the central idea presented in the paragraph, the criteria for approval. The topic sentence introduces the paragraph and the list.

The listing strategy emphasizes the six items that take front and center in the Ashton Lake project. In this situation, listing highlights the rhetorical immediacy of the problem. A traditional paragraph with linear text would obscure the concerns (Figure 4–3).

Avoid excessive use of any writing technique—too many short paragraphs, too much enumeration (first, second, third, etc.), as well as too many lists. Concise paragraphs that begin with topic sentences and well-structured sentences of

---

The City of Ashtonville Economic Development Council presented six main concerns to the City Parks and Recreation Department about the proposed Ashton Lake development proposal:

1. Financial feasibility of the project
2. Ability to raise the necessary funds
3. Project maintenance, including long-term
4. Protection/Enhancement of the shoreline
5. Effect of project on changing lake water levels
6. Erosion of project's materials caused by water/sand

The EDC expressed a major concern that the project may not be as family-friendly as needed because it would require extensive funding to develop the recommended recreation facilities. In addition, the concerns stated suggest that the project, if approved, will have to be funded strictly by private donations rather than city or county parks money.

FIGURE 4–2 Effective Use of a List within a Paragraph

moderate length create clear, readable documents. Why do you find Figure 4–5 easier to read than Figure 4–4?

Note that each paragraph of Figure 4–5 begins with a topic sentence and that a clearly stated subheading introduces each segment of the introduction. (Chapter 5 will explain how to design documents, but this revision also exemplifies excellent document design.)

Basic Principles of Effective Style

Effective writers adjust their style to the needs of their readers: (1) readers' knowledge of the subject, (2) readers' expectations about style based on the specific kind of writing, (3) readers' probable reading level based on the context in which they will read the document, and (4) the writer's relationship to readers—the professional roles of both writer and readers.

Determine your readers' knowledge of the subject. The reader's familiarity with the subject will determine how many specialized terms you can use. If the

---

The City of Ashtonville Economic Development Council presented six main concerns to the City Parks and Recreation Department about the proposed Ashton Lake development proposal: financial feasibility of the project; ability to raise the necessary funds; project maintenance, including long-term; protection/enhancement of the shoreline; effect of project on changing lake water levels; and erosion of project's materials caused by water/sand. The EDC expressed a major concern that the project may not be as family-friendly as needed because it would require extensive funding to develop the recommended recreation facilities. In addition, the concerns stated suggest that the project, if approved, will have to be funded strictly by private donations rather than city or county parks money.

FIGURE 4–3 Ineffective Run-in List

How an MRI Works

Brain perfusion refers to the blood circulation around the brain while exchanging oxygen and nutrients between the blood and the brain tissue. The effectiveness of brain perfusion depends on "blood pressure, blood velocity [...], and diffusion rates of oxygen and nutrients" [1]. An MRI scanner can monitor the perfusion of the brain to determine if a neurological disease is present.

An MRI scan uses a magnetic field to generate enhanced images that show certain parts of the body. The hydrogen atom provides a crucial component in the generation of an MRI scan. The hydrogen atom contains a single proton that interacts with the magnetic field generated by the MRI. The hydrogen atoms in a specific area of the human body will behave differently due to the presence of a magnetic field. The MRI scanner will detect the behavior of these hydrogen atoms to generate an enhanced image of that area.

In many MRI procedures, a contrast agent is used to further enhance the image of the body by increasing the brightness of the tissue being examined. The injection of contrast agents into the body will help the MRI scan retrieve information such as "blood flow [...] and related physiological parameters" [2]. These contrast agents work by altering the local magnetic field in the tissue being examined to generate an illuminated image of that tissue on the MRI scan.

Unfortunately, contrast agents have limited functions when physicians attempt to study the human brain. The brain contains a "membrane structure" known as the blood brain barrier. which protects the brain from foreign chemicals in the blood, while still allowing essential metabolic function [3]. The blood brain barrier filters out any unwanted chemicals that may travel through the brain. Since the blood brain barrier is very effective in its function, the chemicals in the contrast agents will be filtered out by this *membrane structure*. If a neurological disease is present in the brain, it can weaken the blood brain barrier and allow these unwanted chemicals to pass through [3]. The contrast agent would be useful for studying the disease in this situation.

Many methods have been created to penetrate the blood brain barrier. My proposed method will strengthen the interaction between hydrogen atoms with the magnetic field via electrical signals. During an MRI scan, an electrical signal will be emitted around the neck area. This signal will mark the hydrogen atoms that are in the vicinity of the emitted signal. Because no external chemicals are introduced, the blood brain barrier will not detect any foreign chemicals and will let the marked hydrogen atoms into the brain. The hydrogen atoms will travel through the bloodstream around the brain acting as beacons. They will then interact with the magnetic field of the MRI, allowing the MRI to enhance the image of the brain, much like contrast agents [4].

FIGURE 4-4 Original Introduction to a Student Research Proposal

reader has a thorough knowledge of the subject, you can use acronyms, specialized nomenclature, and jargon that readers in a specific discipline regularly read and use. If the reader has little knowledge of the subject, limit the use of specialized vocabulary or perhaps define the terms. Another possibility is to substitute phrases or words that clearly express your meaning.

In a work context, the better you know the people who will read your writing, the better you can design your reports, memos, and e-mails. You can also ask people who know your readers to aid you in finding out preferences in organization, style, and length. When you begin a job assignment, ask your supervisor for his or her preferences in these three areas.

Note: Conciseness does not equal brevity. When you write concisely, you include all that you need to say without extra words and phrases that contribute little to the main idea. Brevity means that you aim only for economy, rather than completeness of thought.

Introduction

Importance of Brain Perfusion

Brain perfusion refers to the blood circulation around the brain while the circulatory system exchanges oxygen and nutrients between the blood and the brain tissue. Brain perfusion depends on "blood pressure, blood velocity [...], and diffusion (dispersal) rates of oxygen and nutrients" [1]. An MRI scanner can monitor the perfusion of the brain to determine if a neurological disease is present.

How an MRI Works

A magnetic resonance imaging (MRI) scan uses a magnetic field (an attractive force) to show enhanced images of certain parts of the body. A dynamic MRI scan displays a fuller and more detailed result, unlike a conventional MRI. The hydrogen atoms in the human body are very important in an MRI procedure. One hydrogen atom contains a single proton that behaves differently when the MRI emits a magnetic field around that area. The MRI scanner will detect the behavior of these hydrogen atoms to generate an enhanced image of that area.

In many MRI procedures, a contrast agent (a compound fluid) is used to further enhance the image of the body by increasing the brightness of the tissue being examined. The injection of contrast agents into the body will help the MRI scan retrieve information such as blood flow and other related physical characteristics [2].

Limits of the MRI

Unfortunately, contrast agents have limited functions when physicians attempt to study the human brain. The brain contains a "membrane structure" known as the blood brain barrier, which protects the brain from foreign chemicals in the blood, while still allowing essential metabolic function [3]. The blood brain barrier filters out any unwanted chemicals that may travel through the brain. Since the blood brain barrier is very effective in its function, the chemicals in the contrast agents will be filtered out by this *membrane structure*. If a neurological disease is present in the brain, it can weaken the blood brain barrier and allow these unwanted chemicals to pass through [3]. The contrast agent would be useful for studying the disease in this situation.

Value of My Method in Improving the MRI

Many methods have been created to penetrate the blood brain barrier. My proposed method will strengthen the interaction between hydrogen atoms with the magnetic field via electrical signals. During a dynamic MRI scan, an electrical signal will be emitted around the neck area. This signal will mark the hydrogen atoms that are in the vicinity of the emitted signal. Because no external chemicals are introduced, the blood brain barrier will not detect any foreign chemicals and will let the marked hydrogen atoms into the brain. The hydrogen atoms will travel through the bloodstream around the brain acting as beacons. They will then interact with the magnetic field of the MRI, allowing the MRI to enhance the image of the brain, much like contrast agents [4].

Subject and Purpose

This research will focus on developing a new method of performing a dynamic MRI scan on the human brain. This new method involves emitting an electrical signal into the body to interact with the hydrogen atoms and performing a dynamic MRI scan. To determine the effectiveness of this new method, I will analyze the MRI scan to determine the rate of perfusion in the brain.

FIGURE 4–5 Revised Introduction to the Student Research Proposal

Determine whether a particular style will be expected. Use the company style sheet and templates. But remember that you still have to adapt what you say to your intended readers.

Adjust the style to the readers, the purpose, and the context. Most business and technical writing should strive for as much conciseness as possible because of the large quantity of information that most readers confront. E-mail messages should have concise paragraphs and concise sentences. Even in complex, highly technical reports, readers value conciseness: the longer the report, the less likely that anyone will read all of it.

KEYS TO BUILDING EFFECTIVE SENTENCES

English sentences work more effectively—they become easier to read, understand, and remember—if they follow basic rules of sentence structure.

Watch sentence length. Documents composed of consistently long sentences can become difficult to read. Sentence length should vary, but consider revising sentences over 15 to 20 words. Even legal documents can benefit from shorter sentences and have improved as a result of plain English laws that now govern insurance policies and many other legal documents in various states. Many government entities want their public documents written in concise, easily understood sentences:

Before:
This Appendix contains a brief discussion of certain economic and demographic characteristics of the Area in which the County is located and does not constitute a part of this Official Statement. Information in this Appendix has been obtained from the sources noted, which are believed to be reliable, although no investigation has been made to verify the accuracy of such information.

Characteristics of Bad and Good Writing

Bad Writing	Good Writing
few verbs/number of words per sentence	many verbs/number of words per sentence
excessive *is/are* verb forms	concrete verbs
abstract nouns	concrete nouns
many prepositional phrases	few prepositional phrases
few clauses	linked clauses
passive voice	active voice
separation of key words: subject-verb, actor-action	clear actor-action relationship
long, rambling sentences	specific, precise sentences
main idea of sentences is difficult to process	meaning of sentences is easy to find and follow
sentences must be read several times	meaning is clear after one reading

After:

This Appendix contains a brief discussion of certain economic and demographic characteristics of the Area in which the County is located. The Appendix does not constitute part of this Official Statement. Information in the Appendix has been obtained from the sources noted. They are believed to be reliable. However, the accuracy of the information has not been verified.

Keep subjects and verbs close together. A recipe for sentence clarity: keep the subject of the sentence and the verb close together, and emphasize verbs. The more verbs in a sentence, the sharper and more direct the sentence. We call this characteristic *verb/word ratio*. For example,

      s   v   s   v

John <u>loves</u> Mary because she <u>inherited</u> money. (verb/word ratio = 2/7)

versus

                                      s   v

Mary's inheritance of money <u>was</u> one of the reasons for John's interest in Mary.

(verb/word ratio = 1/12)

In this simple example, you can see the point: the more verbs, the sharper the sentence.

Let's take this method a step further: lengthy sentences become less distracting to the reader if the writer structures them to enhance clarity and readability. To achieve clarity, build sentences with clauses and as many verbs and verbals (to + verb; verb ending in "ing") as possible. For example, the sentence

When they plan investment portfolios, financial planners recommend a variety of investments because they resist rapid economic changes. (18 words)

develops about three clauses:

When they <u>plan</u> investment portfolios
financial planners <u>recommend</u> a variety of investments
because they <u>resist</u> economic change.

Note that the sentence follows the three guidelines:

Interlocking clauses (three in this sentence)
Specific action verbs: *plan, recommend,* and *resist*
Subject next to the verb in each clause

they plan
planners recommend
they resist

The sentence has a verb/word ratio of 3/17.

Assume that the writer did not follow the guidelines and avoided verbs:

In plans for investment portfolios, a variety of investments is recommended by financial planners because of their resistance to economic changes.

The verb/word ratio is 1/21. The sentence lacks directness and conciseness. Compare the two versions. Can you see the difference? In English sentences, the more verbs and verbals a writer uses, the easier the time the reader has in understanding the sentence(s). In addition, make the actor the subject and follow it with a concrete verb that identifies what the actor does.

For most writing, use specific concrete subjects and verbs. Write to express, not to impress.

Instead of:
There is now no effective existing mechanism for introducing into the beginning initiation and development stages requirements on how to guide employees on how to minimize errors in product development efforts.

(verb/word ratio = 3/31; one *is* verb and two verbals)

Use:
The company has no way to guide employees on how to minimize product development errors during the early development stages.

(verb/word ratio = 3/20) Note that the sentence begins with the actor in the subject position and two verbals.

Note: When a sentence lacks a clear subject/agent doing the action (verb), writers can often drift into the phrases "there is," "there are," "there was," and "there were," which have no meaning and deter conciseness and directness.

Instead of:
Our lack of pertinent data prevented determination of committee action effectiveness in funding target areas that needed assistance the most.

(Note that the sentence incorporates two clauses and two verbs. Verb/word ratio = 2/21.)

Or, even worse:
There was a lack of pertinent data determination of committee action effectiveness in funding target areas that needed assistance the most.

Use, assuming we know who did what action:
Because we lacked pertinent data, our committee could not determine whether we had targeted funds to areas that needed assistance the most.

(Note the revision of the sentence has four interlocking clauses and four verbs. Verb/word ratio = 4/22.) In addition, placing clear subjects close to their verbs makes clear who does what action.

Or use:
We <u>didn't have</u> enough data: we <u>could not decide</u> if the committee <u>had sent</u> funds to areas that <u>needed</u> them most.

When we break the sentence into two sentences, we still have four verbs. In addition, the short sentence followed by the longer, explanatory sentence also clarifies the meaning.

Note that in this revision we place the actor as the subject of the sentence/clause and then follow the subject with the verb, a concrete, descriptive verb that explains what the actor/subject will do.

Remember how direct/indirect words affect tone:

(A) We encourage you to anticipate the amount of correspondence you accumulate and suggest you endeavor to answer it promptly.

(B) Please expect large amounts of e-mail and try to answer it quickly.

Note that you can read (B) more easily and quickly than (A). The tone of (A) sounds pompous.

Write squeaky-clean prose. The following excerpt from *DNA: The Secret of Life* addresses readers interested in science and possessing a basic understanding of genetics. Note the structure of each sentence, the use of topic sentences, and the development of each paragraph:

> The great size of DNA molecules posed a big problem in the early days of molecular biology. To come to grips with a particular gene—a particular stretch of DNA—we would have to devise some way of isolating it from all the rest of the DNA that sprawled around it in either direction. But it was not only a matter of isolating the gene; we also needed some way of "amplifying" it: obtaining a large enough sample of it to work with. In essence we needed a molecular editing system: a pair of molecular scissors that could cut the DNA text into manageable sections; a kind of molecular glue pot that would allow us to manipulate those pieces; and finally a molecular duplicating machine to amplify the pieces that we had cut out and isolated. We wanted to do the equivalent of what a word processor can now achieve: to cut, paste, and copy DNA.
>
> Developing the basic tools to perform these procedures seemed a tall order even after we cracked the genetic code. A number of discoveries made in the late sixties and early seventies, however, serendipitously came together in 1973 to give us so-called "recombinant DNA" technology—the capacity to edit DNA. This was no ordinary advance in lab techniques. Scientists were suddenly able to tailor DNA molecules, creating ones that had never before been seen in nature. We could "play God" with the molecular underpinning of all of life. This was an unsettling idea to many people. Jeremy Rifkin, an alarmist for whom every new genetic technology has about it the whiff of Dr. Frankenstein's monster, had it right when he remarked that recombinant DNA "rivaled the importance of the discovery of fire itself."

Source: Watson, James. *DNA: The Secret of Life*, pp. 87–88. Knopf, 2003. Used by permission.

This excerpt uses a variety of sentences of moderate length, close subject–verb patterns, familiar words, and a description of recombinant DNA in words easily understood by the nonscientific reader: the passage concisely and picturesquely expresses the meaning of recombinant DNA.

Avoid pompous language; write to express, not to impress. The concept of simplicity relates to the concept of naturalness. Writers often believe they must sound learned, aloof, and sophisticated to impress readers. The idea that direct writing lacks sophistication frequently derives from writing done in secondary school. Teachers encourage high school students to expand their vocabularies. Academic writing in college reinforces the importance of using jargon-laden language to convince the professor that the student knows the subject and the nomenclature of the discipline. Instructors may reward students for writing ponderous verbiage in research papers. On the job, however, verbose writing may be ignored or misread by readers who are interested in gleaning information relevant to their job needs.

Remember that writing exists for human beings, and few of us enjoy writing that seems harder to read than it needs to be. What constitutes "difficult" writing depends on the reader, the topic, and the purpose of the document. But direct, concise writing that uses a conversational style will usually be appreciated by your readers. Using shorter rather than longer sentences also helps readers follow your thoughts:

> Please give immediate attention to insure that the pages of all documents prepared for distribution are numbered sequentially and in a place of optimum visibility. This is needed to facilitate our ability to refer to items during meetings.

Versus

> Please correctly number the pages of all documents. Place numbers in the upper right-hand corner. Sequential numbering helps us locate material during meetings.

Or

> Please number all pages sequentially.

Three additional examples:

> 1. It has recently been brought to my attention that only a small percentage of the employees in our division are contributors to the citizens' health research fund supported by this firm. This fund is a major source of money for the encouragement of significant discoveries and innovations made in behalf of research relevant to community health.

Versus

> I have discovered that only a small percentage of employees in our division contribute to the citizens' health research fund. Our firm supports this research because the products of this research improve community health.

> 2. As a result of their expertise, the consulting team is provided with the opportunity to make a reasonable determination of the appropriate direction to proceed regarding their selection of information systems.

Versus

> The consulting team has the expertise to select the best information systems.

> 3. It is our contention that the necessary modifications should be made to make the system operational because its complete replacement is economically prohibitive.

Versus

> We believe that the system should be modified to make it operational. Complete replacement costs too much.

Avoid excessive use of *is/are* verb forms. Choosing specific, concrete verbs for clarity means avoiding forms of the "be" verb, if possible. As the following sentences illustrate, excessive use of "be" verbs often obscures action verbs. Many times, a "be" verb is the best choice (as this sentence exemplifies). However, you can lessen the tendency to use "be" verbs by doing the following:

- Avoid beginning sentences with *there is* or *there are, there was* or *there were*.
- Avoid beginning sentences with phrases such as *it is clear that, it is evident that*, and *it should be noted that*.
- Choose a specific verb rather than *is, are, was*, and *were* verb forms.

"Be" verbs often create a longer, less direct sentence:

> Delegation is a means of lessening the manager's workload.

Versus

> Managers who delegate <u>reduce</u> their workload.

Examine the following three examples:

> 1. Our appraisal system is broken: it is inefficient, it is unfair, and it is costly.

Versus

> Our appraisal system lacks efficiency, fairness, and thrift.

> 2. My decision <u>is based on the assumption</u> that his statement <u>is erroneous</u>.

Versus

> My decision <u>assumed</u> that his statement <u>is erroneous</u>.

> 3. Our office <u>has been provided with the authority to make a determination about the selection</u> of a computing system.

Versus

> Our office was authorized to select a computing system.

As these examples and the ones that follow show, the clearest sentences focus on the agent and the action (the verb):

1. There are two systems presently available for testing job candidates.

Versus

Two available systems can test job candidates.

2. There are several national and global organizations dedicated to promoting environmental sustainability for healthcare facilities.

Versus

Several national and global organizations promote environmental sustainability for healthcare facilities.

Use active voice for clarity. The structure of a sentence—the arrangement of words—affects the clarity of the sentence. In active voice, the agent that does the action occurs next to the verb, the agent and the action both appear in the sentence, and the agent appears as the subject of the sentence.

> *agent*   *verb*
> The department teaches the course every spring term.

> *agent*   *verb*
> Our office submits all travel vouchers within 24 hours of their completion.

The result? Clear, concise, direct sentences.

> *Before:*
> (A) Attempts were made by the division staff to assess the project.

> *After:*
> *agent*   *verb*
> (B) The division staff attempted to assess the project.

Sentence (A) uses passive voice. Sentence (B) uses active voice: the agent (staff) occurs as the subject and appears next to the verb (attempted).

Research to determine the most readable sentence structures indicates that active-voice sentences may be more readable than passive-voice sentences. Readers often need the agent (the actor) placed near the action (the verb) to determine the sentence meaning. The subject and verb contain the essence of the sentence. The following examples illustrate this concept.

> The door is to be locked at 6:00 p.m.

This sentence, which does not specify the agent, could mean either of the following:

> The guard (or some designated person) will lock the door at 6:00 p.m.
> The last person leaving the building at 6:00 p.m. must lock the door.

As both revisions illustrate, to understand a sentence, readers must know the agent and the action carried out by the agent. When you write, be sure your sentences indicate who or what performs the action.

Passive-voice sentences often intentionally do not include the actor or agent doing the action, to hide responsibility. The result may produce a sentence more verbose and less accurate than an active-voice version. Passive-voice sentences often use "there is" and "there are" constructions. Even in engineering writing, such as articles for academic journals, many editors want active-voice sentences because of their increased clarity. As in the examples below, the use of active subjects will usually make an explanation easier to read and easier to understand:

Before:

With the growing request of high-quality multimedia service, especially in portable systems, efficient algorithms for audio and/or video data processing have been developed. These algorithms have the characteristics of high-complexity data-intensive computation. For these applications, there exist two extreme implementations. One is software implementation running on a general-purpose processor and the other is hardware implementation in the form of application-specific integrated circuit (ASIC). In the first case, it is flexible enough to support various applications but may not yield sufficient performance to cope with the complexity of application. In the second case, optimization is better in respect of both power and performance but only for a specific application. A coarse-grained reconfigurable architecture fills the gap between the two approaches, providing higher performance than software implementation and wider applicability than hardware implementation.

After:

(Note: We separate sentences in the original passage here so that you can follow the changes in sentence structure.)

To respond to growing requests for high-quality multimedia services, especially in portable systems, engineers have developed efficient algorithms for audio and/or video data processing.

These algorithms exemplify high-complexity, data-intensive computation.

In addition, these applications use two extreme implementations: (1) a *software implementation* running on a general-purpose processor and (2) a *hardware implementation*, an application-specific integrated circuit (ASIC). The first has flexibility needed to support various applications, but it may not yield sufficient performance to cope with complex applications. In the second, we can optimize by power and performance but only for a specific application. A coarse-grained reconfigurable architecture fills the gap between the two approaches. This architecture provides higher performance than software implementation and wider applicability than hardware implementation.

Breaking long sentences into shorter ones and creating short paragraphs can also produce clarity:

Before:
To ensure dimensional quality of manufactured products, a crucial step is to take coordinate measurements of the geometric features to reconstruct product surface and then to check their compliance with tolerance specifications: my research develops a method to integrate the coordinate measurements from measuring devices of different resolutions for a better reconstruction of the product surface.

After:
To ensure dimensional quality of manufactured products, researchers must take coordinate[d] measurements of the geometric features. Their goal is to reconstruct product surface and then check surface compliance with tolerance specifications.

My research develops a method to integrate the coordinate[d] measurements from measuring devices of different resolutions to better reconstruct the product surface.

WORD CHOICE

To write concise sentences, use clear, concise words and phrases. Avoid using longer words when shorter ones will do just as well. (Write to express, not to impress.)

---

instead of	write	instead of	write
accordingly	so	hence	so
accumulate	gather	implement	carry out
acquaint	tell	initiate	begin
acquire	get	maximum	most
activate	begin	modification	change
aggregate	total	nevertheless	but, however
assist	help	objective	aim
communicate	write, talk, tell	optimum	best
compensation	pay	personnel	people, staff
consequently	so	procure	get
continue	keep up	purchase	buy
demonstrate	show	terminate	end
discontinue	stop	transmit	send
endeavor	try	utilize	use
facilitate	ease, simplify		

Eliminate dead phrases—words that add nothing to the meaning of the sentence.

to the extent that	in view of
with your permission	inasmuch as
hence	as a matter of fact
with reference to	for the purpose of
in connection with	in order
with respect to	
as already stated	

Avoid words that sound knowledgeable without being specific. Many are technical words that have been overused and poorly adapted to nontechnical situations.

parameters	warrants further investigation
logistical interface	broad-based
contact	dynamics
impact	infrastructure
input/output	longitudinal study
conceptualize	matrix
formalize	meaningful
multifaceted	monolithic
systematized	paradigm
prioritize	participatory involvement
time frame	resource utilization
hard date	viability
in-depth study	

Avoid redundant phrases.

absolutely complete	human volunteer
absolutely essential	insist and demand
agreeable and satisfactory	my personal opinion
anxious and eager	necessary essentially
basic fundamentals	past memories
complete absence	point in time
consensus of opinion	right and proper
each and every	sincere and earnest
exactly identical	small in size
example to illustrate	summarize briefly
few in number	thought and consideration

first and foremost   true facts
general consensus   very unique
green in color

Avoid business jargon.

instead of	**write**
consideration was given	I considered
prior to the	before
at the present writing	now
effect an improvement	improve
in the neighborhood of	about
beg to advise	tell
cognizant of	know
thanking you in advance	I would appreciate
endeavor	try
viable alternative	possibility
in regard/reference to	about
send under separate cover	send separately
return same to the above	return to us
needless to say	[omit]
it goes without saying	[omit]
in the normal course of procedure	normally
in this day and age	today
in my opinion	I believe
it is our opinion	we think
on a daily basis	daily
on the grounds that	because
pursuant to our agreement	as we agreed
we are not in a position to	we cannot
without further delay	now
please be advised that	[omit]

STYLE CHECKLIST ✔

Planning

☐ How will you adjust your writing style to accommodate your readers' knowledge of the subject?

☐ How will you meet your readers' expectations about style for the specific kind of document you are writing?

☐ Can you determine the appropriate reading level for the context in which the document will be read?

☐ How will you adjust your style so that it is appropriate for the professional relationship you have to your readers?

Revision

☐ Do paragraphs begin with a topic sentence? Do the following sentences in the paragraph build on the idea in the topic sentence?

☐ Are most sentences 20 words or shorter? Could you make any longer sentences shorter?

☐ Are subjects and verbs close together in your sentences?

☐ Have you used specific nouns and concrete verbs?

☐ Have you avoided ponderous and impersonal language?

☐ Have you avoided *is/are* verb forms whenever possible?

☐ Are most of your sentences written in active voice? Could you change any sentences with passive voice to active voice?

☐ Have you defined everything that might require defining?

☐ Could you write any of your sentences with equal clarity but with fewer words?

EXERCISES

1. Which recommendations in this chapter operate in the revisions of the following sentences? Why do these principles result in clearer, more concise revisions?

> i. A stroke is an increasingly common problem associated with the brain. It is caused when a vein or artery in the brain is obstructed. This results in loss of consciousness from the loss of oxygen in the brain.
>
> > A stroke, an obstruction in a vein or artery in the brain, results in loss of consciousness as the brain does not receive oxygen.
>
> ii. There is another method that is more invasive to the patient but proves to be successful enough to treat stroke.
>
> > Another more invasive method proves successful in treating stroke.

iii. There are many proposed solutions for ridding the United States of oil dependency.

   Many proposed solutions target reduction of US oil dependency.

iv. A significant amount of new research has come out recently about a bodily phenomenon called brown fat. Brown fat is regarded with interest because of its potential use as a new therapy for obese patients. Brown fat is nothing new—it's been known for centuries as "baby fat," the stuff that makes newborn infants so adorably chubby.

   Recent research on human "brown fat" shows that it may help obese patients. Brown fat, known for centuries as "baby fat," makes newborn infants chubby.

v. People in this field work in hospitals and dedicate their time to the mental well-being of children in hospitals as well as to helping the parents be able to support their children and cope with the trauma.

   People in this field work in hospitals. These professionals focus on the mental well-being of hospitalized children and help the parents support their children as they cope with trauma.

vi. Asperger's syndrome, sometimes called high functioning autism disorder, is a disability in young people that is frequently overlooked by educators and medical personnel.

   Asperger's syndrome, a high functioning autistic disorder in young people, is frequently overlooked by educators and medical personnel.

Or

   Educators and medical personnel frequently overlook Asperger's syndrome, a high functioning autism disorder occurring in young people.

vii. Even though Hurricane Gustav left its mark, there was a minimal amount of damage to the town, and finally there was a sigh of relief regarding energy prices.

   Even though Hurricane Gustav left its mark, the town suffered minimal damage and gave a sigh of relief regarding energy prices.

viii. A shoemaking company has signed a deal to share military technology to make their shoes waterproof, at the same time making the fabric breathable.

   A shoemaking company signed a deal to share military technology, making their shoes waterproof and breathable.

ix. There are many issues facing New Orleans related directly to hurricanes and tropical storms. Recently most of the problems caused from these storms have been traced back to the loss of wetlands of New Orleans during the past century. Despite recent pushes by community and political leaders in New Orleans to help restore these very little has been done, leaving much of the city vulnerable to hurricanes and allowing sea level to continue to rise.

New Orleans faces many issues related to hurricanes and tropical storms. Most problems caused by these storms have occurred because of loss of wetlands around New Orleans during the past century. [a. Despite efforts by community and political leaders in New Orleans to restore wetlands, little has been done.] Much of the city becomes increasingly vulnerable to hurricanes as the sea level continues to rise.

[b. Little has been done, despite efforts by community and political leaders in New Orleans to restore wetlands.]

x. Efforts were made on the part of the director's committee for completion of an evaluation of the recommendations of the report.

The director's committee tried to complete their evaluation of the report recommendations.

2. Revise the following:

i. There are several national and global organizations dedicated to promoting environmental sustainability for healthcare facilities.

ii. Though there have been numerous economic anthropology studies in Mexico, they have largely been in three locations.

iii. The intervening steps taken in the site are with the intention to make the site more attractive. This is done by establishing services, heading offices, cultural centers, and commercial firms in the downtown district.

iv. For adequate housing in such circumstances, this paper here presents a model. The model is a step-by-step development process which is aimed at empowering the poor users. This process includes:

- Innovative financing
- Creating employment opportunity
- Make the users aware of sustainable systems for drinking water, sewage disposal, other utilities, and shelter improvement
- Demonstrate the knowledge of the mentioned systems through a construction of a public facility for them.
- Transfer the knowledge by employing the users in the construction and maintenance of the facility.

v. At the same time, these fuels must be easy to burn cleanly. This is typically characteristic of liquid and gaseous fuels. Lastly, it is best that the fuel be a liquid or a gas because of the ability to pump both. This means that the existing infrastructure can be used to distribute the fuel.

vi. Lots of money around the world is poured into cancer research.

vii. There has been success using lymphotropic paramagnetic nanoparticles for imaging prostate cancer.

viii. There is plenty of money and room for big companies to get their foot in and discover new technologies to fight cancer.

ix. Recently there has been a surge in research and development for a material called grapheme. One area grapheme is being applied to is solar cells.

x. Sustainable development is the process of moving human activities into a pattern that can be perpetually sustained. It is an approach that seeks to meet the needs of the present while protecting the resources that will be needed in the future.

3. Figure 4–6 is a memo composed by the manager of The Plaza, a relatively new apartment building in Lubbock, Texas. The manager would like to distribute the memo to all residents as soon as possible. As the regional manager of Rite Properties, which operates The Plaza, you must review the official communications coming from managers to residents. What changes would you advise the manager to make to this memo?

4. Examine Figure 4–7. If you were the author of this document and were given a little extra time to make it more readable, what changes would you make?

5. Examine Figure 4–8. Note the number of stylistic weaknesses highlighted. Revise the report to remove these weaknesses and others you might find.

Rite Properties

7001 South Loop 289
Lubbock, Texas 79425
806-555-1327
manager@theplaza.com

May 15, 2012

TO: ALL PLAZA RESIDENTS
FROM: The Management
SUBJECT: problems

Over recent weeks and months we have had an issue with dogs urinating and defecating in staircases, sidewalks, the courtyard, and in the parking lot. This has required daily cleaning of these areas and is consuming additional janitor hours, which costs money and could soon lead to rent increases. Please be respectful of the buildings and your neighbors and curb your dog from using these public areas as a restroom. If your pet has an accident, it is your responsibility to clean up after it. Okay? If you ever witness this activity, please report it to Management. Monetary penalties will be imposed on the responsible party.

<u>Also effective today, May 15, 2012, Bailey's will tow any Plaza resident vehicles or Plaza visitor vehicles that park in their parking lot.</u> One visitor vehicle from Plaza was parked directly in front of their back door, blocking the way for a big delivery truck coming in this morning, consequently, no one may park over there any more. Residents, we have been over the parking situation over and over, if you or your guests get towed, don't call me. I've put out several newsletters or spoke with individuals about the parking. <u>It is your responsibility to tell your guests where to park.</u> We have had a good relationship with Bailey's for a long time, but when we get in the way, everyone must pay the price. <u>Stop and think where you park! Is it going to be in someone's way in the morning?</u> The next time I am called to clear out the cars to avoid being towed because it is blocking a door or driveway at Bailey's, I promise, I will call the tow truck myself.

We have also been changing the filters and checking the smoke alarm batteries this week and there are those of you that are tampering with your smoke alarms! Repeatedly we have told you not to take the battery out of your smoke alarm. IT IS AGAINST THE LAW. Do you have enough insurance to cover all your home's contents in addition to all the contents of several other homes if a fire breaks out in your apartment and because you have no battery in your smoke alarm, no one was warned before it was too late. When your alarm beeps, it means you need a new battery. We have the batteries in the office, come and get a new battery. <u>There is absolutely no excuse that can be justified for removing or tampering with any part of the smoke alarm that I will accept.</u> If we didn't have an alarm in your apartment and you had a fire, you would immediately blame us, do we get to blame you when you have removed the battery or tampered with the alarm? Each of you that had removed your alarm battery will be fined $25.00 which is clearly stated in your lease and had been explained in several newsletters.

Again, it is your responsibility for telling your guests where to park. You know how it feels when there are no parking spaces because they are filled with guests. I constantly see the same guests parking on the lot, the lot is posted for resident parking with all others towed, I have threatened to tow in the past, but once again, it has gotten out of hand and we are almost full and it takes every space on the lot just so our residents are allowed to park on the lot. <u>Effective today, all non stickered vehicles will be towed with no advance warning. When this happens, perhaps you will want to pay the towing charge to your guest since you did not tell them where to park.</u> When your guests vehicles get towed, please do not contact me. There will be a sticker telling them exactly where to get in touch with the wrecker service to retrieve their vehicle.

continued

FIGURE 4–6 Document for Exercise 3

Once again there is trash being left outside apartments overnight and during the day. Please, we take prospective residents through the courtyard nearly every day and it is very embarrassing to have to pass other people's trash. If you can't carry it out to the dumpster, leave it in your house until you can. It will be greatly appreciated.

Finally, as the weather heats up and you open your windows to cool off, remember that this allows sound to travel. If your windows are open after 10 PM, keep the noise in your apartment down. No loud televisions, parties, or music, please. Your neighbors will thank you. Have a wonderful day!

FIGURE 4-6 *continued*

FACT SHEET

UNITED STATES DEPARTMENT OF AGRICULTURE
FARM SERVICE AGENCY

January 2013

Microloans

Overview

The Farm Service Agency (FSA) developed the Microloan (ML) program to better serve the unique financial operating needs of beginning, niche and the smallest of family farm operations by modifying its Operating Loan (OL) application, eligibility and security requirements. The program will offer more flexible access to credit and will serve as an attractive loan alternative for smaller farming operations like specialty crop producers and operators of community supported agriculture (CSA). These smaller farms, including non-traditional farm operations, often face limited financing options.

Use of Microloans

Microloans can be used for all approved operating expenses as authorized by the FSA Operating Loan Program, including but not limited to:

- Initial start-up expenses;
- Annual expenses such as seed, fertilizer, utilities, land rents;
- Marketing and distribution expenses;
- Family living expenses;
- Purchase of livestock, equipment, and other materials essential to farm operations;
- Minor farm improvements such as wells and coolers;
- Hoop houses to extend the growing season;
- Essential tools;
- Irrigation;
- Delivery vehicles.

Simplified Application Process

The application process for microloans will be simpler, requiring less paperwork to fill out, to coincide with the smaller loan amount that will be associated with microloans. Requirements for managerial experience and loan security have been modified to accommodate smaller farm operations, beginning farmers and those with no farm management experience. FSA understands that there will be applicants for the ML program who want to farm but do not have traditional farm experience or have not been raised on a farm or within a rural community with agriculture-affiliated organizations. ML program applicants will need to have some farm experience; however, FSA will consider an applicant's small business experience as well as any experience with a self-guided apprenticeship as a means to meet the farm management requirement. This will assist applicants who have limited farm skills by providing them with an opportunity to gain farm management experience while working with a mentor during the first production and marketing cycle.

Security Requirements

For annual operating purposes, microloans must be secured by a first lien on a farm property or agricultural products having a security value of at least 100 percent of the microloan amount, and up to 150 percent, when available. Microloans made for purposes other than annual operating expenses must be secured by a first lien on a farm property or agricultural products purchased with loan funds and having a security value of at least 100 percent of the microloan amount.

Rates and Terms

Eligible applicants may obtain a microloan for up to $35,000. The repayment term may vary and will not exceed seven years. Annual operating loans are repaid within 12 months or when the agricultural commodities produced are sold. Interest rates are based on the regular OL rates that are in effect at the time of the microloan approval or microloan closing, whichever is less.

More Information and Eligibility Criteria

Additional information on the FSA microloan program may be obtained at local FSA offices or through the FSA website at www.fsa.usda.gov.

FIGURE 4-7 Document for Exercise 4

Memorandum

TO: Jerry Bradshaw   DATE: March 29, 2011
FROM: Bill Jones
SUBJECT: **The Best Approaches in Alternative Fuel Research**

Purpose

Alternative fuels are becoming an important area of research. There are two main
reasons researchers are looking for a new and reliable fuel source. The first is to combat
the reliance of the U.S on foreign oil. The second is to reduce the amount of harmful
emissions into the atmosphere. The heavy reliance on gasoline is harmful to both our
economy and the environment. In this report I will discuss the best approaches in
alternative fuel research. The three main approaches that will be covered are ethanol,
the hydrogen fuel cell, and biodiesel. For each approach I will explain the process of
producing the fuel, the benefits, costs, and the negative aspects. The information
provided will show what the best alternative fuel will be.

Research Methods

The data and information collected was obtained from several websites and online
databases. I found which alternative fuels were currently in use from the U.S. Department
of Energy website. It had useful information about new alternatives being researched
and also information about fuels currently in use. After determining which three fuels
had the most potential I went into more in-depth research and looked for articles on
the online databases.

Findings

Approach 1—Ethanol

Ethanol is an alcohol produced by the fermentation of the grains or seeds of a plant. The
most common ingredient used in ethanol production is corn. The production process
of ethanol begins with the harvesting of the corn. The kernels of corn are shipped to a
plant where they are first ground and then cooked in a large cooker. After the ground
corn is cooked, the fermentation process can begin. For fermentation to take place a
single celled fungus called yeast is added to the processed corn. The yeast feeds on the
starches in the grain and produces alcohol as a by-product. Ethanol is retrieved from the
fermentation chamber by distilling the fluid inside the chamber. Then trucks are used to

FIGURE 4-8 Document for Exercise 5

transport the ethanol to consumers. Pipes are not used because ethanol is corrosive and would damage the pipes after a certain amount of time.

The main benefit of producing ethanol is that it is renewable. However, most cars cannot use straight ethanol because it is more corrosive than gasoline. Many times ethanol is mixed with gasoline in different ratios so that regular cars are able to function properly without any corrosive damage. The ethanol/gasoline mixture still emits carbon dioxide, but it is cleaner burning than pure gasoline. Plants absorb carbon dioxide, so the carbon dioxide released by the ethanol is recycled by the plants that are used to produce the ethanol.

Some researchers argue that because of all the farm equipment used to produce the corn and the energy needed by the factory, we use more energy producing ethanol than we get out of it. Mass production of corn also takes a heavy toll on the environment. The fertilizers used on the corn release N_2O, which is a very strong green house gas. Farming only corn also takes essential minerals from the soil and can cause erosion. There are efforts being made to negate these problems by finding new processes that will use the whole plant to produce cellulose ethanol instead of just the corn kernels. Using the whole plant would allow ethanol production to become more efficient. It would also allow us to use different grass species. The use of native grass species would require less fertilization and not damage the soil.

Approach 2—Hydrogen Fuel Cells

Fuel cells are able to convert chemical processes into electricity. The type of fuel cell that is best suited to power a vehicle is called a polymer exchange membrane fuel cell (PEMFC). The main reaction that these cells use is converting Hydrogen and Oxygen into water. There are four main components of the PEMFC and each component has a specific job. First is the anode. The anode captures the electrons that are freed during the reaction. These electrons are used in an external circuit that is able to produce work. It also disperses the Hydrogen evenly over the surface of the catalyst. The catalyst is what helps the reaction take place. For this process the catalyst is usually made of platinum nanoparticles that are thinly spread on a thin carbon sheet. The nanoparticles allow for the highest surface as possible, which increases the reaction rate. Third is the electrolyte or proton exchange membrane. This membrane only conducts positively charged ions, while blocking electrons. A downside is that the membrane must remain hydrated in order to function properly. Last is the cathode, it distributes the Oxygen across the catalyst and conducts the electrons from the outer circuit to complete the reaction of Oxygen and Hydrogen to make water. The overall process has four main steps. First the H2 splits when it contacts the catalyst and releases 2 electrons. These

FIGURE 4-8 *continued*

electrons travel to the anode and through the outer circuit to the cathode. Second the O2 also splits forming two highly negative ions. Third, the O- ions attract the H+ ions and draw them through the membrane. Last the Hydrogen and Oxygen pair up with the electrons from the circuit and form water.

This is complicated, but has some excellent benefits. First is the amount of pollution reduction. The only by-product is water. Second is that the process is said to be 64% efficient compared to 20% efficiency of gas combustion engines. However, since the process is complicated the components are expensive, especially the platinum catalysts. The membranes are also not very durable. They tend to degrade under the conditions of turning on and off your car. Very high or low temperatures are also a problem for the membranes. The membranes must stay hydrated to function and in high temperatures they dry out, while in cold temperatures there is the possibility of freezing. There is also a problem with the fuel. To be able to travel a considerable distance, a lot of hydrogen is needed, and we don't have the technology to compress the hydrogen into a small enough tank. We also have no infrastructure to distribute the hydrogen. Therefore this process may need to wait until we have better technology to utilize the fuel cell's capabilities.

Approach 3—Biodiesel

Biodiesel is a form of diesel that is produced from organic renewable oils like vegetable oils and animal fat. Production from these oils is a benefit because they are non-toxic, biodegradable, and renewable. The process of making biodiesel is a simple process. First the recycled oils or fats are filtered and preprocessed to remove any water or contaminants. The oils and fats are then mixed with an alcohol, usually methanol, and a catalyst like sodium hydroxide or potassium hydroxide. The oil molecules are broken apart and reformed into methyl esters and glycerin. Glycerin is a by-product that can be separated out and used in other industries. The methyl esters become the biodiesel.

Using biodiesel has many advantages. Unlike ethanol which is produced from corn, biodiesel can be created from non-food crops or recycled greases. Soybeans are of the major crops that are used in biodiesel production. The oils are extracted from the soybeans and used to make fuel while the rest of the soybean can still be used for livestock feed or other applications, so nothing is wasted. Algae are another source for biodiesel production. Oils can be extracted from the algae and processed. Algae allow for a renewable resource that grows rapidly and is easy to maintain, so it is an excellent option. Some researchers have found that the biodiesel produced contains 3.2 times the amount of energy it takes to produce. This shows that biodiesel production is

FIGURE 4-8 *continued*

an efficient process. Most diesel engines are already compatible with using biodiesel, so fewer changes need to be made. However, biodiesel still needs to be mixed with petrodiesel in order to get the right viscosity, so cars can't run entirely on biodiesel.

Conclusion

There are many reasons why we must take steps to reduce our dependence on fossil fuels. One of the main reasons is to regain energy independence. If we find an alternative fuel that is efficient and easy to produce, we will lessen our dependence on foreign oil and form a more stable economy and country. The three approaches that I researched are some of the best possibilities. Each one has their pros and cons. Although ethanol is renewable it relies too much on food crops and is not efficient enough yet. More research is needed to produce a more efficient procedure. Fuel cells are also a great idea that have no harmful emissions and are very efficient, but we still don't have the technology to make them practical yet. Therefore the most reliable fuel source is biodiesel. The production of biodiesel is simple and efficient. The renewable resources used are abundant, easy to obtain and aren't specifically food crops. Algae are an especially good resource. They would be cheap and easy to mass produce, yet they would provide enough oil for a processing plant. Biodiesel could have an impact on our fossil fuel consumption in the future.

Citations

Kelly, Tom. "Biodiesel Basics." *Beverage World* 129.10 (2010): S20-S21. Web. 29 Mar 2011.

Jozefowics, Chris. "Fuel for Thought." *Current Science* 92.11 (2007): P6-7. Web. 29 Mar 2011.

Lin, Lin. "Opportunities and Challenges for Biodiesel Fuel." *Applied Energy* 88.4 (2011): p. 1020-1031. Web. 29 Mar 2011.

Nice, Karim, and Jonathan Strickland. "How Fuel Cells Work." *How Stuff Works*. Discovery Company, n.d. Web. 29 Mar 2011. <http://auto.howstuffworks.com/fuel-efficiency/alternative-fuels/fuel-ce112.htm>.

"Alternative Fuels and Advanced Vehicles Data Center." *U.S Department of Energy*. N.p., n.d. Web. 29 Mar 2011. <http://www.afdc.energy.gov/afdc/fuels/biodiesel_production.html>.

FIGURE 4-8 *continued*

5

Designing Documents

Effective writing is more than just choosing the right words. And it's more than correct sentences organized in logical paragraphs. To be effective, your document must also look like it will be easy to read and easy to understand.

You have many choices about how a page or screen will appear to your readers. For example, you could integrate illustrations, include audio and video, incorporate links to pages or sites with more information, invite comments from your readers, or encourage readers to share your information through e-mail or social media. It is also easy to overwhelm and confuse your readers with these elements. This chapter will help you make the careful choices that allow for easy reading and navigation of your pages and screens.

Quick Tips

On the job, you will probably write and receive more e-mail messages than any other kind of document. And ordinarily you will be writing your messages and your readers will be receiving your messages on mobile devices. If you design your e-mail messages for quick and easy reading, your recipients will be able to decipher your intended meaning efficiently and respond readily and appropriately to your instructions, questions, and requests regardless of their location. Here are five simple guidelines to follow:

1. **Keep your messages brief**. E-mail is especially effective for brief messages that the recipient will read and reply to quickly. Long, scrolling messages with extensive detail are often better relegated to attachments that could be accessed and studied later.
2. **Use short paragraphs**. Short paragraphs separated by white space encourage quick reading and make it easy for your recipient to perceive and retrieve the chief points of your message.

3. Use the subject line to specify your message. A clear and specific subject line will preview your message for your recipients, making their reading easier and aiding their understanding. You will also be helping recipients to sort and find your messages later, especially if you write a separate message for each topic you address. If your message includes a sizable attachment that might be slow to access from a wireless mobile device, identify it here in brackets—[BIG DOC], [BIG PIC], [AUDIO], or [VIDEO]—so that your recipients know whether to look at it immediately or wait for a more convenient time.

4. Use headings to identify the sections of your message. Headings make it easier for your recipient to skim your message for its chief points and assist in later retrieval of specific information in your message.

5. Ask simple yes/no questions. Make it easy for your recipient to reply to your message with the briefest possible answer. For example, ask, "Should I . . . ?" instead of "How would you like me to handle this situation?" If a series of yes/no questions are necessary, number the questions.

Understanding the Basics of Document Design

Readers judge a document by how it looks as much as by what it contains. In fact, their earliest impression comes from the appearance of the document, not its content. A dense page of long paragraphs will often discourage or annoy readers even before they begin reading. A page designed to help readers locate important information, however, may add to the persuasiveness of your position or convince your readers to put a little more effort into finding what they need and understanding what they find.

Effective document design is built on principles of visual perception—on how human beings perceive and interpret visual information. For example,

- *Contrast*: Different items must be visibly different, especially more important and less important items.
- *Alignment*: Related items must be aligned with each other, and every item must be aligned with some other item (or risk looking misaligned).
- *Proximity*: Related items must be positioned close together.
- *Size*: Greater size implies greater importance.
- *Repetition*: Repetition of design creates unity and builds familiarity.

The following five guidelines will help you adapt the principles of perception to plan your document's visual design:

- Know what decisions are yours to make.
- Choose a design that fits your situation.
- Plan your design from the beginning.

- Reveal your design to your readers.
- Keep your design consistent.

Know what decisions are yours to make. Many companies have a standard format or template for reports, letters, proposals, e-mail messages, and websites. Before you develop your document, determine the pertinent design requirements. Don't change the format arbitrarily just to be different. If you think the template you are supposed to use isn't appropriate for your audience and your message, find out who makes decisions on such design issues and make a case for the changes that you would like to see.

Keep in mind also that you don't have to use exotic or sophisticated software applications to apply basic principles of document design. The typical word processing program has all the functions you will likely need to create a visually effective document.

Choose a design that fits your situation. Don't make your document any more complex than the situation requires. You typically don't need a table of contents or a glossary for reports that are under five pages. Add appendices or multimedia only if this material is necessary and will be useful for your readers.

You'll impress readers most by providing just the information they need in a way that makes it easy for them to find it and understand it. Most people read technical and business documents selectively. They scan the document, looking for sections that are relevant to their needs. They try to grasp the main points quickly because they are busy and have far too much to read. Remember that your readers don't get paid to read documents: they get paid to make decisions and take actions. The more time needed to read your document, the less productive is their paid time on the job. And always keep in mind that they could be reading your document on their smartphones, tablets, or laptops while traveling to and from the job site, while sitting in meetings, and while multitasking in their offices.

For example, users working with a software application are unlikely to read the entire user's manual. They resort to the manual or to online help when they have a specific problem or need instructions for a specific task and can't figure it out by trial and error. They want to get to the right page or screen immediately. They want the instructions to stand out on the page or screen. Look at Figures 5–1 and 5–2. Notice how the numbered steps make for quick reading and easy understanding relative to the long and confusing paragraph of instructions.

Plan your design from the beginning. Before you start writing, carefully consider how you will organize and display your information. Ask questions like the following:

- How will your readers use the document? Will they read it from beginning to end? Will they want to skim it and grab the main points without reading more? Will they want to jump to a specific topic? Even if they read the document through once, will they want to come back later and find a specific point quickly?

- How familiar are your readers with the subject of the document? How much support will they need in understanding and navigating the information you offer?
- How familiar are your readers with the kind of document you are writing? Do they come to the document with certain expectations about how the information will be organized and exhibited?
- Will your readers view this document on paper or on a screen?
- Will your readers be stationary or mobile?
- Will your readers be focused or distracted in their attention to your document?

If, for example, readers are likely to skim your document, adding a table of contents, a subject index, and headings on every page will help them find information quickly. (The rest of this chapter includes techniques for developing effective designs to help people find what they need.)

If readers don't know much about the topic of the document, a glossary of keywords and abbreviations might be necessary. Illustrations or videos could also be important because such readers won't have a store of pictures in their minds on which to call for support. Links to pertinent interactive resources or social media sites might keep these readers attentive and engaged.

If readers are unfamiliar with a particular kind of document, they might benefit from a simple and explicit design that avoids distracting variations or unnecessary decorative elements. Experienced readers, however, might have rigid design expectations: for example, hiring managers assume that letters of application will

All managers must report on the achievement of sales targets by their full-time and part-time salespeople, specifically the number of salespeople meeting weekly, monthly, and quarterly sales targets. These figures are based on the official sales targets issued by the vice president of Merchandising and the number of full-time and part-time salespeople identified by Human Resources as assigned to each department at the beginning of the reporting period. To submit reports, start by logging in to your digital profile in Corporate Measures (http://www.corporatemeasures .com/login/managers). Then, in the Manager section, select Weekly, Monthly, or Quarterly Sales Report, as appropriate. Next, in the Department section, locate each department managed during the reporting period and choose the pencil icon to edit the information for that department. Then, in the list of information for each department, scroll to Sales Targets. Then enter the appropriate figures for both full-time and part-time salespeople. Finally, in the adjacent text box, explain anything new or different during this reporting period that might be related to the achievement of sales targets (e.g., special promotions, changes in aisle displays, suggestive selling techniques).

FIGURE 5-1 Instructions in Ineffective Paragraph Style
This format is difficult to follow, both on screen and on paper.

All managers must report on the achievement of sales targets by their full-time and part-time salespeople, specifically the number of salespeople meeting weekly, monthly, and quarterly sales targets.

These figures are based on

a. official sales targets issued by the vice president of Merchandising
b. the number of full-time and part-time salespeople identified by Human Resources as assigned to each department at the beginning of the reporting period

To submit reports,

1. Log in to your digital profile in CorporateMeasures (http://www.corporatemeasures.com/login/managers).

2. In the Manager section, select Weekly, Monthly, or Quarterly Sales Report as appropriate.

3. In the Department section, locate each department you managed during the reporting period and choose the pencil icon to edit the information for that department.

4. In the list of information for each department, scroll to Sales Targets.

5. Enter the appropriate figures for both full-time and part-time salespeople.

6. In the adjacent text box, explain anything new or different during this reporting period that might be related to the achievement of sales targets (e.g., special promotions, changes in aisle displays, suggestive selling techniques).

FIGURE 5-2 Instructions in Effective List Style

adopt a standard format of business correspondence (with return address, salutation, signature, etc.) and are usually annoyed with applications that violate their expectations.

If the document is going to be read on a screen, you may have both more constraints and more choices than if the document were printed on paper. Illustrations and color will be easier and less expensive to include in a digital document than in a paper document. Readers can also typically adjust the size of text on a screen to their level of comfort, which could easily disrupt your intended spatial relationship of words and images on the screen.

If readers are mobile, your document could be competing for their attention with all the distractions in their changing environment. In this case, headings, lists, illustrations, and white space will help readers to keep track of their location in your document and to recognize key items of information.

Reveal your design to your readers. Research on how people read and process information shows that readers must see how information is organized in order to make sense of it. That is, as you read, you try to do two things at the same time: you try to make sense of the passage you're reading and you try to make sense of how this passage fits with previous passages and what it contributes to the entire document. The more difficult it is to do one or the other, the more difficult the document is to read.

Tables of contents and headings reveal the organization, scope, and direction of your document and give readers a clear overview or map with which to proceed. Using headings (at least one on every page or screen) in a memo, e-mail message, or brief report will show the structure and logic of the discussion and help readers recognize, remember, and retrieve your major points. Longer documents definitely need headings and a table of contents that lists the headings. In digital documents, a table of contents typically links to the pages listed.

Keep your design consistent. Consistency in design makes for easy reading. When you have considered your audiences, the information you have to deliver, and the ways that people will read and use your document, you must develop a page or screen design that will work for your situation. Once you have decided on the appropriate design, don't change it for arbitrary reasons. You want your readers to proceed confidently and comfortably through the document. You want them to know immediately when they are beginning a new section or when they are in another part of the same section because they recognize the differences in the design of the headings at each level.

To achieve this consistency, identify the different types of information in your document and use the styles function of your word processing program to duplicate the design. First, think about all the types of information you will need to display, such as paragraphs, quotations, lists, examples, equations, formulas, and various levels of headings. Second, plan a design that always shows the same type of information in the same way throughout your document. The design could include the type size, the typeface (or font), the placement of an element on the page or screen, whether the text has a border (also called a "line" or a "rule") over or under it, whether the text and headings are bold or italic, the amount of space that comes before and after a heading, the style of the text that follows each kind of heading, and so forth. Third, use the style function of your word processing program to label and fix the design of each type of information.

Figure 5–3 shows a letter using document design to reveal the content and the relationships among the sections. This example shows how any document, including routine letters and memos, will benefit from the use of a consistent design.

DEPARTMENT OF HEALTH & HUMAN SERVICES   Public Health Service

---

Centers for Disease Control
and Prevention

March 18, 2011

Dear Colleague,

April is STD Awareness Month, an annual observance to raise public awareness about
the impact of sexually transmitted diseases (STDs) on the lives of Americans and the
importance of discussing sexual health with healthcare providers and sex partners. This
letter summarizes CDC activities and resources that we hope will complement and
support your local activities.

CDC's Public Health Grand Rounds on Chlamydia
The Division of STD Prevention (DSTDP) participated in a series of presentations for the
CDC Public Health Grand Rounds on Chlamydia last May. A forthcoming article in the
Morbidity and Mortality Weekly Report (MMWR) will reprise these presentations.

The *GYT: Get Yourself Tested* Campaign
CDC is again partnering with MTV, the Kaiser Family Foundation, Planned Parenthood
Federation of America, and others to bring attention to the epidemic of STDs in the
United States. This year the GYT website is offering resources for providers to help them
better serve teen and young adult patients. The GYT provider site addresses young
people's common STD-related misconceptions and offers simple answers to patients'
questions. The website features CDC's 2010 STD Treatment Guidelines, reference guides
and charts (e.g. STD basics, who should be tested for what, where, and when); resources
for talking to patients about sexual history and STD testing; training resources; and GYT
materials for your office or clinic. You are invited to a March 24th webinar to learn about
this year's updates to the GYT campaign (prior registration is required).

National HIV and STD Testing Resources
CDC continues to update its interactive STD and HIV testing locator on the National HIV
and STD Testing Resource website www.findSTDtest.org. If you have not already done
so, please check to be sure your clinic(s) are included in the database and that the
information about them is current. Since the findSTDtest.org website will be promoted
throughout April as a source for STD testing locations, we want to ensure that the
information is accurate. Please contact Rachel Kachur (rlk4@cdc.gov) if you need
assistance updating your information.

STD Awareness Resource Site
CDC's STD Awareness Resource Site (formerly the STD Awareness Month website) has
been updated for 2011 and provides our STD prevention partners with information and
tools to support your local STD Awareness Month activities all year round. Included on
this site are resources for developing STD-related health communication and social
marketing campaigns. Additionally, there is a media kit that includes sample press
releases, public service announcement scripts, and tips for media outreach. Widgets to

FIGURE 5-3 Letter That Uses Document Design Principles

provide zip-code-based locator information for STD testing sites are available for download.

STD Prevention Materials
CDC is providing approximately 5,000 GYT clinic kits to state and local health departments. These kits will arrive by the end of March to the addresses provided by those who ordered them. If you have not ordered kits or need additional kits, you can download the kit materials and additional electronic resources now at the GYT Toolkit tab on the GYT website.

CDC STD Fact Sheets are available for download and local printing and are designed to print in either black and white or in color.

The following printed materials are available for shipment to you without charge through our online ordering page:

- STD "The Facts" brochure series. This award-winning series of brochures is available in English or Spanish, and is written in plain language. The brochures are available to health departments and other partners and may be ordered in quantities up to 500 per brochure. Electronic copies in PDF format can also be found on the STD Awareness Resource Site.
- *A Guide to Taking a Sexual Health History*. This practical guide is a good resource to share with healthcare providers in your community, especially if you are making them aware of STD Awareness Month activities. Five copies of this guide are included in each GYT clinic kit.
- The *2010 STD Treatment Guidelines* will be available soon for ordering in hard copy, wall chart, and pocket guide formats.

The Division of STD Prevention's Twitter account
The Twitter account @CDCSTD now has over 2,700 followers, is following more than 200 individuals and organizations, and is included in nearly 200 lists. It is now verified and listed as an official HHS Twitter account.

I look forward to hearing from you about your successes with your STD Awareness activities, not only in April, but throughout the year. Many thanks for your hard work and persistent efforts toward reducing the burden of sexually transmitted diseases on the American people.

Sincerely,

/Gail Bolan/

Gail Bolan, MD
Director, Division of STD Prevention
National Center for HIV/AIDS, Viral Hepatitis, STD and TB Prevention

FIGURE 5-3 *continued*

Designing Effective Pages and Screens

Visually effective pages and screens are designed on a grid so that readers know where to look for information. Using space inside the text, around the graphics, and in the margins will keep pages uncluttered and information easy to locate. The right line length and margins will also help people read easily. The following suggestions will help you develop visually effective pages and screens:

- Use blank space to frame and group information.
- Set the spacing for easy reading.
- Use a medium line length.
- Use a ragged right margin.
- Position words and illustrations in a complementary relationship.

Use blank space to frame and group information. Don't think of blank space as wasted or empty space. Space is a critical element in design for both paper and screens because it makes information easier to find and read. Look at Figures 5–4 and 5–5. Which do you think is easier to read?

You can incorporate blank space into documents in several ways. A critical location for blank space is at the margins. Here, blank space serves to enclose and contain the information and keep the page or screen from looking crowded and chaotic. Clear and generous margins make your information look organized and coherent.

If your document will be read on paper, also think about how it will be bound. If you are putting your work in a binder, be sure to leave room for the binding so that holes don't punch through the text. Similarly, think about whether a reader will want to punch holes in a copy later or put the work in a binder. On a standard 8½ by 11–inch page, use the following margins:

top margin	1 inch
bottom margin	1 inch
left margin	1 inch, if material is not being bound
	1½ inches, if material is being bound
right margin	1 inch

If you are going to photocopy on both the front and back of the page, leave space for the binding in the left margin of odd-numbered pages and in the right margin of even-numbered pages. Word processing programs allow you to choose mirror margins so that they alternate for right-hand (odd-numbered) pages and left-hand (even-numbered) pages. If you cannot set alternating margins, set both the right and the left margins at about 1½ inches to allow for binding two-sided copies.

The space in the margins is important, but it's not enough. Graphic designers call margins *passive space* because margins only define the block of the page or screen for readers. Graphic designers know that *active space*—the space inside the

TO: All Department Heads

SUBJECT: New Media Request Procedures

A recent study of our media center request procedures indicates that we are not fulfilling media requests as efficiently as possible. A number of problems surfaced in the study. First, many requests, and particularly special requests (e.g., poster printing, laminating, close captioning of audio/videocasts, audio/video editing) are submitted in the evening after the media center has closed for the day or in the early morning before it has opened. As a result, the media center has an enormous backlog of orders to fill before it can begin servicing the orders submitted after 8:30 A.M., when it officially opens. This backlog may throw the center two or three hours behind schedule. Media requests submitted throughout the day cannot be completed promptly. By 2:00 P.M., requests submitted may not be filled that same day. If special orders arrive unexpectedly, even a routine copy request may take two business days to complete.

To remedy the situation, we will change to the following media request procedure beginning Monday, February 3. The media center will close at 3:00 every afternoon. Two employees will work at the center from 3:00 until 5:00 to complete all routine orders by 5:00. If you submit requests by 3:00, the center will have them ready by 5:00. In short, all requests will be filled the day they are submitted. However, do not leave requests after 3:00, as these will not be processed until the following day. However, we guarantee that if you submit your request between 8:00 and 3:00, you will have it fulfilled that day.

Special requests—including over 100 copies of one item, single/multiple copies of any document over 50 pages, front/back copying of one item up to 50 copies, poster (large-scale) printing, laminating, jobs involving trimming or folding, close captioning of audio/videocasts, and audio/video editing—will require that notice be given to the media center at least one day in advance (more time for unusual projects such as audio/videocasts in excess of five minutes or close captioning in multiple languages). That way, the center can prepare for your request and be sure to have it ready for you.

A copy of the special request form is attached to this message. Please complete this form and e-mail it to Lydia Schlosser at the media center (at mediarequest@crsu.edu) so that she can schedule all special jobs. If you submit a special request without having completed the form, your project will be completed after all other special requests are completed.

Allow plenty of time for routine jobs—at least two hours, and three if possible. Beginning February 3, give all media requests to the receptionist in the media center. Be sure you attach complete instructions. Give your name, employee number, e-mail address, and department. Describe the project in detail, such as the number of copies required, paper color, collating and binding for multipage copies, color or black and white printing.

Pick-up procedures also change February 3. All jobs, after they are complete, will be placed in each department's mail box. No jobs will be left outside the media center after closing time. No jobs will be left with the receptionist. Large orders that will not fit mail boxes will be delivered to your office. Electronic projects will be delivered to the e-mail address listed on the request form.

If you have questions about this new procedure, please contact Lydia Schlosser (by voice or text) at 742-2500.

FIGURE 5–4 Memo That Violates Format Guidelines

TO: All Department Heads
DATE: January 23, 2014
FROM: Lydia Schlosser
SUBJECT: New Procedures for Ordering Jobs from the Media Center

EFFECTIVE DATE: MONDAY, FEBRUARY 3, 2014

To handle orders more quickly and efficiently, the Media Center is changing its procedures. Please inform everyone in your department and ask them to follow these new procedures.

Special Orders versus Routine Requests
Decide if you have a special order or a routine request. A special order is
- over 100 copies of one item
- single/multiple copies of any document over 50 pages
- front/back copying of one item up to 50 copies
- poster (large-scale) printing
- laminating
- jobs involving trimming or folding
- close captioning of audio/videocasts
- audio/video editing

Procedure for a Routine Request
1. Attach complete instructions to your request. Include
   - your name, employee number, e-mail address, and department
   - project details such as number of copies
   - special instructions such as collating and binding for multipage copies

2. Give all routine requests to the receptionist in the Media Center.

3. Allow 2 hours for your order to be filled.

NOTE: Routine requests submitted between 8:00 A.M. and 3:00 P.M. will be processed by 5:00 P.M. on the same day.

Procedure for a Special Order
1. Fill out the attached Special Order form.
2. E-mail the completed form to mediarequest@crsu.edu at least one day in advance of the day you need the job completed (two days in advance for unusual projects such as audio/videocasts in excess of five minutes or close captioning in multiple languages).

NOTE: This will allow us to schedule your job with the appropriate staff and equipment and complete your job promptly.

The Copy Center will close at 3:00 P.M. Orders submitted after that time will be processed the next business day.

Copy Pick-Up Procedures
Jobs will be delivered to your department's mailbox. If the order is too large for the mailbox, it will be delivered to your office. Electronic jobs will be delivered to your e-mail address.

If you have questions, call or text Lydia Schlosser at 742-2500.

FIGURE 5-5 A Revision of Figure 5-4

text—makes the real difference in designing effective pages or screens. Blank space helps readers to find information quickly, keep track of the location of information, process the information in identifiable chunks, and retrieve the information later.

Here are three techniques to bring active space to your pages and screens:

- Use headings frequently (and at least once per page or screen). Put each heading above the text or to the left of the text with an extra line of space before the heading to separate it visually from the preceding text.
- Use bulleted lists for three or more parallel points. Use numbered lists for steps in instructions. Lists are often indented inside the text, and each item may be separated from the others by blank space.
- Separate paragraphs with extra blank space, or indent the first line of each paragraph. In digital documents, make your paragraphs even shorter than you would in paper documents so that there is notable space even on the small screens of mobile devices. On a smartphone or tablet, one instruction or one short sentence may make an appropriate paragraph.

Space the lines of text for easy reading. It's customary in paper documents to use single spacing. (For a brief letter or memo of a single paragraph, double spacing is appropriate.) When you use single spacing, insert an extra line between paragraphs.

Drafts of documents submitted for review and editing are often double-spaced to give writers and editors more room in which to write corrections and notes. When you use double spacing in drafts, you need to have some way to show where new paragraphs begin. Either indent the first line of each paragraph or add an extra line between paragraphs.

For documents that will be read on screens, use single spacing, with an extra line inserted between paragraphs. Double spacing is rarely used for continuous text on the screen because it increases the need for scrolling through the document: a typical screen displays only a portion of what a paper page does, especially the screens on mobile devices.

Adjust the line length to the size of the page or screen. The number of words that fit on a line depends in part on the size and style of type that you are using. Long lines of text fatigue readers and make them lose their place in moving from the right margin back to the left margin of the next line. Short lines are also difficult to read because readers are almost continuously shifting their eyes from the right margin to the left margin of the next line with little time for moving across each line. Figure 5–6 illustrates the problems with both long and short lines of text.

Use a ragged right margin. Although text is almost always lined up on the left margin, it is sometimes also aligned on the right margin, creating a tidy rectangle of text. The text of this book, for example, aligns on both the left and the right

Long lines of type often prove challenging for people to read. The problem is that readers find it difficult to get back to the correct place at the left margin. They may find themselves inadvertently skipping a line or rereading the same line. If they make this mistake more than once or twice, they may get tired or annoyed and stop reading altogether. The smaller the type, the harder it is for most people to read long lines of type.

Very short lines
of type can look
choppy on a
page and make
comprehension
difficult because
readers must
work harder to
put the words
together in
coherent groups.

FIGURE 5-6 Line Length
Very long lines and very short lines are hard to read.

margins. Most of the examples in the figures in this chapter, however, align on the left but not on the right. The technique of making all the text align exactly on both the left and the right margin is called *justifying the text*. If the text aligns on the left but not on the right, it has a *ragged right margin*.

Justified type gives a document a formal appearance, whereas ragged right type gives a document a friendly, informal feeling. Justified text is often more difficult for readers because every line is the same length, thus eliminating a visual signal that helps readers both to keep track of each line and to locate the next line of text. Digital documents ordinarily have ragged right margins in order to make reading on screens of various sizes as easy as possible.

Position words and illustrations in a complementary relationship. The visual and verbal information in your document must reinforce each other, working together to communicate your message. Position each illustration as close as possible to the text that relates or refers to it. Keep in mind also that readers will typically examine the illustrations on a page or screen before giving their attention to the words in captions, headings, and paragraphs. Make sure the text adjacent to each illustration supports the message of the illustration and doesn't lead to mis-interpretation. Notice, for example, how the position of the diagram in Figure 5–7 could cause readers to think it displays the kind of extension cord that ought to be discarded instead of the kind with desired safety features. A quick fix here would be to change the neighboring heading to "Look for Key Safety Features" so that the words truly complement the diagram.

CPSC Safety Alert

Household Extension Cords Can Cause Fires

Choose the right extension cord for the job, and use it as specified. Extension cords can overheat and cause fires when used improperly. Overheating is usually caused by overloading or connecting appliances that consume more watts than the cord can handle. Damaged extension cords can also cause fires. Extension cords should only be used temporarily. Protect young children by keeping them away from extension cords and unplugging the cords when not in use.

Overheating or Damage
- Overheating can occur at the plug, at the socket, or over the entire length of the cord. Hot plugs and sockets are often caused by deteriorated connections to the cord's wires.
- Look for visible signs of excessive wear or damage to the plug, sockets or insulation. Replace damaged extension cords.

Prevent Overloading
- If any part of the extension cord is hot while in use, it is a warning sign that it may be overloaded. Check if the extension cord is properly rated for the products that are plugged into it. Also, inspect the cord along its entire length to ensure it has not been damaged.
- Do not overload your extension cord by using it to power appliances beyond its capacity. You can check its capacity, or rating, by looking at the tag on the cord or its packaging.

Protect Extension Cords from Damage
- Do not run extension cords under carpets, through doorways or under furniture.
- Only use an extension cord outdoors if it is marked for outdoor use.
- Never alter a cord to change its length or perform inadequate repairs such as taping up damaged insulation. Do not trim, cut or alter the plug blades in any way.
- Unplug an extension cord when it is not in use. The cord is energized when it is plugged in and can overheat if shorted.

Discard Older Extension Cords
- Discard cords that are old and/or are missing important safety features, including safety closures, polarized blades and a large plug face that covers the outlet's slots and is easy to grasp to unplug.
- Extension cords should be at least 16 AWG, unless they are 18 AWG with fuse protection. AWG refers to the size of the wires in the cord. The wire size is imprinted on the cord's surface.

POLARIZED PLUG BLADES (one wider blade)
LARGE PLUG FACE
EASY GRIP
AWG 16 CORD
SAFETY CLOSURE PREVENTS SHOCKS AND MOUTH BURNS

Check cords to make sure they have been listed by a recognized national testing laboratory, such as Underwriters Laboratories (UL), Intertek (ETL) or CSA-International (CSA).

Publication 5032 009810 042012

U.S. CONSUMER PRODUCT SAFETY COMMISSION • (800) 638-2772 • www.cpsc.gov • www.SaferProducts.gov

FIGURE 5–7 Ineffective Word–Illustration Relationship
The diagram and adjacent heading communicate opposite messages instead of reinforcing each other.

HELPING READERS LOCATE INFORMATION

To help your readers find what they need and make sense of what they find, you must plan a useful structure for your document and make that structure evident to your readers. In the previous sections of this chapter, we showed you how to use page layout and fonts to make your document clear and easy to use. In this section, we explain how to give your readers signs of your document's overall structure.

On the job, readers of technical and business documents rarely examine every page; instead, they read selectively. They may glance over the table of contents to see what the document is about and then pick and choose the sections to read by finding headings that match their needs and interests. They may go straight to the search function for digital documents (or the index of paper documents) to locate a specific topic. They may skim through the document, stopping only when a heading or example or illustration strikes them as important. They may go back to the document later to retrieve or verify specific facts. Long reports or proposals may be divided among several readers, with each reviewing the section pertinent only to his or her field (e.g., accountants checking the budget, engineers examining the schematics, managers assessing the timeline) before coming together to discuss the subject of the document or sharing comments and questions through a wiki or e-mail list.

Following are four ways to help your readers find information easily:

- Use frequent headings.
- Write descriptive headings.
- Design distinctive headings.
- Use page numbers and headers or footers.

Use frequent headings. Frequent headings help readers know where they are in a document at all times. In a report, for example, you want a heading for every section and subsection, with at least one heading on every page or screen. You want to keep each topic short (one to three paragraphs) and give each topic a heading.

Write descriptive headings. *Headings* are the short titles that you use to label each section and subsection of your document. Even brief documents, such as letters and e-mail messages, can benefit from headings. Compare Figures 5–8 and 5–9 to see how useful headings are in a brief message.

Headings are the roadmap to your document, identifying the key topics and revealing the direction of thought.

These five suggestions will help you write useful headings:

- Use concrete language.
- Use questions, verb phrases, and sentences instead of nouns alone.
- Use standard keywords if readers expect them.
- Make the headings at a given level parallel.
- Make sure the headings match any list or table of contents in the document.

Use concrete language. Generic headings such as "Introduction" or "Conclusion" give no indication of the topic you are discussing. Make your headings specific to your document. Make your headings reveal the subject and claims of your document. Readers should be able to read only your headings, without any of the accompanying text, to get a clear sense of your overall message.

UNITED STATES OF AMERICA
FEDERAL TRADE COMMISSION
WASHINGTON, D.C. 20580

Division of Advertising Practices

Mary K. Engle
Associate Director

February 19, 2013

Petty Tsay Rader, Esq.
Munchkin, Inc.
16689 Schoenborn Street
North Hills, CA 91343

      Re: *Munchkin, Inc.*, FTC File No. 122-3165

Dear Ms. Rader:

    As you are aware, the staff of the Division of Advertising Practices of the Federal Trade Commission investigated Munchkin, Inc. for possible violations of Sections 5 and 12 of the Federal Trade Commission Act, 15 U.S.C. §§ 45, 52. Our investigation concerned Munchkin's advertising (including labeling) for the Sesame Street® Chilly Ring Teether, which is filled with a blend of water and propylene glycol. Our inquiry focused on Munchkin's statement that the product is "water-filled," among others. Upon careful review of the matter, including non-public information submitted to the FTC, we have determined not to recommend enforcement action at this time. The factors we considered in making this determination included the limited nature of the advertising containing the claims at issue and Munchkin's voluntary action to modify all advertising (including labeling) to ensure compliance with the FTC Act. In particular, the company has stopped characterizing its product as "water-filled."

    We appreciate Munchkin's cooperation in resolving this matter. The staff encourages Munchkin to exercise caution in future advertising when describing the contents or otherwise characterizing the safety of a product.

    This action is not to be construed as a determination that a violation of law did not occur, just as the pendency of an investigation should not be construed as a determination that a violation has occurred. The Commission reserves the right to take such further action as the public interest may require.

                        Very truly yours,

                        Mary K. Engle
                        Associate Director
                        Division of Advertising Practices

FIGURE 5-8 Letter That Lacks Headings

UNITED STATES OF AMERICA
FEDERAL TRADE COMMISSION
WASHINGTON, D.C. 20580

Division of Advertising Practices

Mary K. Engle
Associate Director

February 19, 2013

Re: *Munchkin, Inc.*, FTC File No. 122-3165, Closing of Investigation

Dear Ms. Rader:

As you are aware, the staff of the Division of Advertising Practices of the Federal Trade Commission (FTC) investigated Munchkin, Inc., for possible violations of Sections 5 and 12 of the Federal Trade Commission Act, 15 U.S.C. §§ 45, 52. Our investigation concerned Munchkin's advertising (including labeling) for the Sesame Street® Chilly Ring Teether, which is filled with a blend of water and propylene glycol. Our inquiry focused on Munchkin's statement that the product is "water-filled," among others.

FTC will take no action now
Upon careful review of the matter, including nonpublic information submitted to the FTC, we have determined not to recommend enforcement action at this time. The factors we considered in making this determination included the limited nature of the advertising containing the claims at issue and Munchkin's voluntary action to modify all advertising (including labeling) to ensure compliance with the FTC act. In particular, the company has stopped characterizing its product as "water-filled."

Munchkin will exercise caution in future advertising
We appreciate Munchkin's cooperation in resolving this matter. The staff encourages Munchkin to exercise caution in future advertising when describing the contents or otherwise characterizing the safety of a product.

FTC could take action later
This action is not to be construed as a determination that a violation of law did not occur, just as the pendency of an investigation should not be construed as a determination that a violation has occurred. The commission reserves the right to take such further action as the public interest may require.

Very truly yours,

Mary K. Engle
Associate Director
Division of Advertising Practices

FIGURE 5–9 Revision of Figure 5–8

Use questions, verb phrases, and sentences instead of nouns alone. The best way to write headings is to put yourself in your readers' place. Will readers come to your document with questions? If so, then questions will make good headings. Will they come wanting instructions for doing tasks? If so, then verb phrases that match the actions they need to take will make good headings. Will they come seeking knowledge about a situation? If so, then statements of fact about that situation will make good headings.

In addition, avoid headings that are individual nouns or strings of nouns: such headings are often perceived as ambiguous. For example, a heading such as "Evaluation Questionnaire Completion" makes it impossible to predict the kind of information that this section will offer. Much clearer would be headings such as "How Do I Complete the Evaluation Questionnaire?" or "What Is the Deadline for Completing the Evaluation Questionnaires?" or "Who Must Complete the Evaluation Questionnaire?" Figure 5–10 shows how effective it can be to use questions, verb phrases, and statements as headings.

Use standard keywords if readers expect them. You may be working on a document for which readers expect to see a certain set of headings in a certain order,

---

Questions are useful as headings in a brochure.
>   What does the gypsy moth look like?
>   How can we protect trees from gypsy moths?
>   How often must we spray?
>   What is the chemical composition of the pesticide?

Verb phrases are useful in instruction manuals.
>   Verb phrases can be gerunds, like these:
>> Choosing the right kind of graph
>> Importing the data
>> Preparing the image
>> Adding a title
>
>   Verb phrases can be imperatives, like these:
>> Make your attendance policy clear.
>> Explain your grading scale.
>> Announce your office hours.
>> Supply titles of required textbooks.
>> Review assignments and due dates.

Short sentences are useful in memos and reports.
>   Our workload has doubled in the past year.
>   We are also being asked to do new tasks.
>   We have logged 500 hours of overtime this year.
>   We need three more staff positions.

FIGURE 5–10 Different Structures You Can Use for Effective Headings

Project Summary	Facilities and Equipment
Project Description	Personnel
Rationale and Significance	Budget
Plan of Work	

FIGURE 5–11 Keywords as Headings in a Proposal

as in a standard proposal format. In that case, organize your material in the order and with the headings that your readers expect. Figure 5–11 shows the headings you might use in a standard proposal format.

Make the headings at a given level parallel. Like the items in a list, headings at any given level in a document should be parallel. Parallelism is a very powerful tool in writing because it helps readers to recognize the similarity among the listed items. See for yourself the difference parallelism makes by comparing the two sets of headings in Figure 5–12.

Make sure the headings match any list or table of contents in the document. To check how well your headings tell your story and how well you've maintained parallel structure in headings, use your word processing program to create an outline view or a table of contents for your draft document. Readers can use the table of contents to locate a particular section. They know they're in the right place if the heading for that section matches the wording in the table of contents. This function is especially important in digital documents, where readers almost always navigate by jumping directly from a heading in the table of contents to a screen of information. If the heading on the screen they come to doesn't match the heading that they clicked on in the table of contents, they may be unsure of their location. Their confusion will quickly lead to irritation and a loss of trust in your attention to accuracy.

Design distinctive headings. Headings do more than outline your document. They also help readers find specific parts quickly, and they show the relationship among the parts. To help readers, headings have to be easily distinguished

Nonparallel Headings	**Parallel Headings**
Graph Modifications	Modifying a graph
Make data changes	Changing the data
To insert or delete columns	Inserting or deleting columns
How to adjust color or gradients	Adjusting the color or gradients
Titles and legends can be added	Adding titles and legends

FIGURE 5–12 Nonparallel and Parallel Headings
Headings that use the same sentence structure—parallel headings—are easier for users to follow.

from the text and each level of heading has to be easily distinguished from all other levels. Figure 5–13 is a good example of a document with four levels of headings. You can see how the writer uses boldface to distinguish all headings from the text and then uses type size, capitalization, and position on the page to distinguish each level of heading from the other levels.

These seven suggestions will help you design distinctive headings:

- Limit the number of heading levels.
- Create a pattern for the headings and stick to it.
- Match size to importance.
- Put more space above a heading than below it.
- Keep each heading with the section it covers.
- Use headings frequently.
- Consider using numbers with your headings.

Limit the number of heading levels. Don't make the hierarchy of levels more complicated than it needs to be. Headings are supposed to be an aid to reading, not an obstacle. The more levels of headings you use, the more that readers must do to keep the hierarchy straight in their minds. Paper documents don't need more than four levels of headings. If you have more than four levels, consider dividing the material into two major sections or two separate documents. Digital documents don't need more than two levels of headings: readers scrolling through documents displayed on various sizes of screens will easily lose track of different levels of headings.

Create a pattern for the headings and stick to it. Although your choices depend in part on the technology you are using, you almost certainly have several options for showing levels of headings. Figure 5–13 demonstrates a variety of ways to show different levels of headings. You can combine these to create the pattern for your headings. For example, you can change size, position, and capitalization to show the different levels of headings.

Match size to importance. Changing the type size is one way to indicate levels of headings. If you use different type sizes, make sure that you match the size to the level of importance. If the headings are different sizes, readers expect first-level headings to be larger than second-level headings, second-level headings to be larger than third-level headings, and so on, as shown in Figure 5–13. The lower-level headings can be the same size as the text, but no level of heading should be smaller than the text. That would violate readers' expectations. If you use different type sizes for different heading levels, don't make the differences in type sizes excessive.

Put more space above a heading than below it. Headings announce the topic that is coming next in your document. Therefore, you want the heading to lead the reader's eye down the page or screen into the text that follows. One way to do that is to have more space, on the page or screen, above the heading rather than below it. In this way, the heading and its accompanying text constitute a visible chunk of information.

Controlling Soil-Borne Pathogens in Tree Nurseries

Types of Soil-Borne Pathogens and Their Effects on Trees
Simply stated, the effects of soil-borne pathogens . . .

Soil-borne fungi
At one time, it was thought that soil-borne fungi . . .

    Basiodiomycetes. The Basiodiomycetes are a class of fungi whose species . . .

    Phycomycetes. The class of Phycomycetes is a highly diversified type of fungus. It is the . . .

Plant parasitic nematodes
Nematodes are small, unsegmented . . .

Treatments and Controls for Soil-Borne Pathogens

FIGURE 5-13 Four Levels of Headings in a Report

If you are going to use a rule with the heading, consider putting it *above* the heading rather than below it. A rule above the heading creates a "chunk" that includes both the heading and the text that it covers. A rule above the heading also draws the reader's eye down into the text that follows instead of up and away from that text.

Keep each heading with the section it covers. Don't leave a heading at the bottom of a page when the text appears on the next page. Make sure you have at least two lines of the first paragraph on the page with the heading. In some cases, you may want each topic to be on a separate page so that the heading and all the text of a topic appear together. Most word processing programs have functions that help you keep headings from being stranded at the bottom of a page and that allow you to set up your document so that all headings of a certain level start on a new page.

Consider using numbers with your headings. In many companies and agencies, the standard for organizing reports and manuals is to use a numbering system with headings. Figure 5–14 shows two such numbering systems: the alphanumeric system and the decimal system.

A numbering system allows you to refer precisely and concisely to a section of the report by the number of its heading (e.g., Section II.A.3.b or Section 4.3.7). This function could be especially important if the document is subject to review by multiple readers from different divisions or remote locations. If you add or delete a section later, however, you must be sure to revise the numbering of the report. In addition, a lot of readers find numbering systems confusing and, for example, would fail to recognize immediately that a section numbered 6.8.23 would come after a section numbered 6.8.2.3. If you use a numbering system with your headings, include the numbers before the entries in your table of contents.

TITLE
I. FIRST-LEVEL HEADING   **alphanumeric system**
   A. Second-Level Heading
     1. Third-level heading
     2. Third-level heading
   B. Second-Level Heading
II. FIRST-LEVEL HEADING
   A. Second-Level Heading
     1. Third-level heading
       a. Fourth-level heading
       b. Fourth-level heading
     2. Third-level heading
   B. Second-Level Heading

TITLE
1 FIRST-LEVEL HEADING   **decimal system**
   1.1 Second-Level Heading
     1.1.1 Third-level heading
     1.1.2 Third-level heading
   1.2 Second-Level Heading
2 FIRST-LEVEL HEADING
   2.1 Second-Level Heading
     2.1.1 Third-level heading
       2.1.1.1. Fourth-level heading
       2.1.1.2. Fourth-level heading
     2.1.2 Third-level heading
   2.2 Second-Level Heading

FIGURE 5-14 Two Types of Numbering Systems

Use page numbers and headers or footers. In addition to clearly worded and visually accessible headings, page numbers and running headers and footers are important aids to efficient reading.

Number the pages. Page numbers help readers keep track of where they are and provide easy reference points for talking about a document. Always number the pages of drafts and final documents that people are going to read on paper.

If the document is going to be read on screens, inserting page numbers could be unnecessary. In websites, for example, readers jump from topic to topic and page to page instead of proceeding in a specified order from start to finish. In addition, word processing programs keep track automatically of the number of pages and typically display this information in the bottom margin of the document window as a reference point for writers and readers. If readers are likely to print your digital document, however, they will certainly appreciate your inclusion of page numbers. In slides prepared for oral presentations, you will help your audience to track your progress and to stay attentive if you identify the slides with a numbering notation like this:

Slide 1 of 10

Short manuscripts and reports that have little prefatory material almost always use Arabic numerals (1, 2, 3). The common convention is to center the page number below the text at the bottom of the page or to put it in the upper outside corner (upper left corner for left-hand pages, upper right corner for right-hand pages). Always leave at least two lines of space between the text and the page number. Put the page number in the same place on each page. Page numbers at the bottom of the page often have a hyphen on each side, like this:

-17-

As reports grow longer and more complicated, the page-numbering system also may need to be more complex. If you have a preface or other material that comes before the main part of the report, it is customary to use lowercase Roman numerals (i, ii, iii) for that material and then to change to Arabic numerals for the body of the report.

In addition, the title page doesn't show the number but is counted as the first page. The page following the title page is number 2 or ii.

You also have to know whether the document will be printed on one side of the paper or both. If both sides will carry printing, you may have to number several otherwise blank pages in word processing files. New chapters or major sections usually start on a right-hand page. The right-hand page always has an odd number. If the last page of your first chapter is page 9, for example, and your document will be printed double-sided, you have to include a blank page 10 so that the first page of your second chapter will be a right-hand page 11 when the document is printed, copied, and bound.

The body of a report is usually paginated continuously, from page 1 to the last page. For the appendices, you may continue the same series of numbers, or you may change to a letter-plus-number system. In that system, the pages in Appendix A are numbered A-1, A-2, and so on. The pages in Appendix B are numbered B-1, B-2, and so forth. If your report is part of a series or if your company has a standard report format, you will need to make your page numbering match that of the series or standard format.

Numbering appendices with the letter-plus-number system has several advantages:

- It separates the appendices from the body. Readers can tell how long the body of the report is and how long each appendix is.
- It indicates that a page is part of an appendix and identifies which appendix. It makes pages in the appendices easier to locate.
- It allows the appendices to be printed separately from the body of the report. Sometimes the appendices are ready before the body of the report has been completed, and being able to print the appendices first may save time and help you meet a deadline.
- It allows the pagination of either an appendix or the body to be changed without requiring changes in the other parts.

Include headers or footers. In long documents, it helps readers if you give identifying information at the top or bottom of each page. Information at the top of the page is a *header*; information at the bottom of the page is a *footer*. Organizations often have standard practices regarding the information to be displayed in headers and footers. A typical header for a report might show the author's name, the title of the report, and the date. It might look like this:

Lisa Miller   Feasibility Study   November 2014

In this case, the page numbers would likely appear in the footer.

A typical header for a letter might show the name of the person receiving the letter, the page number, and the date. It might look like this:

Dr. Emilia Rodriguez   -2-   November 16, 2014

or

Dr. Emilia Rodriguez
Page 2
November 16, 2014

Note that headers and footers rarely appear on the first pages of documents because first pages already carry identifying information like the title, author, recipient, and date. Word processing programs allow you to start headers and footers on the second page.

DOCUMENT DESIGN CHECKLIST

- ☐ Is your document clean, organized, and attractive?
- ☐ Is your text easy to read?
- ☐ If your document is supposed to conform to a standard template, does it?
- ☐ Have you left adequate margins? (If necessary, have you left extra room for binding?)
- ☐ Is the spacing between the lines and paragraphs consistent and appropriate?
- ☐ Can the reader tell easily where sections and paragraphs begin?
- ☐ Are words and illustrations positioned appropriately?
- ☐ Have you checked the headings? Are the headings informative? Unambiguous? Consistent? Parallel?
- ☐ Will readers get an overall picture of the document by reading the headings?
- ☐ Is the hierarchy of the headings obvious?
- ☐ Can readers tell at a glance what is heading and what is text?
- ☐ If readers want to find a particular section quickly, will the size and placement of the heading help them?

DOCUMENT DESIGN CHECKLIST ✔   *continued*

☐ Have you checked the page breaks to be sure that you do not have a heading by itself at the bottom of a page?
☐ Are the pages of a paper document numbered?
☐ Are there appropriate headers or footers?
☐ If you are using a numbering system, is it consistent and correct?
☐ Did you test representative readers for their ability to locate information easily?

EXERCISES

1. Visit the websites of two government agencies in your city or state that might employ majors in your field (e.g., the Department of Commerce and the Department of Transportation). What kinds of documents do you find at each site? Do the agencies seem to have a standard template for the design of their documents? From looking at a number of the documents at each site, how would you describe their design? Which agency does a better job of designing documents for its audience and purpose? How would you make the design of each agency's documents more effective?

2. Revise the headings used in Figure 5–3, substituting questions for the original phrases. Revise again, using short sentences for the headings. Which revision do you consider more appropriate for the audience and purpose of this letter?

3. Figure 5–15 is a draft of the instructions for the GrillWizard.

   You are a friend and neighbor of the owner and founder of Fierce Products, a new company in Lubbock that manufactures the GrillWizard. Fierce Products is a family-owned company with 45 employees. The owner (whose spouse composed this draft version) has hired you to revise the design of the instructions in order to ensure safe and efficient operation of the product as well as to convey a positive impression of the quality of Fierce Products.

   After you revise the design of this document, you will return it to the company owner for final approval. Fierce Products is ready to release the GrillWizard to market. As soon as the instructions are ready, the product will be boxed and shipped. The sooner you submit your revised design, the sooner Fierce Products will start making money.

4. Examine Figure 5–16. How would you revise the design of this document to make it more effective? Which changes will have the greatest impact on the user's ability to read and remember the safety tips?

GrillWizard

Fast and efficient frying and cooking with propane gas. Light the burner and instantly you have a hot 100,000 BTU continuous flame.

The **GrillWizard** will help you eliminate fish and other lingering cooking odors from your home.

The **GrillWizard** is used for fast frying of fish potatoes, onion rings, chicken, vegetables, and donuts. Substitute water for oil and it's also great for shrimp, crab, and lobster boils as well as steaming clams.

This cooker is completely portable with all parts easily assembled and disassembled for compact transportation and storage, yet it weighs only 40 lbs.

When the control valve is open, a full 100,000 BTUs of powerful heat prepares cooking oil in 3 minutes for frying. Adjust the heat with a touch of the control valve.

Operating Instructions

Place grill on level ground. Insert tapered end of tubing into the hole in the base of the grill. (CAUTION: Make sure the cooker is level and the burner is facing up.)

Attach grill connector to propane cylinder.

Completely open propane valve.

Slightly open control valve at grill connector and light cooker at top of tube immediately (CAUTION: **DO NOT stand directly over cooker when lighting burner**).

Adjust control valve for desired flame height.

When finished cooking, always close both grill connector valve and propane cylinder valve completely.

The **GrillWizard** works with any size of propane tank cylinder and all will give off the same amount of heat. A 20-pound cylinder will provide approximately 6 hours of cooking time if valves are completely opened.

The intense heat produced by the **GrillWizard** allows you to fast fry all foods. The cooking oils of conventional fryers drop in temperature as food is added, but the **GrillWizard** maintains its temperature with just a quick touch of the control valve.

The **GrillWizard** has been designed for easy care. However, keep all dirt and foreign objects out of connectors, hose, valves, and openings. Failure to do so could cause obstruction of gas and greatly diminish the effectiveness of the **GrillWizard.**

CAUTION: If you suspect leaks **DO NOT** light unit before checking.

For outdoor use only.

After washing pan with soap and water, dry thoroughly and coat the entire pan with cooking oil on paper towel to prevent rusting.

FIGURE 5–15 Document for Exercise 3

Child Safety
Keep Your Little One Safe When You Fly

The Federal Aviation Administration (FAA) recommends that parents secure a child in a child restraint system (CRS) or device based on the child's weight. Parents are encouraged to take this tips sheet with them when they travel by air with small children.

Tips for Parents

- Make sure your CRS or device is government-approved for use on airplanes and has "This restraint is certified for use in motor vehicles and aircraft" printed on it. Otherwise, you may be asked to check the CRS as baggage.
- While booster seats and harness vests enhance safety in vehicles, the FAA prohibits passengers from using these types of restraints on airplanes during taxi, take-off and landing. These restraints should be checked as baggage. Supplemental lap restraints or "belly belts" are not approved for use in both airplanes and vehicles in the United States.
- The CARES Child Safety Device is the only FAA-approved harness-type restraint for children weighing between 22 and 44 pounds. It will have "FAA Approved in Accordance with 14CFR 21.305(d), Approved for Aircraft Use Only" on it.

- **If your child weighs...**   **Use a...**
  Less than 20 pounds   Rear-facing CRS
  20 to 40 pounds   Forward-facing CRS
  22 to 44 pounds   CARES child safety device
  More than 40 pounds   Airplane seat belt

- Measure the width of your CRS. It should fit in most airplane seats if it is no wider than 16 inches.
- Ask your airline for a discounted fare. Buying a ticket for your child is the only way to guarantee that you will be able to use a CRS.
- Reserve adjoining seats. A CRS should be placed in a window seat so it will not block the escape path in an emergency. Do not place a CRS in an exit row.
- If you do not buy a ticket for your child, ask if your airline will allow you to use an empty seat. If your airline's policy allows this, avoid the busiest days and times to increase the likelihood of finding an empty seat next to you.
- Arrange for your airline to help you if you need help making a connecting flight. Carrying a CRS, a child, and luggage through a busy airport can be challenging.
- Pack a bag of toys and snacks to keep your child occupied during the flight.
- Always use a CRS when driving to and from the airport.
- Wear your seat belt at all times.

Federal Aviation Administration

FIGURE 5-16 Document for Exercise 4

Source: US Federal Aviation Administration. *Child Safety: Keep Your Little One Safe When You Fly.* Washington, DC: GPO, 2012. http://www.faa.gov/passengers/fly_children/media/child-safety-tips.pdf

Designing Illustrations

In communicating technical information, you will often need to use illustrations either in addition to words or instead of words to convey your message. How do you determine whether illustrations are really desirable? Ask yourself two questions:

- What do you want the reader to do or think after reading your document?
- How will illustrations help you to achieve your objective?

Quick Tips

Illustrations divide into two categories: tables and figures. Tables display numbers and words in columns and rows. Every other kind of illustration is a figure.

If you want to summarize information to make it easier to remember or retrieve, use a table.

If you want to emphasize information to clarify or reinforce its meaning, use a figure.

CREATING ILLUSTRATIONS

In deciding when and how to use illustrations, remember the following guidelines:

- **Simplify your illustrations.** Keep your illustrations as simple as possible so that your reader has no difficulty understanding your message. Avoid distracting your reader with unnecessary details or decorative flourishes.
- **Use computer applications critically.** Graphics applications and clip art allow you to include a wide variety of illustrations in your document. To make

sure your tables and figures are effective, however, you must evaluate the choices available. Graphics applications, for example, might create artistic but misleading graphs, and clip art might exhibit a pictorial style that isn't quite serious enough or detailed enough to do justice to your subject. It is your job to choose illustrations that display your information clearly and correctly.

- **Consider size and cost.** Calculate the impact of illustrations on the expected length of your document. Illustrations will often increase the size of a document and, if printed, add to the cost of production and distribution.
- **Title your illustrations.** Give each table and figure a title that clearly indicates the message you wish it to convey.
- **Number your illustrations.** If you use several illustrations in your report, number them. Number the tables and figures separately (e.g., Table 1, Table 2, Figure 1, Figure 2).
- **Alert your readers.** Always alert your readers to illustrations by referring to them in the text. Every time you refer to the illustration, use the table or figure number (e.g., see Table 1). Announce the illustration—what it is or shows— then add any verbal explanation your reader will need to fully understand it. Don't lead readers through a complicated explanation and only later refer to the illustration. Point them to the illustration immediately so that they can shift back and forth between the explanation and the illustration as necessary.
- **Position your illustrations strategically.** Place each illustration as close to the passage it explains as possible.
- **Identify your sources.** If you borrow or adapt a table or figure from another source, identify that source (and, if necessary, permission to use it) below the illustration (e.g., *Source: A Practitioner's Guide to Statistical Illustrations*, p. 157, by Xiling Xiao. Copyright 2014 by Oxford University Press. Reprinted by permission.).

Tables.

- Note that the purpose of a table is to summarize information in specific categories to assist the viewer in accessing and retrieving this information.
- Every column in a table should have a heading that identifies the information below it. In a table of numbers, include the unit of measurement, such as "miles per hour." For large numbers, add a designation such as "in thousands" or "in millions" to the column heading (and delete the corresponding zeros from the numeric data). Headings should be brief. If headings need more explanation, include this information in a footnote below the table. Use lowercase letters, numbers, or symbols (e.g., a, *, or 1) to indicate footnoted material.
- If possible, box your table to separate it from surrounding paragraphs.
- Keep tables as simple as possible. Include only data relevant to your purpose.
- Consider omitting lines between rows and columns to avoid giving your table a crowded appearance. If possible, use white space to separate rows and columns.

Figures 6–1 through 6–3 show three ways of presenting tabular data.

Table 4.

Refugee Arrivals by Age, Sex, and Marital Status: Fiscal Years 2010 to 2012

Characteristic	2012		2011		2010	
	Number	**Percent**	**Number**	**Percent**	**Number**	**Percent**
AGE						
Total	58,179	100.0	56,384	100.0	73,293	100.0
0 to 17 years.	18,876	32.4	19,232	34.1	25,373	34.6
18 to 24 years.	9,700	16.7	9,588	17.0	11,853	16.2
25 to 34 years.	13,491	23.2	11,802	20.9	14,954	20.4
35 to 44 years.	7,446	12.8	7,124	12.6	9,587	13.1
45 to 54 years.	4,409	7.6	4,230	7.5	5,727	7.8
55 to 64 years.	2,441	4.2	2,438	4.3	3,218	4.4
65 years and over	1,816	3.1	1,970	3.5	2,581	3.5
SEX						
Total	58,179	100.0	56,384	100.0	73,293	100.0
Male	31,380	53.9	29,436	52.2	38,624	52.7
Female	26,799	46.1	26,948	47.8	34,669	47.3
MARITAL STATUS						
Total	58,179	100.0	56,384	100.0	73,293	100.0
Married	22,322	38.4	22,095	39.2	28,567	39.0
Single	32,608	56.0	31,324	55.6	41,022	56.0
Other*.	3,249	5.6	2,965	5.3	3,704	5.1

*Includes persons who were divorced, separated, widowed, or of unknown marital status.

Source: U.S. Department of State, Bureau of Population, Refugees, and Migration (PRM), Worldwide Refugee Admissions Processing System (WRAPS).

FIGURE 6–1 Table This table is effectively designed using thin vertical lines and generous white space to separate the columns and thin horizontal lines to separate the three categories of information, making the entire table look easy to navigate. Note that the row labels are aligned on the left for easy reading, while the numerical data in the rows are aligned on the right (or on the decimal point) to allow the easiest possible comparison of information across the rows and columns. The column headings are centered, while the subheadings ("Number" and "Percent") are aligned on the right with their corresponding rows of numerical data. The column headings and subheadings are displayed in bold, making the use of color here nice but unnecessary. The use of all capital letters for the headings of the three major categories is also unnecessary but consistent and unlikely to disrupt reading. Note that the title of this table is "Fiscal Years 2010 to 2012" but the columns are arranged left to right from 2012 to 2010, creating a possibility for misinterpretation. Changing the subtitle to "Fiscal Years 2012, 2011, and 2010" or putting the columns in chronological order from left to right would avoid this potential confusion.

Source: US Department of Homeland Security, Office of Immigration Statistics. *Refugees and Asylees; 2012.* Washington, DC: GPO, 2013. p. 4. http://www.dhs.gov/publication/refugees-and-asylees-2012

Table 3. Bus driver's preaccident activities (March 9–12, 2011).

Time	Activities	Location
Wednesday, March 9		
9:00 p.m.	Arrive for duty, World Wide Travel terminal	Brooklyn, NY
9:15 p.m.	Depart terminal	Brooklyn
10:00 p.m.	Pick up passengers; depart en route to Mohegan Sun	Flushing, NY
Thursday, March 10		
1:00 a.m.	Arrive Mohegan Sun casino, drop off passengers, and move motorcoach to parking lot	Uncasville, CT
1:20–3:15 a.m.	Sleep in bus (self-reported)	Parking lot
5:15 a.m.	Pick up passengers	Bus lobby entrance
5:30 a.m.	Depart Mohegan Sun en route to New York	I-95 south
8:58 a.m.	Return World Wide Travel terminal	Brooklyn
10:00 a.m.–4:00 p.m.	Sleep (self-reported)[a]	Residence
9:30 p.m.	Arrive for duty, World Wide Travel terminal	Brooklyn
9:45 p.m.	Depart terminal	Brooklyn
10:30 p.m.	Pick up passengers; depart en route to Mohegan Sun	Flushing
Friday, March 11		
1:56 a.m.	Arrive Mohegan Sun, drop off passengers	Uncasville
~2:00 a.m.–6:00 a.m.	Sleep in bus (self-reported)	Parking lot
~6:30 a.m.	Depart Mohegan Sun en route to New York	I-95 south
~9:50 a.m.	Arrive World Wide Travel terminal	Brooklyn
11:00 a.m.–4:00 p.m.	Sleep (self-reported)[b]	Residence
6:15 p.m.	Arrive for duty, World Wide Travel terminal	Brooklyn
7:40 p.m.	Pick up passengers; depart en route to Mohegan Sun	Bowery, New York City, NY
10:36 p.m.	Arrive Mohegan Sun, drop off passengers	Uncasville
10:54 p.m.	Drive to casino parking lot	Parking lot
11:46 p.m.	Drive from parking lot to casino for lost item	Bus lobby entrance
11:55 p.m.	Return to casino parking lot from bus lobby entrance	Parking lot
~11:55 p.m.–3:17 a.m.	Sleep on bus	Parking lot
Saturday, March 12		
3:17 a.m.	Receive incoming cell phone call (32 seconds)	Parking lot
3:19 a.m.	Pick up passengers	Bus lobby entrance
3:48 a.m.	Depart Mohegan Sun en route to New York	I-95 south
5:38 a.m.	Accident	I-95, near New York City

[a]The driver's cell phone was in use multiple times between 10:15 a.m.–12:03 p.m. Additionally, a 24-second incoming call occurred at 3:39 p.m.
[b]The driver's cell phone was in use numerous times during this period, with the longest period of nonuse between 11:38 a.m.–12:17 p.m. (39 minutes).

FIGURE 6-2 Table This table is easy to read with effective spacing to separate the columns and rows of information. Major headings are centered and displayed in distinctive white type on black horizontal bars. The subheadings are consistently centered and displayed in bold black type on gray horizontal bars. Two notes are used to clarify important abbreviations. The result is a table that viewers could readily navigate to compare and contrast information or to retrieve specific pieces of evidence. Times in the left column would be easier to read if aligned on the colon.

Source: National Transportation Safety Board. *Motorcoach Run-Off-the-Road and Collision with Vertical Highway Signpost, Interstate 95 Southbound, New York City, New York, March 12, 2011.* Highway Accident Report NTSB/HAR-12/01. Washington, DC: GPO, 2012. p. 22.

Table 4

Pedalcyclists Killed and Injured and Fatality and Injury Rates by Age and Sex, 2011

Age (Years)	Male			Female			Total		
	Killed	Population (thousands)	Fatality Rate*	Killed	Population (thousands)	Fatality Rate*	Killed	Population (thousands)	Fatality Rate*
<5	3	10,300	0.29	2	9,863	0.20	5	20,162	0.25
5–9	17	10,384	1.64	9	9,950	0.90	26	20,334	1.28
10–15	27	12,717	2.12	8	12,145	0.66	35	24,862	1.41
16–20	44	11,339	3.88	6	10,745	0.56	50	22,083	2.26
21–24	42	8,963	4.69	11	8,594	1.28	53	17,558	3.02
25–34	59	21,044	2.80	12	20,746	0.58	71	41,790	1.70
35–44	72	20,223	3.56	6	20,404	0.29	78	40,628	1.92
45–54	130	22,019	5.90	27	22,699	1.19	157	44,718	3.51
55–64	98	18,358	5.34	10	19,704	0.51	108	38,062	2.84
65–74	50	10,476	4.77	8	12,005	0.67	58	22,482	2.58
75–84	23	5,573	4.13	0	7,602	0.00	23	13,175	1.75
85+	9	1,894	4.75	0	3,843	0.00	9	5,737	1.57
Total‡	**578**	**153,291**	**3.77**	**99**	**158,301**	**0.63**	**677**	**311,592**	**2.17**
Age (Years)	Male			Female			Total		
	Injured	Population (thousands)	Injury Rate*	Injured	Population (thousands)	Injury Rate*	Injured	Population (thousands)	Injury Rate*
<5	**	10,300	**	**	9,863	**	**	20,162	**
5–9	2,000	10,384	175	1,000	9,950	57	2,000	20,334	117
10–15	5,000	12,717	409	2,000	12,145	137	7,000	24,862	276
16–20	6,000	11,339	497	3,000	10,745	257	8,000	22,083	380
21–24	3,000	8,963	360	1,000	8,594	129	4,000	17,558	247
25–34	5,000	21,044	252	2,000	20,746	83	7,000	41,790	168
35–44	4,000	20,223	213	**	20,404	**	5,000	40,628	116
45–54	5,000	22,019	236	1,000	22,699	57	7,000	44,718	146
55–64	4,000	18,358	243	1,000	19,704	41	5,000	38,062	138
65–74	2,000	10,476	155	**	12,005	**	2,000	22,482	80
75–84	1,000	5,573	110	**	7,602	**	1,000	13,175	48
85+	**	1,894	**	**	3,843	**	**	5,737	**
Total	**38,000**	**153,291**	**245**	**11,000**	**158,301**	**67**	**48,000**	**311,592**	**154**

* Rate per million population.
** Less than 500 injured, injury rate not shown.
‡ Total includes 4 males killed of unknown age.
Source: Fatalities — Fatality Analysis Reporting System, NHTSA.   Injured — General Estimates System, NHTSA.   Population — Bureau of the Census projections.

FIGURE 6–3 Ineffective Table This table tries to offer a lot of information in a single location but makes reading and retrieval difficult as a consequence. Notice that bold, centered type displayed in horizontal blue bars, used for both first-level and second-level headings, occurs both at the top of the table (which is to be expected) and again in the middle of the table (which would be unexpected). Similarly, a row of totals in a lighter blue bar occurs at the bottom of the table (which is to be expected) but also in the middle of the table (which would be unexpected). Unnecessary vertical grid lines and cells with double asterisks only exacerbate the visual confusion. At least two separate tables would make for easier reading: one for numbers and one for percents and/or one for fatalities and one for injuries.

Source: US Department of Transportation, National Highway Traffic Safety Administration. *Bicyclists and Other Cyclists.* Traffic Safety Facts DOT HS 811 743. Washington, DC: GPO, 2013. p. 3.

Bar and column graphs.

- Note that the purpose of a bar graph is to compare and contrast two or more subjects at the same point in time, whereas the purpose of a column graph is to reveal change in a subject at regular intervals of time.

- Avoid putting excessive information in a bar or column graph and thereby complicating the reader's ability to decipher it. Consider using a separate graph to communicate each point.

- Be sure to label the horizontal or, x, axis and the vertical or, y, axis—what each measures and the units in which each is calibrated. Readers can't understand your graph if they don't know what you are measuring or how it is measured.

- For bar graphs, start the x axis at zero and equally space the intervals on the x axis to avoid distorting the length of the bars. For column graphs, start the y axis at zero and equally space the intervals on the y axis to avoid distorting the height of the columns. In a bar graph, neither axis is a measure of time; in a column graph, the x axis is ordinarily a measure of time.

- Color can enhance the effect of a graph, but excessive color can reduce comprehension and distort information. Use the same color for all bars or columns that represent the same categories of information. Avoid using color as simple decoration.

- Make the graph accurate. Typical graphics applications allow you to add a range of special effects. However, artistic graphs are not always either effective or accurate. Three-dimensional bars and columns are often deceptive because readers have difficulty visually comparing the relative lengths of the bars and the relative heights of the columns. If you choose three-dimensional bar or column graphs, watch for possible distortion.

- Limit the number of divisions. The more a bar or column is divided, the more difficult it is to understand. Bars or columns with more than four or five divisions are especially challenging for readers and could lead to confusion or misinterpretation.

- For divided bar and column graphs, use color or solid shading to distinguish divisions instead of textured patterns that may create distracting optical illusions.

- Put labels on or near the bars or columns. Keep in mind that explanatory legends (or keys) will slow comprehension because the reader is required to look back and forth between the bars and the legend. Placing labels on bars can be difficult, however, especially if you use colored bars. Even black text will often be difficult to read on any colors but very light ones. In short, use a legend only if you can't use labels. And if you do use a legend, locate it as close to the bars or columns as possible.

- Avoid crowding the bars or columns within a graph. Such visual clutter makes a graph look difficult to interpret. Using three-dimensional bars or columns will also reduce the number that will fit in a given space. Effective and inviting graphs leave generous space between the bars or columns.

See Figures 6–4 through 6–7 for examples of bar and column graphs.

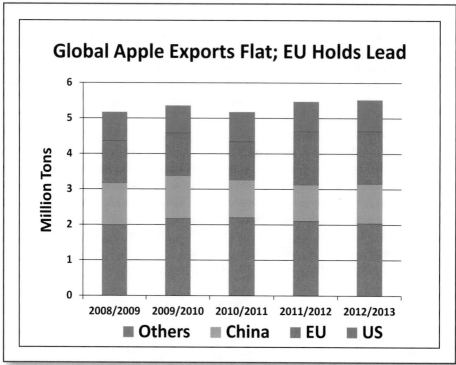

FIGURE 6–4 Column Graph This divided column graph is clean and concise and makes its two key points explicit in the title. The columns are labeled clearly for each period, arranged in chronological order from left to right, and spaced generously. The colors used for the sections of each column are bright and distinctive from each other. Numerical labeling of each section and a total for each column would offer more specific information, but the horizontal grid lines here are sufficient to support the claim in the title. Instead of the inefficient use of a separate legend to identify the meaning of each color (as well as the potentially confusing position of this legend along the horizontal axis), the sections of the last column could be labeled "US," "EU," "China," and "Other" for quicker and easier access to the same information.

Source: US Department of Agriculture, Foreign Agricultural Service. *Fresh Deciduous Fruit (Apples, Grapes, & Pears): World Markets and Trade.* Washington, DC: GPO, 2013. p. 1.

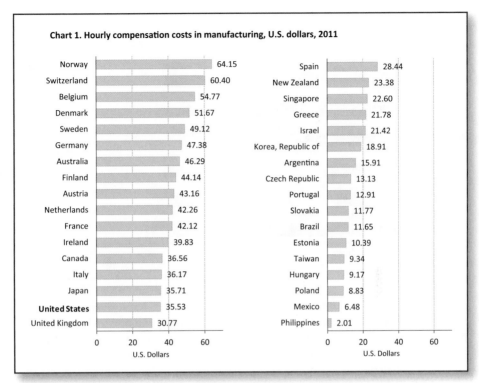

Chart 1. Hourly compensation costs in manufacturing, U.S. dollars, 2011

	U.S. Dollars		U.S. Dollars
Norway	64.15	Spain	28.44
Switzerland	60.40	New Zealand	23.38
Belgium	54.77	Singapore	22.60
Denmark	51.67	Greece	21.78
Sweden	49.12	Israel	21.42
Germany	47.38	Korea, Republic of	18.91
Australia	46.29	Argentina	15.91
Finland	44.14	Czech Republic	13.13
Austria	43.16	Portugal	12.91
Netherlands	42.26	Slovakia	11.77
France	42.12	Brazil	11.65
Ireland	39.83	Estonia	10.39
Canada	36.56	Taiwan	9.34
Italy	36.17	Hungary	9.17
Japan	35.71	Poland	8.83
United States	35.53	Mexico	6.48
United Kingdom	30.77	Philippines	2.01

FIGURE 6-5 Bar Graph This bar graph is a coherent display of the same issue at different locations at the same point in time. The ordering of the bars from largest to smallest assists the viewer in navigating the 34-nation list and making rapid sense of it. The United States is listed in bold type to emphasize its position on the list and to facilitate comparison and contrast with nations higher and lower on the list. Each bar carries its numerical identity, making the vertical grid lines unnecessary. The blue coloring of the bars is also unnecessary but at least uniform, indicating clearly that it is the same information (i.e., labor costs per hour in US dollars) that is being displayed about each nation on the list: dark blue instead of light blue would offer better contrast and easier viewing against the white background.

Source: US Department of Labor, Bureau of Labor Statistics. *International Comparisons of Hourly Compensation Costs in Manufacturing, 2011*. News Release USDL-12-2460. Washington, DC: GPO, 2012. p. 1.

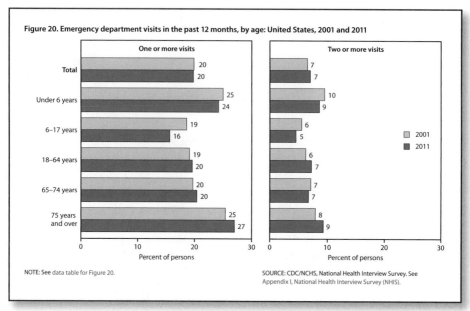

Figure 20. Emergency department visits in the past 12 months, by age: United States, 2001 and 2011

NOTE: See data table for Figure 20.

SOURCE: CDC/NCHS, National Health Interview Survey. See Appendix I, National Health Interview Survey (NHIS).

FIGURE 6-6 Bar Graph This bar graph makes a clear and vivid point regarding the frequency of emergency room visits of people of various ages at two different points in time. The two years, 2001 and 2011, are easily identified by obviously different colors, as explained in a legend conveniently positioned inside the graph but visibly separated from the bars. The age categories are all crisply labeled for easy reading and understanding. The categories are positioned on the graph according to age. The major distraction of the graph is the separation into two grids, "One or more visits" and "Two or more visits," instead of a single display with divided bars. The separated display might disrupt or delay recognition that "two or more" is included in "one or more" and thus lead to misinterpretation. Each bar is numerically labeled (making grid lines unnecessary), but it would be wise to specify the numbers to at least one decimal point, instead of rounding, to eliminate the likely confusion created by the different size bars both labeled "20" and the different size bars both labeled "7" in the "Total" and "65–74 years" categories.

Source: US Department of Health and Human Services, National Center for Health Statistics. *Health, United States, 2012: With Special Feature on Emergency Care.* DHHS Publication No. 2013-1232. Washington, DC: GPO, 2013. p. 22.

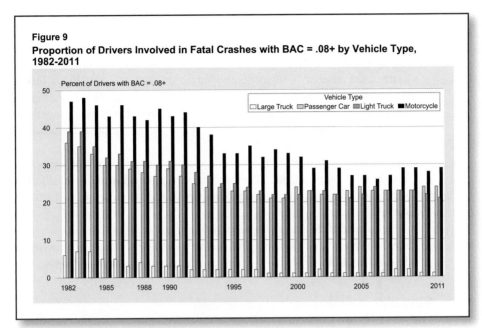

Figure 9
Proportion of Drivers Involved in Fatal Crashes with BAC = .08+ by Vehicle Type, 1982-2011

FIGURE 6-7 Ineffective Column Graph This column graph is difficult to decipher because it displays too many columns at too many points in time in too narrow a space. The distracting visual effect makes the data look crowded and cramped instead of readily accessible. The graph would be more effective, for example, if only the odd-numbered years were displayed. The alternative, of course, is to allow the entire graph more horizontal space. In addition, the color of the columns for the large truck category is almost indistinguishable from the white background of the graph, making these particularly short columns easy to overlook in the visual clutter: gray would have been a better color choice for this category of vehicle. At least the explanatory legend is conveniently positioned inside the graph and sticks to simple solid colors instead of using dots and diagonal lines.

Source: US Department of Transportation, National Highway Traffic Safety Administration. *Traffic Safety Facts 2011.* DOT HS 811 754. Washington, DC: GPO, 2013. p. 37.

Circle graphs.

- Note that the purpose of a circle graph (also referred to as a "pie chart") is to display the number and relative size of the divisions of a subject.
- Restrict the number of segments in a circle graph to seven or eight. This is the limit on the number of segments into which a circle can be divided before comprehension of the relative sizes is jeopardized. If necessary, combine several smaller segments and create a second circle graph to display the composition of that combined segment.
- Watch for possible distortion when you use three-dimensional circle graphs.
- Clearly label all segments. Whether they are placed inside or outside the circle, labels should be horizontal for easier reading. Avoid using explanatory legends (or keys).
- As you segment the graph, begin with the largest section in the upper right-hand quadrant. The remaining segments should be arranged clockwise, in descending order. Color the sections clockwise from darker to lighter so that color reinforces size.

See Figure 6–8 for an example of a circle graph.

Line graphs.

- Note that the purpose of a line graph is to show the degree and direction of change relative to two variables.
- Label each axis clearly. Like bar and column graphs, line graphs must have clearly labeled scales to show the variables you are measuring. Ordinarily, the independent variable is placed on the horizontal, or x, axis and the dependent variables are placed on the vertical, or y, axis (and, in a three-dimensional graph, on the diagonal, or z, axis).
- Choose the scale of each axis to show the appropriate steepness of the slope of the line. Typically, the scales start at zero, with the intervals equally spaced on each axis.
- The major difficulty in designing line graphs lies in choosing the spacing for each axis so that the steepness (slope) of the line accurately measures the actual trend suggested by the data. Typical graphics applications will allow you to adjust the intervals on the x axis and the y axis, but your job is to decide whether the slope of the graph accurately depicts your data or gives a distorted impression.
- Avoid using more than three data lines in one graph unless they are separated by visible space. Graphs with lots of overlapping and intersecting data lines are usually difficult for readers to interpret.
- Keep the data lines in your graph distinctive by using either different colors or different styles for each line.
- If possible, label each data line. Avoid explanatory legends (or keys) because these will slow reader comprehension.

See Figures 6–9 and 6–10 for examples of line graphs.

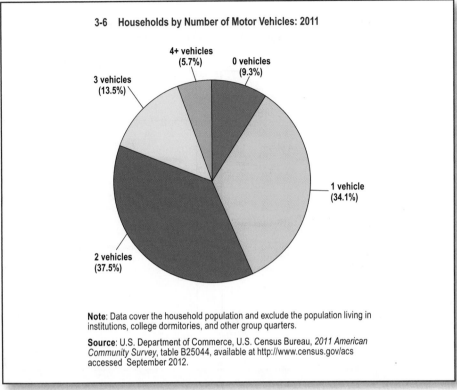

3-6   **Households by Number of Motor Vehicles: 2011**

4+ vehicles
(5.7%)

0 vehicles
(9.3%)

3 vehicles
(13.5%)

1 vehicle
(34.1%)

2 vehicles
(37.5%)

Note: Data cover the household population and exclude the population living in institutions, college dormitories, and other group quarters.

Source: U.S. Department of Commerce, U.S. Census Bureau, *2011 American Community Survey*, table B25044, available at http://www.census.gov/acs accessed September 2012.

FIGURE 6–8 Circle Graph This circle graph uses clear labels for each of its five sections and consistently positions the labels outside the circle adjacent to each respective section. While effective circle graphs typically order their sections from largest to smallest, this circle graph is nevertheless logically organized (proceeding clockwise from 0 vehicles to 4+ vehicles). This logical ordering of the sections makes the graph easier to decipher. The coloring of the sections, however, is entirely decorative. For example, instead of the darkest color used for the largest section and the lightest color used for the smallest section, color here is arbitrarily assigned instead of systematically applied. And given the wide variety of available colors, it is curious (and potentially confusing) that two shades of yellow are adopted.

Source: US Department or Transportation, Bureau of Transportation Statistics. *Pocket Guide to Transportation.* Washington, DC: GPO, 2013. p. 19.

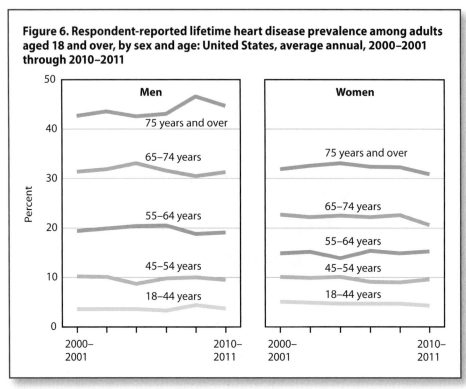

Figure 6. Respondent-reported lifetime heart disease prevalence among adults aged 18 and over, by sex and age: United States, average annual, 2000–2001 through 2010–2011

FIGURE 6-9 **Line Graph** This pair of line graphs uses clear labeling to identify and differentiate five data lines. It also uses color to allow easy comparison and contrast of the data lines across the two graphs, making evident that men and women have similar rates of heart disease until about age 55. Horizontal grid lines assist viewers in deciphering the relative positions of the paired data lines, while vertical grid lines prove unnecessary for the essentially static 10-year period. Unlike Figure 6-6, the two separate grids here are appropriate for the two separate and mutually exclusive sets of data.

Source: US Department of Health and Human Services, National Center for Health Statistics. *Health, United States, 2012: With Special Feature on Emergency Care.* DHHS Publication No. 2013-1232. Washington, DC: GPO, 2013. p. 10.

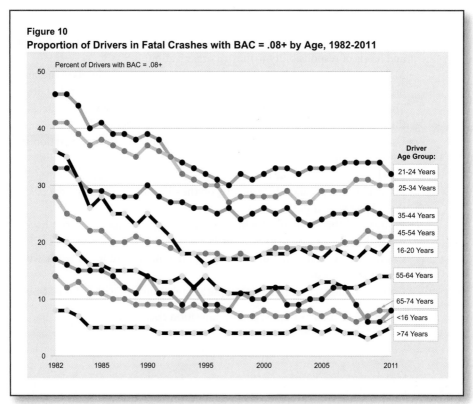

Figure 10
Proportion of Drivers in Fatal Crashes with BAC = .08+ by Age, 1982-2011

FIGURE 6-10 Ineffective Line Graph This could be the least effective line graph ever designed because it focuses the viewer's attention on all the unnecessary circles instead of the data lines themselves. The graph does include a label for each line, but the nine labels are located on the right side of the graph, making it necessary for readers to read each data line from right to left instead of left to right. The graph also uses three different color combinations to differentiate the nine data lines—itself a confusing practice—but the use of color is also arbitrary: the key point of the graph is that alcohol-related driver fatalities in all age groups dropped from 1982 to 2008 but most precipitously in the 16–20 age group. Greater emphasis on this point would be achieved if this were the only group displayed with a black data line.

Source: US Department of Transportation, National Highway Traffic Safety Administration. *Traffic Safety Facts 2011.* DOT HS 811 754. Washington, DC: GPO, 2013. p. 39.

Organization charts.

- Note that the purpose of an organization chart is to map the various divisions and levels of responsibility within an organization.
- Make the chart as simple as the organization itself, with the levels of hierarchy organized highest to lowest and positioned on the chart from top to bottom. The more levels of hierarchy in the organization, the more vertical the chart; the more divisions in the organization, the more horizontal the chart.
- Use the same shape in the same size for all divisions of the organization that are at the same level in the hierarchy.
- Label each division of the organization.
- If space allows, put the labels directly on the division; if not, attach the label to the subject with thin rules (never arrows).
- Position all the labels on the horizontal so that the viewer doesn't have to rotate the page or screen to read the labels.
- Connect each level of the organization to the higher and lower levels with a clear line (never arrows).
- Connect optional, informal, or temporary relationships with a dotted line.

See Figure 6–11 for an example of an organization chart.

FIGURE 6–11 Organization Chart (see facing page) This organization chart effectively depicts the multiple levels of authority at the US Nuclear Regulatory Commission: for example, regional administrators report to a deputy executive director, who reports to the executive director, who reports to the chair of the commission. Notice also that certain officials in the organization report to the chair, while others report to the entire commission. Instead of the usual boxes with position title only, here photographs and names of the individuals in each position, diminishing in size according to their level in the hierarchy, bring a sense of humanity to the organization.

Source: US Nuclear Regulatory Commission. NRC Organization Chart. Washington, DC: GPO, 2013. http://www.nrc.gov/about-nrc/organization/nrcorg.pdf

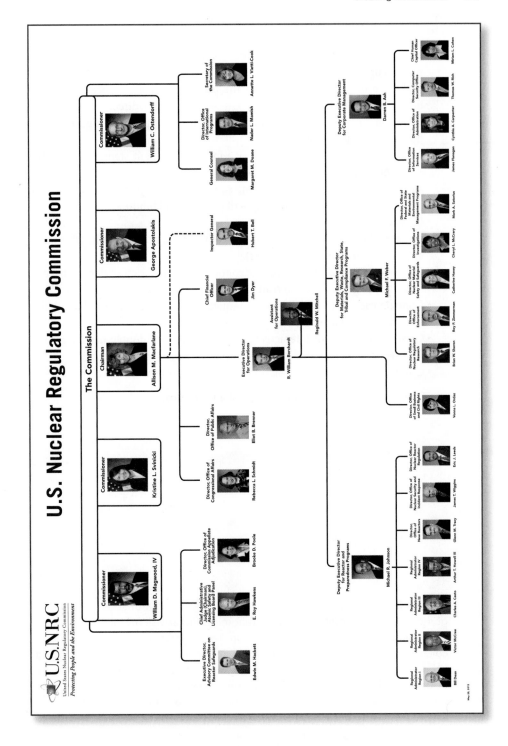

Flow charts.

- Note that the purpose of a flow chart is to show the sequence of steps in a process or procedure.
- Make the flow chart as simple as the process itself. If a process is simple, design the flow chart so that it progresses in a single direction, usually top to bottom or left to right. Complicated designs that spiral and zigzag imply a more complicated process.
- Use the same shape in the same size for all equivalent steps or phases but different shapes for steps or phases of a different kind (e.g., circles for the stages in researching a document, squares for the stages of writing, and diamonds for the production stages of printing and binding).
- Label each of the steps or phases.
- If space allows, put the labels directly on the step; if not, attach the label to the step with thin rules (never arrows).
- Position all the labels on the horizontal so that the viewer doesn't have to rotate the page or screen to read the labels.
- Connect each step or phase in the sequence to the next step or phase with a clear directional arrow.
- Connect reversible or interactive steps or phases with double-headed arrows.
- Connect recursive or cyclical steps or phases with circular arrows.
- Connect optional steps or phases with dotted-line arrows.

See Figures 6–12 and 6–13 for examples of a flow chart.

FIGURE 6–12 Flow Chart (see facing page) This flow chart uses several levels of redundancy to simplify a process that might otherwise mystify or intimidate a prospective filer of a patent application. The process is organized in three columns: the number of each step, the agent of action, and the corresponding action itself. Actions are displayed in blue rectangles, while decision points are displayed in white diamonds. Arrows direct the viewer to the next step in the process and reinforce the numbering. Links to more information are included in almost every step but always opening as new pages in the viewer's browser to make it easy to preserve this overview of the entire process.

Source: US Department of Commerce, Patent and Trademark Office. *Process for Obtaining a Utility Patent.* 2013. http://www.uspto.gov/patents/process/index.jsp

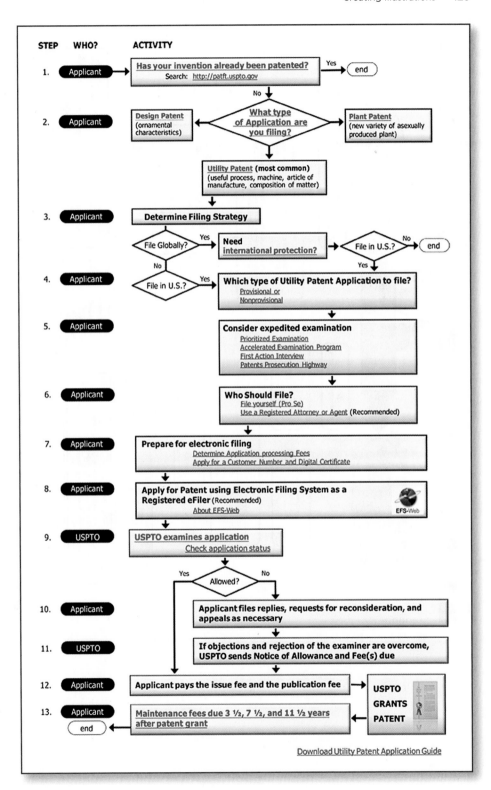

STEP   WHO?   ACTIVITY

1.   Applicant   **Has your invention already been patented?**   Yes → end
Search: http://patft.uspto.gov

No ↓

2.   Applicant   **Design Patent** (ornamental characteristics) ← **What type of Application are you filing?** → **Plant Patent** (new variety of asexually produced plant)

↓

Utility Patent (most common) (useful process, machine, article of manufacture, composition of matter)

↓

3.   Applicant   **Determine Filing Strategy**

File Globally?   Yes →   **Need international protection?**   →   File in U.S.?   No → end
No ↓   Yes ↓

4.   Applicant   File in U.S.?   Yes →   **Which type of Utility Patent Application to file?** Provisional or Nonprovisional

↓

5.   Applicant   **Consider expedited examination** Prioritized Examination / Accelerated Examination Program / First Action Interview / Patents Prosecution Highway

↓

6.   Applicant   **Who Should File?** File yourself (Pro Se) / Use a Registered Attorney or Agent (Recommended)

↓

7.   Applicant   **Prepare for electronic filing** Determine Application processing Fees / Apply for a Customer Number and Digital Certificate

↓

8.   Applicant   **Apply for Patent using Electronic Filing System as a Registered eFiler** (Recommended) / About EFS-Web   EFS-Web

↓

9.   USPTO   **USPTO examines application** Check application status

↓

Yes ←   Allowed?   → No

10.   Applicant   **Applicant files replies, requests for reconsideration, and appeals as necessary**

↓

11.   USPTO   **If objections and rejection of the examiner are overcome, USPTO sends Notice of Allowance and Fee(s) due**

↓

12.   Applicant   **Applicant pays the issue fee and the publication fee** →   **USPTO GRANTS PATENT**

13.   Applicant   **Maintenance fees due 3 ½, 7 ½, and 11 ½ years after patent grant** ←
end

Download Utility Patent Application Guide

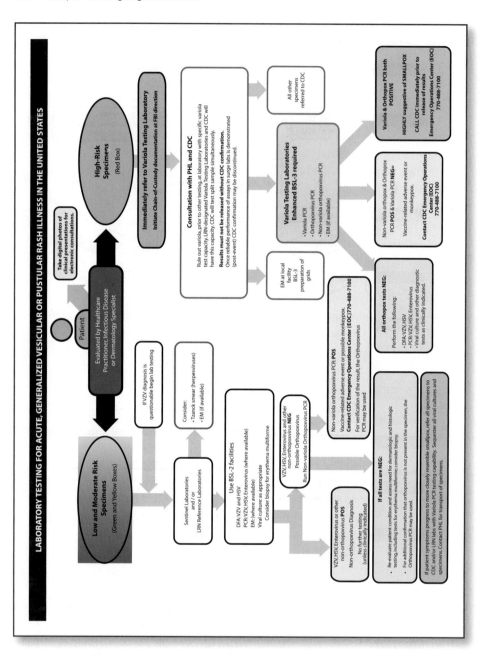

LABORATORY TESTING FOR ACUTE, GENERALIZED VESICULAR OR PUSTULAR RASH ILLNESS IN THE UNITED STATES

FIGURE 6-13 Ineffective Flow Chart (see facing page) This flow chart uses a confusing array of rectangles in various sizes and colors, but the figure has a fairly straightforward organization: left or right then top to bottom. Some of the boxes carry clear labels in bold type, others have only bold type, and still others have only plain type or combinations of bold and plain type. Some of the boxes are outlined in red, one is outlined in green, and most are outlined in black. The significance of these typographical and color variations is not immediately clear. Neither is it evident why some of the boxes are white, while others display a variety of colors. The unexplained variations make deciphering the process unnecessarily challenging.

Source: US Department of Health and Human Services, Centers for Disease Control and Prevention. *Acute, Generalized Vesicular or Pustular Rash Illness Testing Protocol in the United States.* 2012. http://www.emergency.cdc.gov/agent/smallpox/diagnosis/pdf/poxalgorithm1-5-12.pdf

Project schedule charts.

- Note that the purpose of a project schedule chart (also referred to as a "Gantt chart," after its creator, Henry Gantt) is to display the start and stop dates for all consecutive and simultaneous tasks related to a project. It offers a quick picture of tasks in progress at any given point in time, as well as how much work has been completed on each task and how much work remains.
- Make the project schedule chart as simple as the project itself, with one horizontal bar for every major task or activity and vertical grid lines for each reporting period (e.g., daily, weekly, monthly). Excessive division of the project into a multitude of minor tasks or milestones will create a chart that is difficult to decipher. Also avoid abbreviations and symbols and their explanatory legends (or keys).
- List the tasks from top to bottom in order of their start date.
- Extend a horizontal bar for each task from its start date to its anticipated date of completion or specified deadline.
- Fill in (or color) each bar from left to right as work on the task proceeds.
- If you don't have access to applications that will generate a project schedule chart, create a table in your word processing program with a row for each task and a column for each reporting period and fill in cells from left to right as progress is made.
- Incorporate additional information for each task as necessary and appropriate (e.g., names of task managers, facilities and resources assigned for each task, funds budgeted and expended in each reporting period), but realize that every extra level of detail complicates the illustration and slows interpretation.

See Figure 6–14 for an example of a project schedule chart.

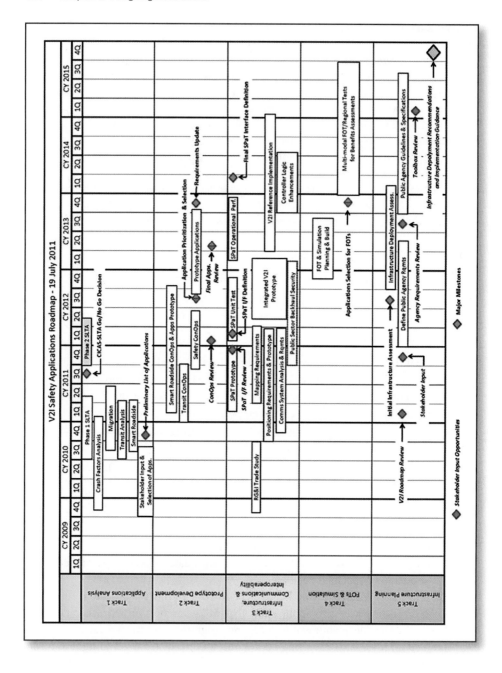

FIGURE 6-14 **Project Schedule Chart** (see facing page) This project schedule covers five tasks across 7 years, with each year divided in quarters. The five tasks are displayed in cells of different colors, a nice but unnecessary decoration devoid of meaning. Color is used more productively to identify opportunities for stakeholder input and major milestones in the schedule: the purple and green diamonds are readily visible among all the white rectangles and horizontal and vertical grid lines, and each diamond is clearly labeled with arrows running from label to diamond. Putting white rectangles on a white grid does create a visual challenge and make for slower reading: light coloring of the grid background would make it easier for viewers to recognize the borders of the rectangles and thus the start/stop times of tasks in the schedule.
Source: US Department of Transportation, Research and Innovative Technology Administration. Vehicle-to-Infrastructure Safety Applications Roadmap. 2011. http://www.its.dot.gov/roadmaps/v2i_roadmap.htm

Diagrams.

- Note that a diagram should identify parts of a subject and their spatial or functional relationship.
- Keep the diagram as simple as possible, avoiding unnecessary details or distracting decorations and focusing the viewer's attention on the key features or parts of the subject.
- Label each of the pertinent parts of the subject.
- If space allows, put each label directly on its corresponding part or adjacent to it; if not, connect the label to the part with a thin rule or arrow.
- Position all the labels on the horizontal so that the viewer doesn't have to rotate the page or screen to read the labels.

See Figure 6–15 for an example of a diagram.

FIGURE 6-15 Diagram (see facing page) This diagram of the calcium carbide process focuses on the details pertinent to the investigation of the explosion and fire that killed two and injured two at a five-floor facility. As the diagram makes clear with its concise labels and simplified drawings, coke is unloaded, dried, mixed with lime, and brought by conveyor to a furnace fueled by electricity; exhaust gases are scrubbed before release into the atmosphere, and the resulting product is transported away by automatic conveyor. Note the inconsistent use of color: while red (implying heat) might be appropriate for the coke-drying equipment as well as for the furnace, the red stripes on the vehicle unloading lime are entirely decorative. It might also be appropriate to add human icons to the control room: this is where operators are located during the process, and this was the location of the individuals killed and injured during the accident in question.

Source: US Chemical Safety and Hazard Investigation Board. *Carbide Industries, LLC, Louisville, Kentucky, Electric Arc Furnace Explosion, March 21, 2011.* Case Study No. 2011-05-I-KY. Washington, DC: GPO, 2013. p. 16.

Photographs.

- Note that photographs show what a subject looks like in realistic detail.
- Keep the photograph as simple as possible, focusing the viewer's attention on the key features or parts of the subject.
- Exercise caution in editing or polishing the photograph. Never insert or delete images. Viewers typically expect a photograph to be a representation of reality, and ethical communicators strive to meet that expectation. Cropping a photograph in order to close in on a subject and eliminate distractions in the background is a standard practice, but never insert or delete objects or change the size or color of objects in a photograph.
- If appropriate or necessary for the viewer's understanding, apply labels for each of the pertinent parts of the subject to direct the viewer's attention.
- If space allows, put each label directly on its corresponding part or adjacent to it; if not, connect the label to the part with a thin rule or arrow.
- Position all the labels on the horizontal so that the viewer doesn't have to rotate the page or screen to read the labels.

See Figures 6–16 and 6–17 for examples of photographs.

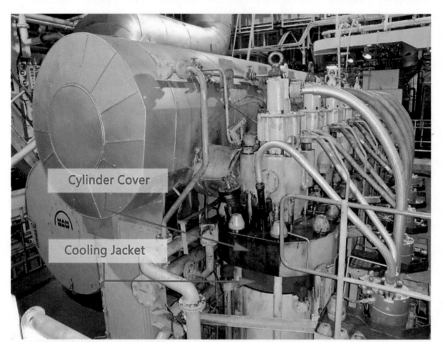

The *FR8 Pride*'s main propulsion engine, with the location of the failed engine component (cooling jacket for cylinder No. 5) indicated. Photo by the Coast Guard.

FIGURE 6-16 Photograph This photograph carries clear labels to identify the pertinent components of a ship engine that failed and caused the collision of two vessels (with no injuries but $16–17 million in damages). The photograph puts viewers in the engine room and makes visible (and almost tangible) the complexity of the technology, while the arrows direct attention across all the pipes, wires, hoses, and valves to the source of the engine failure.

Source: National Transportation Safety Board. *Marine Accident Brief: Collision of Oil Tanker* FR8 Pride *with MODU* Rowan EXL I. NTSB/MAB-13/13. Washington, DC: GPO, 2013. p. 5.

Figure 5. Suppression of the fire.

FIGURE 6–17 Ineffective Photograph This photograph is designed to display the suppression of fire following the horrific collision of two trains, but viewers are likely to focus their initial attention at least on the four firefighters standing in the foreground and watching the suppression effort. The photograph thus does a disservice both to the severity of the collision, which is here backgrounded, and to the heroic efforts of the firefighters, who are here depicted as passive observers. A more effective photograph would either exclude the distraction of the inactive firefighters and direct attention to the burning trains being doused with water or include active firefighters engaged in the fire-suppression effort.

Source: National Transportation Safety Board. *Head-On Collision of Two Union Pacific Railroad Freight Trains Near Goodwell, Oklahoma, June 24 2012.* Railroad Accident Report NTSB/RAR-13/02. Washington, DC: GPO, 2013. p. 6.

Infographics.

- Note that infographics make information easier to understand and easier to remember.
- Keep the pictorial images of the infographic as simple and straightforward as possible, avoiding unnecessary details or distracting decorations and focusing the viewer's attention on key traits.
- Choose pictorial images appropriate to your likely viewers (especially their age, level of education, and culture). Make sure the image clarifies without insulting or irritating your viewers.
- Choose pictorial images appropriate to the gravity of your subject (especially for issues related to safety). Make sure the image neither trivializes nor distorts the subject.
- Reinforce the pictorial image with a minimum of explanatory text, labels, or captions.

See Figure 6–18 for an example of an infographic.

FIGURE 6-18 Infographic (see facing page) This series of infographics composes a vivid warning about the relative dangers of fireworks. The images of the various kinds of fireworks, from sparklers to public displays, are simplistic but concisely identified. The images of the human beings, however, are almost comical in the ungainly positioning of the bodies and could keep viewers from taking seriously the risk of injuries on which the infographic is reporting.
Source: US Consumer Product Safety Commission. *Fireworks Injuries.* 2013. http://www.cpsc.gov/en/ Safety-Education/Safety-Education-Centers/Fireworks/

Fireworks Injuries

Fireworks. They are synonymous with our celebration of Independence Day. Yet, the thrill of fireworks can also bring pain, and even death. In 2012, CPSC staff conducted a study of fireworks injuries from June 22 through July 22. Here's what we learned.

★ **200** people on average go to the emergency room every day with fireworks-related injuries in the month around the July 4th holiday.

★ **60%** of these fireworks injuries in 2012 occurred during the month surrounding July 4th.

★ **Illegal** and **homemade fireworks** were involved in all **6** fireworks-related **deaths** reported to CPSC in 2012.

Most Injured Body Parts

12% Eyes

19% Heads, faces and ears

41% Hands and fingers

1% Arms

15% Trunks

13% Legs

More than **HALF** the injuries were burns.

Injuries by Fireworks Type*

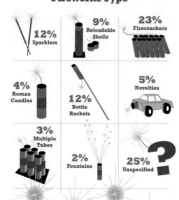

12% Sparklers

9% Reloadable Shells

23% Firecrackers

4% Roman Candles

12% Bottle Rockets

5% Novelties

3% Multiple Tubes

2% Fountains

25% Unspecified

2% Public Display

*Those percents do not account for how many products are used.

Injuries by Age

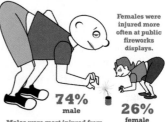

23% 25-44   **15%** 45-64   **10%** 5-9   **9%** 10-14

16% 20-24   **15%** 15-19   **10%** 0-4   **1%** 65+

Injuries by Gender

Females were injured more often at public fireworks displays.

74% male

26% female

Males were most injured from firecrackers, sparklers, bottle rockets, novelty devices, Roman candles and reloadable shells.

Fireworks Safety Tips

★ Never allow **children** to play with or ignite fireworks.

★ **Never** try to **re-light** or **pick up** fireworks that have not ignited fully.

★ Keep a **bucket of water** or a **garden hose** handy in case of fire or other mishap.

★ Make sure fireworks are **legal** in your area before buying or using them.

★ Light fireworks **one at a time**, then **move back** quickly.

★ More Fireworks Safety Tips – **www.cpsc.gov/fireworks**

Source: U.S. Consumer Product Safety Commission 2012 Fireworks Annual Report

U.S. Consumer Product Safety Commission
CPSC Hotline: (800) 638-2772
www.cpsc.gov

Video Clips.

- Note that video clips illustrate a process, operation, or incident. Film clips depict activity in realistic detail. Animation clips simulate activity that would be impossible, imprecise, or impractical to film.
- Keep the video clip as simple and straightforward as possible, avoiding unnecessary details or distracting decorations and focusing the viewer's attention on the key actors, equipment, and steps or stages in the process, operation, or incident.
- If appropriate or necessary for the viewer's understanding, insert audio narrative or titles and captions that direct the viewer's attention to pertinent information.
- Choose animation images appropriate to your likely viewers (especially their age, level of education, and culture) as well as the gravity of your subject

FIGURE 6-19 Film Clip This 10-minute video explains how workers who sometimes must wear a respirator on the job should store and clean this important safety equipment. The film covers a variety of workers in different kinds of jobs using a variety of respirators. Titles and diagrams are inserted periodically during the film to reinforce points made in the audio narration and in the activities depicted. The use of film for this video serves to reinforce the reality of hazardous work environments and the importance of proper care and maintenance of respirators.

Source: US Department of Labor, Occupational Safety and Health Administration. *Maintenance and Care of Respirators.* Video. 2012. https://www.osha.gov/video/respiratory_protection/maintenance.html

(especially for issues related to safety). Make sure the animation clarifies without insulting or irritating your viewers and neither trivializes nor distorts your subject.

- Exercise caution in editing or polishing a film clip. Viewers typically expect a film clip to be a representation of reality, and ethical communicators strive to meet that expectation. Editing a film for extraneous material is a standard practice, but never change the speed or sequence of the process, operation, or incident (unless the change is altogether obvious), and always make sure that your editing creates no distortions or misrepresentations. In addition, treat each frame of the film as you would a photograph: never insert or delete objects or actors and never alter the size or color of the items in a frame.

See Figures 6–19 and 6–20 for screen shots of sample video clips.

FIGURE 6–20 Animation Clip This 23-second video simulates how flood debris during storms can destroy buildings and create more debris. The use of animation offers a quick, simple, and coherent picture of a process that would be expensive and dangerous to film and that would be neither clear nor vivid in static diagrams or photographs.

Source: US Department of Homeland Security, Federal Emergency Management Agency. *Coastal Animation— Debris Impact Forces.* Video. 2012. http://www.fema.gov/medialibrary/media_records/10513

DESIGNING ILLUSTRATIONS ETHICALLY

Displaying information ethically requires that you make careful choices about the design of your illustrations.

For example, the scale of the x and y axes on a line graph has a significant impact on the data display. In designing a graph, you ordinarily start the x and y axes at zero. Exceptions are possible if beginning at some other point ("suppressing the zero") will not distort information. If several graphs are positioned side by side on a page, thus inviting comparison and contrast of the data, it would be unethical to use differing vertical scales because readers would likely come to incorrect conclusions.

Using distorted graphs, however, isn't the only error that will result in the creation of unethical illustrations. It is unethical to create a drawing that puts features on a product it doesn't really have. It is unethical to design a flowchart that disguises the complexity of a process by making it look relatively simple. It is unethical to stage or doctor a photograph to create a positive or negative impression of your subject that isn't fully justified. It is unethical to edit a film so that a 5-minute operation seems to take only 4 minutes. It is unethical to incorporate illustrations that might be easily misinterpreted.

In addition, if you illustrate information regarding people, you must strive to be sensitive to their humanity. A thoughtless or insensitive illustration could be widely and quickly circulated on the Internet, jeopardizing your organization's reputation for caring about people, especially the people in the communities in which it operates. For example, if you choose a conventional bar graph to depict the human beings killed in mining explosions, you make simple objects of real people who have surviving families and friends. In mapping the damage from a tornado, if you ignore the locations of the people who were killed and focus only on demolished buildings, you both diminish the loss of life and give a distorted picture of the disaster. Always consider ways that you might bring a sense of humanity (and thus greater accuracy) to your illustrations. Notice, for example, how the human dimension of a tragic highway accident is made evident in Figure 6–21 by the inclusion of brief demographic details and icons of the passengers.

Figure 4. Motorcoach occupant diagram, showing seating location, injury, and ejection information.

FIGURE 6–21 Diagram of Injuries and Fatalities Sensitive to Victims This diagram of passenger locations on a bus involved in a highway accident is sensitive to the humanity of the victims. Implicit in the demographic details of age and sex are the relationships among the passengers (e.g., row 10: F-19, F-46, M-01 On-lap). The human icons (displayed in red for the four passengers killed) make the truly tragic consequences of this accident vivid and undisguised.

Source: National Transportation Safety Board. *Motorcoach Roadway Departure and Overturn on Interstate 95 Near Doswell, Virginia, May 31, 2011.* Highway Accident Report NTSB/HAR-12/02. Washington, DC, 2012. p. 17.

ILLUSTRATION CHECKLIST ✔

Planning

☐ What kinds of illustrations is your audience familiar with?

☐ Do you have information that could be more easily or quickly communicated to your audience visually or in a combination of words and graphics?

☐ Do you have definitions that could be displayed visually in whole or in part?

☐ Do you have any processes or procedures that could be depicted in a flow chart?

☐ Do you have information on trends or relationships that could be displayed in tables and graphs?

☐ Do you have masses of statistics that could be summarized in tables?

☐ Do you need to depict objects? If so, what do you need to display about the objects? Do you need to focus attention on specific aspects of the objects with diagrams? Do you require the realistic detail of photographs?

☐ Do you need to depict incidents or operations? If so, what do you need to display? Do you need to focus attention on specific aspects through animation? Do you require the realistic detail of film?

☐ What are the design conventions of your illustrations?

☐ Are there suitable illustrations you could borrow or adapt? Or will you need to create them yourself?

Revising

☐ Are your illustrations suited to your purpose and audience?

☐ Do your illustrations communicate information ethically?

☐ Are your illustrations effectively located and easy to find?

☐ Are your illustrations numbered and labeled?

☐ Do your verbal and visual elements reinforce each other?

☐ Are your illustrations genuinely informative instead of simply decorative?

☐ When necessary, have you helped your readers to interpret your illustrations with commentary or annotations?

☐ Have you acknowledged the sources for borrowed or adapted tables and figures?

EXERCISES

1. Examine the illustrations used in five journals and magazines in your major field of study. Which types of illustration do you ordinarily find? Which types don't you find? Which occur most often? Which occur least often? What are the conventions for tables and figures in your field? Summarize your findings in a brief oral presentation with slides that display representative illustrations.

2. Zeitgeist Corporation designs games for mobile devices and is looking for individuals with the experience and training to conceive and design state-of-the-art products. It has been a small private corporation for 5 years and hopes to achieve a major expansion. The president of Zeitgeist, Miyuki Sasaki, has asked you to develop recruiting materials. You've compiled the following information that you believe will be pertinent:

History	Founded in 2010, in Long Beach, California, by Miyuki Sasaki, BS in Computer Science, 2007, University of Ohio; MBA, 2009, University of Los Angeles
Sales	2010: $250,000; 2011: $750,000; 2012: $1 million; 2013: $4 million; 2014: $7 million
Employees	With specialists in design (3), graphics and animation (4), programming (5), video editing (3), production (2), quality assurance (1), documentation (1), sound (1), and music (1)
Products	5 current products, looking to diversify, especially in action–adventure, newest product: Street Maniac (driving game); biggest seller: Trojan War (fantasy role-playing)
Facilities	State-of-the-art equipment, serves both Android and iOS systems
Location	1221 Alamitos Avenue, Long Beach, California: new environment-friendly building, spacious offices
Salary/benefits	Competitive with industry; exceptional medical coverage, 4 weeks vacation each year, 2 weeks paid community-service time each year

Design recruiting materials for Zeitgeist that incorporate as wide a variety of illustrations as possible.

3. In designing your recruiting materials for Zeitgeist, you are surprised by a dilemma. Miyuki Sasaki is willing to hire people with disabilities and would like you to include that information. At this time, Zeitgeist has no employees with disabilities, but Jacqueline Bailey, a programmer, injured her back in a skiing accident and is temporarily restricted to a wheelchair. Sasaki proposes including a photograph of various Zeitgeist employees (including Bailey in the wheelchair)

to indicate to prospective employees that people with disabilities would find a supportive environment at Zeitgeist. You mention the idea to staff photographer Kishor Mitra, but he objects, claiming such a picture would be deceptive. As project leader, you have to decide who is right and compose one of two e-mail messages: either to convince Sasaki that such a photograph would be unethical or to direct Mitra to take the photograph in spite of his objections. How do you proceed?

4. Examine Figure 6–22. If you were a collaborator on this document and were given 15 minutes to design more effective illustrations, what changes would you make? You likely won't have time to fix everything but must decide quickly which changes will have the greatest impact.

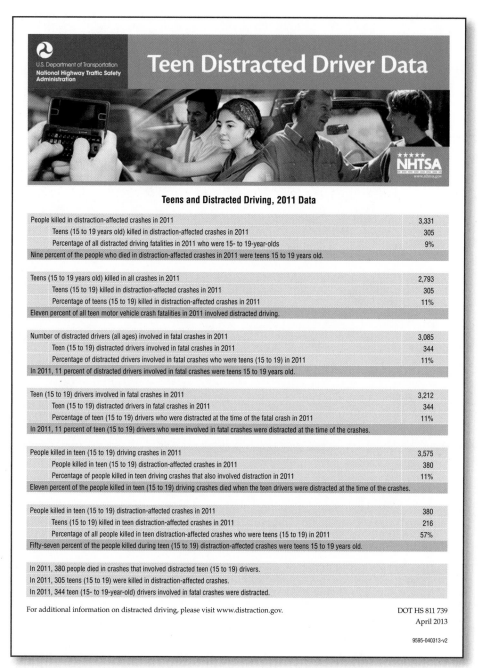

Teens and Distracted Driving, 2011 Data

People killed in distraction-affected crashes in 2011	3,331
Teens (15 to 19 years old) killed in distraction-affected crashes in 2011	305
Percentage of all distracted driving fatalities in 2011 who were 15- to 19-year-olds	9%
Nine percent of the people who died in distraction-affected crashes in 2011 were teens 15 to 19 years old.	
Teens (15 to 19 years old) killed in all crashes in 2011	2,793
Teens (15 to 19) killed in distraction-affected crashes in 2011	305
Percentage of teens (15 to 19) killed in distraction-affected crashes in 2011	11%
Eleven percent of all teen motor vehicle crash fatalities in 2011 involved distracted driving.	
Number of distracted drivers (all ages) involved in fatal crashes in 2011	3,085
Teen (15 to 19) distracted drivers involved in fatal crashes in 2011	344
Percentage of distracted drivers involved in fatal crashes who were teens (15 to 19) in 2011	11%
In 2011, 11 percent of distracted drivers involved in fatal crashes were teens 15 to 19 years old.	
Teen (15 to 19) drivers involved in fatal crashes in 2011	3,212
Teen (15 to 19) distracted drivers in fatal crashes in 2011	344
Percentage of teen (15 to 19) drivers who were distracted at the time of the fatal crash in 2011	11%
In 2011, 11 percent of teen (15 to 19) drivers who were involved in fatal crashes were distracted at the time of the crashes.	
People killed in teen (15 to 19) driving crashes in 2011	3,575
People killed in teen (15 to 19) distraction-affected crashes in 2011	380
Percentage of people killed in teen driving crashes that also involved distraction in 2011	11%
Eleven percent of the people killed in teen (15 to 19) driving crashes died when the teen drivers were distracted at the time of the crashes.	
People killed in teen (15 to 19) distraction-affected crashes in 2011	380
Teens (15 to 19) killed in teen distraction-affected crashes in 2011	216
Percentage of all people killed in teen distraction-affected crashes who were teens (15 to 19) in 2011	57%
Fifty-seven percent of the people killed during teen (15 to 19) distraction-affected crashes were teens 15 to 19 years old.	
In 2011, 380 people died in crashes that involved distracted teen (15 to 19) drivers.	
In 2011, 305 teens (15 to 19) were killed in distraction-affected crashes.	
In 2011, 344 teen (15- to 19-year-old) drivers involved in fatal crashes were distracted.	

For additional information on distracted driving, please visit www.distraction.gov.

DOT HS 811 739
April 2013

9595-040313-v2

FIGURE 6–22 Document for Exercise 4

Source: US Department of Transportation, National Highway Traffic Safety Administration. *Teen Distracted Driver Data.* 2013. http://www-nrd.nhtsa.dot.gov/Pubs/811739.pdf

PART **TWO**

Applications

E-mails, Texts, Memos, and Letters

E-mails, text messages, memos, and letters are among the most frequently written business documents. Because people receive so much correspondence, they can often overlook, forget, or misread important messages.

Quick Tips

- Prepare readers for the key information in your message by using a clearly worded subject line or opening sentence.
- Put the most important information in the opening paragraph.
- Use formatting techniques to help readers proceed rapidly through your message. The format should help readers recognize the organization and major points of the message.
- Always be concise. Keep sentences and paragraphs short. Readers should not have to read your message more than once to grasp your meaning.
- Try to keep your message to a single page or screen. If your message exceeds a single page or screen, use headings to guide your reader through it.
- Keep a copy of every message you send and every message you receive for at least 2 years. When others forget what they wrote to you or what you wrote to them (or the fact that you wrote at all), your ability to retrieve the message in question will save time, avoid arguments, correct mistakes, and reinforce your reputation for efficiency.

E-MAIL AND TEXT MESSAGES

E-mail and text messages constitute the majority of job-related communication. You will likely find yourself reading or writing at least 100 e-mail and text messages every day. Text messages are by definition brief and, thus, ideal for urgent notifications and quick acknowledgments—information that can't wait for e-mail. Appropriate e-mail and texting etiquette continuously evolve, but both kinds of messages require conciseness and readability. Keep in mind that typical mobile devices display the list of incoming e-mail messages with the opening five lines of each message and the list of incoming text messages with the opening two lines. You will assist your recipients in navigating their incoming messages if you make sure that your key point is located in the opening lines.

Consider, for example, the series of informative and encouraging text messages prepared by the Department of Health and Human Services in conjunction with the American Academy of Pediatrics (see Figure 7–1). Addressed to mothers and fathers of young children and sensitive to the strains of family life, these once-a-week parenting tips average 145 characters (including spaces) to make for quick and convenient reading. Which do you think are effective? Notice that the messages written with "be" verbs and punctuated as exclamations sound unconvincing and condescending: for example, "Lowfat milk & water are the best drink choices for your child, but we all need variety! Limit sugar-sweetened drinks like soda to special occasions." Choose action verbs whenever possible, keep your messages brief, and leave the exclamations for emergency warnings only: for example, "Lowfat milk & water make the best drink choices for your child. Limit juice and soda to special occasions."

Note also that a growing problem involves people texting carelessly and without thinking of issues that could arise if their messages were later forwarded to others or made public. Always exercise discretion in texting. Keep in mind that everything digital is infinitely available. As hundreds of government and corporate officials across the world have discovered, if you or anyone you know is ever investigated for any kind of impropriety, all of your e-mail and text messages could be exposed for humiliating scrutiny.

MEMOS AND LETTERS

Memos, the relatively short and informal documents that circulate within organizations, address internal decisions and operations. Memos can be posted to the organization's website (including bulletin boards, blogs, and wikis), sent as attachments to e-mail messages, or printed and distributed.

Letters differ from memos in that letters are usually written to individuals outside the writer's organization, but they can be used as official documents within an organization. Like memos, letters can be sent as e-mail attachments. Memos serve

TXT4Tots: Parenting is a busy job! Take time when you grocery shop to pick out easy to eat on-the-go healthy snacks like a banana or whole grain cereal.

TXT4Tots: Lowfat milk & water are the best drink choices for your child, but we all need variety! Limit sugar-sweetened drinks like soda to special occasions.

TXT4Tots: Toddlers like to copy everything you do. Demonstrate healthy habits you want them to learn like eating fruit for snacks & walking after dinner.

TXT4Tots: Sometimes it seems harder to be active inside, but you can do it! TXT4Tot tips: 1) Play hide and seek 2) Create an obstacle course with pillows.

TXT4Tots: Snacking on the run? Keep lowfat cheese sticks, bananas, and small, whole grain crackers on hand. 2–3 snacks a day can help prevent temper tantrums.

TXT4Tots: Create an activity chart with your child. Track active play time at home or childcare with stickers for each activity. Think of a fun reward!

TXT4Tots: Eating directly out of a box or container can allow for overeating. Make sure to serve food from a plate, bowl, or dish.

TXT4Tots: Get moving! Make a list of your family's favorite ways to be active. Each Saturday pick an activity out of a hat to do together.

TXT4Tots: Try to set screen time limits for the entire family. After the time is up, play board games or work on a puzzle together.

TXT4Tots: Eat as a family whenever you can. Keep meal time relaxed to nourish the body & help your family stay connected.

FIGURE 7-1 Sample Text Messages from the Txt4Tots Series.
Source: US Department of Health and Human Services, Health Resources and Services Administration. "TXT4Tots." http://www.hrsa.gov/healthit/txt4tots/txt4totsfiles.html. This TXT4Tots Library of Messages was developed under a cooperative agreement between the American Academy of Pediatrics and the Health Resources and Services Administration of the U.S. Department of Health and Human Services. This version of the Library represents a derivative work.

as intraorganizational communications only and are not accepted channels for official business with other companies and clients.

Both memos and letters must be worded and formatted for quick skimming and rapid reading. Consider the letters in Figures 7–2 and 7–3. Which would you be more likely to read? Why?

DEPARTMENT OF HEALTH & HUMAN SERVICES   Public Health Service

                                                    Centers for Disease Control and
                                                    Prevention (CDC)

                                                    March 18, 2013

Dear Colleagues,

World TB Day is March 24. This annual event commemorates the date in 1882 when Dr. Robert Koch announced his discovery of *Mycobacterium tuberculosis*, the bacillus that causes tuberculosis (TB). This day is observed each year to raise awareness about TB-related problems and solutions and to support worldwide TB-control efforts. For the second year, the Centers for Disease Control and Prevention (CDC) joins the global Stop TB Partnership in adopting the slogan *Stop TB in my lifetime,* which echoes the theme of calling for a world free of TB. The slogan and theme encourage people worldwide to make an individual call to stop TB and say what changes they expect to take place in their lifetimes.

One of the goals of this year's campaign is to highlight personal stories from TB patients and their families, focusing on the services and support that public health programs provide to successfully diagnose and treat people with TB disease and latent TB infection. The stories are available on the CDC TB website. I hope you will visit this page and share these stories with other colleagues, both in the United States and globally. DTBE also has developed resources in English and Spanish to assist partners in developing messages for World TB Day. I hope that you find these resources useful and incorporate them into your World TB Day events and activities.

Lastly, on March 22, 2013, CDC will host a World TB Day event that will be available via live webcast. The event will highlight 2012 U.S. TB surveillance data, as well as a keynote address by Dr. Susan Ray, Professor, Emory University School of Medicine, Atlanta, Ga. For more information about this event, please visit the event webpage.

Let's continue to reach out to those at highest risk for TB and to identify and implement innovative strategies to improve testing and treatment among high-risk populations.

Thank you for all the work you do to Stop TB in our lifetime!

Sincerely,

Rima F. Khabbaz, MD
Deputy Director for Infectious Diseases
Director, Office of Infectious Diseases
Acting Director, National Center for HIV/AIDS, Viral Hepatitis, STD, and TB Prevention
Centers for Disease Control and Prevention
http://www.cdc.gov/nchhstp

RADM Kenneth G. Castro, M.D.
Assistant Surgeon General, U.S. Public Health Services
Commanding Flag Officer, CDC/ATSDR Commissioned Corps
Director, Division of Tuberculosis Elimination
National Center for HIV/AIDS, Viral Hepatitis, STD, and TB Prevention
Centers for Disease Control and Prevention
http://www.cdc.gov/tb

FIGURE 7-2 Letter

DEPARTMENT OF HEALTH & HUMAN SERVICES   Public Health Service

Centers for Disease Control and
Prevention (CDC)

March 18, 2013

Dear Colleagues,

World TB Day is March 24. This annual event commemorates the date in 1882 when Dr. Robert Koch announced his discovery of *Mycobacterium tuberculosis*, the bacillus that causes tuberculosis (TB). This day is observed each year to raise awareness about TB-related problems and solutions and to support worldwide TB-control efforts.

Campaign Slogan: Stop TB in my lifetime
For the second year, the CDC joins the global Stop TB Partnership in adopting the slogan *Stop TB in my lifetime*. This slogan echoes the campaign theme of calling for a world free of TB.

Campaign Goal: Highlight Patient and Family Stories
One of the goals of this year's campaign is to highlight personal stories from TB patients and their families, especially about the services and support that public health programs provide to successfully diagnose and treat people with TB disease and latent TB infection. The stories are available on the CDC TB website. Please visit this page and share these stories with colleagues both in the United States and globally.

Campaign Resources: Help with Developing Messages
CDC's Division of TB Elimination has developed resources in English and Spanish to assist partners in developing messages for World TB Day. I hope that you find these resources useful and incorporate them into your World TB Day events and activities.

Campaign Event: Live Webcast on March 22
CDC will host a World TB Day event that will be available via live webcast. The event will highlight 2012 U.S. TB surveillance data and include a keynote address by Dr. Susan Ray, Professor, Emory University School of Medicine in Atlanta. For more information, please visit the event webpage.

Let's continue to reach out to those at highest risk for TB and to identify and implement innovative strategies to improve testing and treatment among high-risk populations.

Thank you for all the work you do to Stop TB in our lifetime!

Sincerely,

Rima F. Khabbaz, MD
Deputy Director for Infectious Diseases
Director, Office of Infectious Diseases
Acting Director, National Center for HIV/AIDS, Viral Hepatitis, STD, and TB Prevention
Centers for Disease Control and Prevention
http://www.cdc.gov/nchhstp

RADM Kenneth G. Castro, M.D.
Assistant Surgeon General, U.S. Public Health Services
Commanding Flag Officer, CDC/ATSDR Commissioned Corps Director,
Division of Tuberculosis Elimination
National Center for HIV/AIDS, Viral Hepatitis, STD, and TB Prevention
Centers for Disease Control and Prevention
http://www.cdc.gov/tb

FIGURE 7–3 Revised Letter

GUIDELINES FOR ENSURING QUALITY

- Make sure you look and sound professional as you choose words and put together sentences. Review and proofread messages carefully before you send them. Note especially that every digital message you send will exist indefinitely and could easily be forwarded to additional readers across the world or obtained by unexpected audiences for years to come. Note also that every message you send or receive through your company e-mail account and every text message you send or receive through your company's message service is subject to review at any time by officials of your company and their legal representatives. Your only assured privacy covers messages sent or received through your private accounts on your personal equipment.

- Keep the six *C*'s in mind as you develop e-mails, texts, memos, and letters: be *concise*, *concrete*, *complete*, *correct*, *courteous*, and *clear*. Be attentive to how your words can be perceived by your reader. Rephrase anything that you think might be construed in ways you do not intend.

- Be aware of how your message looks. A letter or memo with typographical or formatting errors makes an unfavorable statement about you and your organization. Design e-mail messages with equal care because these may be printed and copies made and distributed.

- Watch for spelling errors as well as errors in sentence structure and word usage. For especially important e-mail messages, consider printing copies before you send them. Editing printed material often reveals errors you may miss as you compose or proofread on a screen. Also use the spellcheck function available on your e-mail application to help you find errors, but keep in mind that spellcheck is no substitute for careful reading.

- For clarity in international communication, spell months of the year instead of using numbers: that is, 3/11/14 would likely be read as March 11 in the United States but as November 3 in Europe and Asia, and readers won't necessarily know whether you are using your practice or trying to be sensitive to cultural differences and adopting theirs.

- For text messages on the job, avoid slang and abbreviations unless you know your recipient personally. Always keep your messages on point and professional. Always assume that your message could be forwarded to your boss, published in the newspaper, or distributed on Twitter.

- For memos, letters, and e-mail messages, use the following structure:

First Paragraph:	State the purpose of the message or the main information the reader needs. If your message delivers unfavorable news, you may want to cushion the reader's disappointment with a gentle and supportive opening.
Middle Paragraphs:	Support or develop the main topic stated in the first paragraph. Limit each paragraph to one idea.

Final Paragraph: Tell the reader what to do or what position to take. If
                            appropriate, include closing comments (e.g., offer to
                            explain in greater detail or answer questions).

APPROPRIATE TONE IN E-MAILS, TEXTS, MEMOS, AND LETTERS

Be careful of how you come across to your readers. What you write always carries
with it a perceptible attitude toward the audience and the subject. As they absorb
your message, readers often vocalize your sentences and phrases. Anticipate the
intonation that your readers will attach to what you write. Adopt a respectful, positive perspective toward your reader—a "you" attitude that addresses the subject
from your reader's point of view—and avoid wording that might sound rude, sar-
castic, or irritable. Remember: You want to be clear, but you also want to maintain
a favorable relationship with your readers.

Avoid phrases that suggest that the reader is careless or unintelligent:

You neglected to . . .
You failed to . . .
I fail to understand how you could . . .

Also avoid phrases that imply that the reader might be lying or exaggerating:

You claim that . . .

Avoid writing what your reader might interpret as excessive flattery. Many people
are sensitive to insincerity and suspicious of any attempt to induce them to respond in a certain way:

It is indeed a profound privilege for us to work with you on your immensely
important project.

We look forward to the extraordinary opportunity to submit to you our full
proposal. Your firm is known worldwide for its commitment to excellence and
value, which in all humility we are confident we offer.

Attempting to sound formal and objective, however, can often produce impersonal writing:

The changes agreed upon per our conversation are herein included.

Your immediate reply will be greatly appreciated.

Instead, write with a conversational tone that lets your reader know that you are
grateful for his or her attention:

The changes included here were the ones we discussed during our telephone
conversations.

Please let us know your decision as soon as you can.

Many times, in an effort to be clear, writers may sound tactless:

> If your employees had actually read the procedures before trying to install the pump, they would have seen that the installation requires that the sealant be allowed to set for 1 hour before beginning the second stage of the process. Not following this procedure causes the diagnostic to shut down because the sealant has not dried enough to absorb the test stress. To avoid future installation problems, please see to it that your employees read the procedures and read them thoroughly.

Thinking about the emotion that your writing may convey or provoke will help you to present the same idea in a less caustic way:

> Please note in the procedure manual that the second stage of the installation process cannot begin until the sealant has set for 1 hour. This amount of time is necessary for the sealant to dry enough to absorb the test stress. If the second stage of the installation process is attempted before the sealant has dried, the diagnostic will shut down.

Following are additional examples of tactless phrasing and their revisions:

> Regrettably, equipment expense reimbursements are not allowed.

Versus

> My office can reimburse you for travel-related expenses only. Your invoice for equipment may be submitted to the VP of Operations.

> We cannot fulfill your request for a list of company positions and annual salaries as none but company executives are allowed to have access to this information.

Versus

> The list of positions and salaries you requested is confidential information limited to use by our company personnel only. For your research, however, we can send you related information on industry-wide trends in similar positions.

GUIDELINES FOR DEALING WITH TONE

As a writer, you cannot anticipate every nuance of meaning that your message will have for your reader. But you can achieve clarity and maintain goodwill with your reader if you keep the following guidelines in mind as you compose any letter, memo, text, or e-mail:

- Allow more time for designing messages that arise from sensitive issues and that may be subject to misreading. Try to avoid sending bad news by text or e-mail.

- Anticipate the effects of messages that will convey negative or unwelcome news. These can be misinterpreted by readers, who can also misjudge your attitude toward them and the information you are transmitting. People often take bad news personally and may be more inclined to take offense or respond with anger.
- Always analyze your reader as carefully as possible, particularly the reader's frame of reference and attitude toward both you and the subject. Always be courteous.
- Read aloud what you have written. When you hear what you have written, you may often detect words and phrases that don't convey the attitude you intend.
- For any especially difficult message, compose the message, save the draft, and then let it cool for a while. Later, read the message aloud to check for clarity and courtesy.

PLANNING AND WRITING CORRESPONDENCE

The following cases (7–1 through 7–6) introduce typical situations on the job in which writers must write a letter, memo, or e-mail message.

Review of Principles

As you plan your e-mails, texts, memos, and letters . . .

10 Questions to Ask Yourself

1. How well do I know the readers?
2. How much do they know about the topic discussed in the message?
3. How will they respond to what I will need to say? In what ways can I use this communication to build rapport for my organization?
4. What exactly am I trying to accomplish with this message?
5. What is their level of knowledge about the subjects discussed?
6. What is their attitude toward me and my organization?
7. What previous business dealings have I/we had with them?
8. How much and what kind of information should I include, based on their profile?
9. How technical can I be in presenting my message?
10. What strategies can I use to make this message easy to read and understand?

Case 7–1: Informational E-mail Message

David Reilly, an administrative assistant to HCI's chief financial officer (CFO), must notify 14 division managers of a meeting to deal with looming budget cuts. David decides to attach three documents that those attending will need to have read by the meeting. The e-mail message explains the reason for the meeting, the documents attached, and the proposed agenda. As David explains, everyone should bring the attached documents to the meeting.

This e-mail message exemplifies effective design (Case Document 7–1). Note the clearly phrased subject line, the action-required statement in bold type, and the placement of the main information—the meeting and required attendance—in the opening sentence. The meeting agenda appears as a list, and the managers are told exactly what they must do with the attached materials. The message is concise and visually accessible.

CASE DOCUMENT 7–1

From: David Reilly [david.reilly@hci.com]
Date: February 27, 2014
To: Operations Support Staff
Subject: **Agenda for March 12 Meeting on Proposed Budget Cuts**

Attachments: CFO 1-31.pdf, NSP bid.pdf, OSHA oxygn proc rev.pdf, SHR mntnc cost.pdf

ACTION REQUIRED: Prepare New Cost Figures for Your Projects by March 12

All division managers are required to attend a March 12 meeting to decide how to cut the third and fourth quarter budgets. Please refer to the CFO's memo of January 31 (see attached) for guidelines on cuts.

You will receive a separate e-mail message later today asking you to put this mandatory meeting on your calendar. Please clear your calendar now of any conflicting appointments.

Time and Location: March 12, 8:00–11:00 a.m., Third Floor Conference Room

Please have four copies of your budget prepared with the 15% cut suggested by the CFO's 1/31 memo.

Meeting Agenda
1. Budget presentations and discussion
2. Proposed bid development on the North Shore Power project (see attached)
3. Revision of Procedures to Meet OSHA guidelines on oxygen tanks (see attached)
4. Cost overrun problem with maintenance contractors on the Signal Hill Reclamation project (see attached)

Please review all attached documents and be prepared to discuss viable solutions.

If you have problems or questions, please let me know.

DR

David Reilly
Administrative Assistant
Office of the CFO
HCI, Inc.
889–742–2500

Case 7–2: Instructional Memo

The following memo (Case Document 7–2) announces Quaker Pharmaceutical's new agreement with a local rental car dealership and provides instructions for renting cars for business use. The writer, Luning Jiang, suggests that readers may want to keep the memo for reference.

CASE DOCUMENT 7–2

MEMORANDUM

TO: Quaker Pharmaceutical sales representatives
FROM: Luning Jiang
SUBJECT: New Benefits from Horizons Rent-A-Car
DATE: 10/17/14

Quaker Pharmaceutical signed an agreement today with Horizons Rent-A-Car offering a 20% discount for all business-related travel. Please retain this memo for your files.

Business Vehicle Guidelines
- 1 or 2 employees traveling: use compact car; 3 or more employees traveling: use sedan or minivan.
- Luxurious vehicles for transporting clients or potential clients must be approved by your supervisor.
- Please lease the most economical vehicle available.

Regulations
- Employees are expected to follow all traffic laws.
- Employees under 25 cannot drive unless accompanied by a fellow employee over 25.
- Vehicle cannot be used to transport persons or property for hire.
- Vehicle cannot be used to tow anything.
- Horizons Rent-A-Car and **Quaker Pharmaceutical** are not responsible for loss or theft of personal belongings left unattended in the vehicle.

Reservations

To reserve your car for business travel, go online at www.you-go.com or call 1–800–RENTCAR.

- Valid driver's license needed for reservation
- No deposit required

Online Instructions:

1. Go to http://www.you-go.com
2. Locate login box for Gold Corporate Customers (GCC)
3. Enter username: QPrental (case-sensitive)
4. Enter password: your last name (all lowercase)
5. Follow remaining steps as prompted

For questions, contact Shirley Kimball, the Office Services Manager, at 898–757–3478 by call or text.

Case 7-3: Letter Requesting Information

Juanita Márquez serves as local arrangements chairperson for the American Society of Engineering Education's (ASEE's) regional meeting. One of her first responsibilities is to locate a hotel in which the regional meeting can be held. She decides to send a letter to seven hotels near the sports arena so that conferees can enjoy sports events during the evenings of the 3-day conference. Juanita writes the same letter to all seven hotels requesting information about their facilities.

When Juanita writes to each of the seven hotels, she has a number of questions she needs to be answered about availability, facilities, and equipment. Juanita needs each hotel to respond by a specific deadline, and she wants to be sure that every hotel supplies all of the information she requests. With these answers, she can compare costs and services of hotels that are interested in hosting the convention. She uses a *block letter format* with all elements aligned on the left margin. She chooses to conduct this correspondence by formal business letter instead of e-mail, to make sure she obtains signed paper copies of all promises made by the hotel representatives.

CASE DOCUMENT 7-3

Pittsburg City College
200 Rosser Hall
Department of Civil Engineering
Pittsburg, KS 66760

(date)

(name of convention manager)
(name of hotel)
(mailing address)
(city, state, zip)

Dear (   ):

The South Central Chapter of the American Society of Engineering Education will have its annual meeting in Kansas City, April 22–26, 2016. With the convention 15 months away, the local arrangements committee is seeking a hotel that will serve as convention headquarters. Our committee would like to consider your hotel as a possible site because of its location and its reputation in handling conventions. By March 1, 2015, I will need answers to the following questions, if you are interested in hosting our convention. I will contact you by telephone no later than March 15, 2015, to arrange an appointment to discuss our needs and your facilities in further detail.

Number of Conferees Expected
Based on registration from past conventions, we expect approximately 500 people to attend. Of that number, approximately 450 will need rooms. Of that number, approximately 250 will require double occupancy rooms.

Conference Accommodations Needed
We need you to provide answers to the following questions:

1. Can your hotel accommodate 500–600 people, April 22–26, 2016?
2. Since the conference will feature 3 days of concurrent sessions, do you have available conference rooms in the following configurations?

   - three rooms located in the same general area that will hold 100 people
   - three additional rooms that will hold 50 participants, with six persons per round table

   We will have three concurrent sessions at each time slot—two in the morning and two in the afternoon.
3. Can your hotel provide the following multimedia equipment for each room?

   - screen
   - projector that can be attached to a variety of mobile devices
   - microphones
   - lectern
   - high-speed Internet connection (please specify wired or wireless)

4. Can the seating arrangements be altered in each room between sessions?
5. Can the hotel provide refreshments during the morning break and afternoon break? Can these refreshments be made available at a location that is convenient to the meeting rooms?

continued

Response Deadline

I will need a written response to each question by *March 1, 2015*. Also, please include in your reply a price list and menu for refreshments available for conferences.

If you have any questions, please contact me at 620-231–7000 Monday–Friday. If I am away from my desk, just leave a message on my voice mail and a time when I may return your call.

Sincerely,

Juanita Márquez

Juanita Márquez
ASEE Convention Arrangements Chair

Case 7-4: Unfavorable News Letter

Nancy Krueger is a highly experienced safety engineer who was interviewed by Novel Engineering. Ms. Krueger was impressive during the visit to corporate headquarters, but the interview team recommended another finalist for the position. Despite Ms. Krueger's exceptional credentials, the second candidate had qualifications that more clearly matched the company's needs. Ritu Srivastava, director of personnel, has to write to Ms. Krueger to explain that a job offer will not be forthcoming.

In this situation, Ritu may call Ms. Krueger to give her the disappointing news, but she will also follow the call with a formal letter. Avoid using e-mail or text messages to deliver bad news. Note that the letter gives reasons for Ms. Krueger not being selected, but the news is not explained in a harsh, critical way. It is intended to announce the bad news in as positive and supportive a way as possible. Notice also that Ritu cushions the bad news with a friendly opening and appropriate compliments. Because Ritu was heavily involved with Ms. Krueger during the application and interview process, she addresses her as "Nancy." Do not use first names unless you really know the person you are addressing. Ritu uses a *modified block format*, which uses paragraph indentions and shifts the date, closing, and signature line to the right side of the page (Case Document 7-4), because she thinks this format looks a little less stiff and formal.

CASE DOCUMENT 7-4

NOVEL ENGINEERING, INC. ———————— ESTABLISHED 1954
    4415 JUNIPERO STREET
    SYRACUSE, NY 13233

                   May 14, 2014

Ms. Nancy Krueger
1248 Orange Avenue
Valley View, OH 44125

Dear Nancy:

    All of us here enjoyed the two days you visited with us two weeks ago. Your perception of our clients' needs indicates that you have a firm understanding of the role safety engineering plays in the local contracts we win and manage.

continued

While your background would be invaluable to us, we have only one position available, and we have selected an applicant with international project experience.

As we discussed during your visit, our range of clients has expanded. We now receive RFPs from countries in Europe, South America, and the Pacific Rim. Developing responses to their needs requires us to expand our team, particularly when we can do so with professionals with international experience.

Thank you for considering us. We were genuinely impressed with your professionalism, your analytical skills, and your record as a team player. And I was immensely pleased to have the opportunity to meet you.

Sincerely,

Ritu Srivastava

Ritu Srivastava
Manager, Personnel

Case 7-5: **Claim Letter**

Catherine Leigh, a senior at Astoria College, discovers that she has been denied a tuition refund for a course she thought she dropped. As though the monetary penalty weren't enough, she also discovers that she has been given a grade of WF (Withdraw Failing) for the course. This withdrawal on her transcript will hurt the exceptional grade point average (GPA) she has been building for 3 years and makes admission to veterinary medicine almost impossible. She takes her complaint to the dean of Academic Affairs and follows up the face-to-face meeting with a formal letter that explains the problem in detail as well as the solution she is proposing.

Catherine knows that the dean of Academic Affairs has a lot to do and may give the letter to an assistant to investigate the complaint. She therefore doesn't assume that her reader remembers (or even knows) anything more about the case than what she provides in her letter. And she uses headings to help the reader find information quickly. She adopts a block format to make herself look judicious and methodical, and she proofreads carefully so that she doesn't come across as the kind of student who easily makes mistakes.

CASE DOCUMENT 7-5

Catherine Leigh
601 Ocean Avenue
Astoria, FL 34103

September 8, 2014

Dr. Neville Hodges
Dean of Academic Affairs
Astoria College
Astoria, FL 34103

Dear Dean Hodges:

I am writing to give you a written record of the situation I discussed in your office on Friday, September 5. I enrolled in the summer session of Math 3325, in section 001, under the instruction of Dr. Voskovic. I dropped the course before classes started but nevertheless was denied a tuition refund and given a grade of WF.

continued

Description of Events

I registered for Math 3325 in February, but later my plans for the summer changed. I believed that all I had to do was go online and drop the course. I dropped the course online and did not attend any of the class sessions. During the first week of June, I called the registrar for information about my refund check. I asked the woman who answered my call if I had taken care of everything I needed to in order to get my refund check. She asked for my ID number, which I gave her. She then informed me that I should receive my check in the mail shortly. I waited for two months.

I returned to the registration office in person during the first week of August to inquire about the refund check. It was then that I was informed that I was dropped from the course due to excessive absences and I could not receive a refund. However, I was assured that I would get a W on my transcript. I was upset that my tuition for the course would not be refunded, but I was relieved that the incident would not hurt my GPA.

This relief, however, was short-lived. The following day, while I was working on my application to the Florida College of Veterinary Medicine, I discovered that my unofficial transcript online has a grade of WF for Math 3325. This grade will seriously hurt my chances of admission.

My Request

I request that the WF on my transcript be changed to a W and that my tuition for the course be refunded. I sincerely believed that I had dropped the course. My telephone conversation with the registration office only confirmed this belief.

I am a senior honors student with a GPA of 3.63. I fully understand that once a student enrolls in a course, he or she must attend class. I came to Astoria to prepare for a career in veterinary medicine: I would never enroll in a course and not attend it. I hope you will agree that I am conscientious about my studies. Please do not allow this one error to ruin my chances of acceptance to the Florida College of Veterinary Medicine.

I know that the registrar's office is hectic, and I regret that this error occurred. However, given the situation I have described, I do not know how I could have proceeded in a different way. I believe that I did everything I could to drop the course correctly.

Sincerely,

Catherine Leigh

Catherine Leigh
Astoria ID: 418556

Case 7-6: Letter of Reply

Gabriel Dominguez owns an agricultural equipment business. He has just received an order and a check from Bruce Pickering, who wants to replace an irrigation pump.

Gabriel needs to notify Mr. Pickering that the pump he has ordered is being discontinued and replaced by two new pumps, both of which are more expensive but are designed to offer better performance. Gabriel wants to be sure that Mr. Pickering doesn't think that Gabriel is simply trying to sell him a more expensive pump when the current model will do the job.

In designing this letter, Gabriel wants to be sure that he does not suggest to Mr. Pickering that he is the object of a bait and switch. Thus, he opens the letter by telling Mr. Pickering that he has the pump and can ship it immediately (Case Document 7–5). However, he wants Mr. Pickering to know that several improved models are available and that he may wish to choose one of the new models. Gabriel then presents all the information he believes Mr. Pickering will need to make his decision and then invites Mr. Pickering to contact him. Gabriel could call Mr. Pickering and talk to him, but in this case a business letter that expresses concern for Mr. Pickering's investment and presents factual information about the alternative pumps will make Mr. Pickering's decision easier. Note that Gabriel chooses the block letter format: he thinks its formal and straightforward appearance will reinforce his credibility and indicate respect for his customer and their business relationship. To give Mr. Pickering the necessary information as soon as possible, Gabriel will prepare the letter with his electronic signature and e-mail it as an attachment; for this important sale, he will also put a paper copy of the letter with his ink signature in 2-day delivery so that Mr. Pickering is confident that everything is official and legal. (In the e-mail message, Gabriel advises Mr. Pickering of the attached electronic copy and the forthcoming paper copy of the letter.)

CASE DOCUMENT 7-6

---Dominguez Irrigation Manufacturing & Supply Co.---

2101 Cerritos Drive
Altair, TX 77412
(817) 569-3766

(date)

Mr. Bruce Pickering
Route 1, Box 41
Silsbee, TX 77656

Dear Mr. Pickering:

We have received your order and check for $698 as payment for our 15-hp Model XM 21 auxiliary pump. Although we do have several in stock and we can mail one to you immediately, manufacture of this model has ended. We are holding your check and order until we hear from you regarding the following alternatives.

Here is the situation. Purchasers of the 15-hp model have indicated that the low output of this pump is generally insufficient to run the new drip irrigation systems currently being manufactured. This particular pump is not designed to handle extensive operation of these larger systems. As a result, we have discontinued the model. We have reduced the price to $548 due to our inability to warranty the motor for more than 1 year or supply replacement parts for more than 2 years.

We now manufacture two larger pumps, a 50-hp model and a 75-hp model. The 50-hp model will power a #70 system. The 75-hp model will power a #90 system. Here is a comparison of these three models.

Hypo	Pumps			
Motor	Model	Rating	Warranty	Price
15 hp	21	30	1 yr = 2 pts	$ 448
50 hp	31	70	full + 5 yr	$1,795*
75 hp	41	90		$2,350
*Additional warranty available for $125/yr.				

While we believe that you will be happier with the reliability and efficiency of either of the larger pumps, if you choose to purchase the 15-hp model, we will ship it immediately and refund the $150 difference in price. I am sending brochures on both new motors. Please look them over and call me at our 800 number, extension 145, if you have any questions. As soon as we hear from you, we will ship you the motor of your choice within 2 working days.

Sincerely,

Gabriel Dominguez

Gabriel Dominguez
Owner

CORRESPONDENCE CHECKLIST ✔

Planning

☐ What is your subject, and what is your purpose?

☐ What do you want to have happen as a result of your correspondence? What will you do to achieve your objective?

☐ Who are your primary readers? Who are your secondary readers? Do your primary and secondary readers have different needs? How will you satisfy the different needs of all your readers?

☐ Why will your readers read your correspondence?

☐ What is the attitude of your readers toward you, toward your subject, toward your purpose?

☐ If you are addressing international readers, do you understand their cultural practices? What adjustments in your correspondence will their cultural practices require?

☐ Will you write an e-mail, text, memo, or letter?

☐ Will you choose a direct or an indirect style?

Revision

☐ Are your topic and purpose clearly identified?

☐ Have you satisfied your readers' purpose in reading?

☐ Have you adopted a style suitable to your readers' culture?

☐ Have you avoided jargon and clichés?

☐ Is your message clear, concise, complete, and courteous?

EXERCISES

1. Civil Engineering Associates (CEA) of Lubbock, Texas, was established in 2009 by Robert Davidson and Walter F. Posey, both graduates of the University of Amarillo. Business for the company was good, and CEA took on six additional partners between 2010 and 2014: Alvin T. Bennett, Wayne Cook, Frank Reynolds, John W. Castrop, George Ramirez, and Richard M. Burke—all graduates of the University of Amarillo.

   For the last 5 years, the partners of CEA have met every Friday for a working lunch at Coasters, a local restaurant and bar that features attractive young waitresses wearing provocative swimsuits. The customers are almost exclusively men, and the interaction between customers and waitresses often appears flirtatious.

This year CEA hired Elizabeth Grider, a Missouri State University graduate, as a new partner in the firm. She has attended two of the working lunches at Coasters and finds herself uncomfortable in this environment. She does not feel, however, that she can just skip the lunches because these are the only regular times at which all of the partners gather. Projects are often discussed and work assignments decided at these Friday meetings. In addition, the lunches offer the opportunity to establish a cordial and friendly working relationship with the partners. She realizes that "Friday at Coasters" has become a long-standing tradition at the firm, and she hesitates to upset the status quo. However, she really wishes they could find a better place for their meetings.

As Elizabeth Grider, write a memo to the senior partner, Robert Davidson, explaining the issue with Coasters and recommending one or more solutions. Of course, Grider will speak to Davidson directly but thinks that writing a memo will help her organize her thoughts, anticipate objections, and prepare for a likely difficult conversation. And after their meeting, she will leave the memo with Davidson as a written record of her recommendations.

2. You are the manager of Activ8 Sports in Cleveland, Ohio. Activ8 Sports, a regional chain of 19 athletic equipment stores, has headquarters in Detroit. This morning you received a call from a vice president of the company. The corporate office's social media specialist was doing the usual daily search on Twitter, Yelp, Facebook, etc., for customer comments about Activ8 Sports. She discovered that several of the salespeople at the Cleveland location were using Twitter to make derogatory comments about a coach at a local high school. The coach, a regular and frequent customer at the store, sometimes appears abrupt and aggressive with the salespeople. The vice president directed you to "fix this Twitter thing immediately" and implied your job was at risk. You decide to reprimand each of the salespeople in question. You will also issue a memo for their file as an official warning with a reminder of the corporate policy that proscribes discussing company business in public, including social media.

   As the manager of Activ8 Sports, write this memo: try to keep your salespeople cooperative and productive but also make clear that this is their only warning.

   You also consider writing a letter of apology to the high school coach before he finds out about the indiscretion of your salespeople. You know that sooner or later he will find out: the comments could easily go viral at the high school. The coach has been a loyal customer, purchasing all uniforms and equipment through your store year after year. You can't afford to lose his business.

   As the manager of Activ8 Sports, write to the coach to explain the situation and apologize for the comments of your salespeople on Twitter.

3. Examine Figure 7–4. After Clarion News Service issued this memo to the employees at its Portland location, the hostile reaction was swift. The company was widely criticized in local newspapers for insensitivity to people losing their jobs.

Several city officials accused the company of deceptive communication because, it was claimed, the memo buried important and time-sensitive information in a long paragraph, hoping it would go unnoticed. The criticism of Clarion was picked up by its competitors in the news business and distributed nationally.

If Clarion News Service were to issue a revised letter, what changes would you advise it to make? And, at this point, what other audiences should Clarion address and what other messages should it compose to restore its reputation for ethical business practices?

4. You are special assistant to Jordan Williams, president of Phoenix Engineering Services. He asks you to "check this e-mail message for errors and revise as necessary before distributing," a routine activity for you in your administrative position. This time, however, the message leaves you worried about its practical and ethical implications: it aggressively tries to push more experienced and higher-paid engineers out of their jobs in order to make room for less experienced and lower-paid substitutes (see Figure 7–5). Will this message demoralize and antagonize its recipients? Could it cause a "brain drain" and loss of tacit knowledge at Phoenix? Does this message constitute age discrimination?

What revisions could you make to the wording of this message that would alleviate your anxieties about it? How might you raise your questions about this message without insulting your boss?

5. Adapt the information in Figure 7–6 as a series of 10 one-a-day text messages from the National Park Service to individuals who have made appointments to visit this important site. In addition to a message with directions to the location, compose messages with interesting historical and architectural details that will prepare visitors for a productive and enjoyable tour. To allow the text messages to be recycled later for Twitter, limit each message to 140 characters (including spaces).

CLARION **News Service**

MEMORANDUM

TO: Clarion News Service Employees

FROM: Elsa White, VP Operations

DATE: March 16, 2014

RE: Clarion Communications
            702 East Burnside Street
            Portland, OR 97214

In order to increase the efficiencies of our operations and better serve our clients, Clarion will be closing its Portland location on May 15, 2014.

Business considerations and realignments have compelled us to arrive at this carefully considered and irrevocable decision. Please be assured that we are making every effort to provide opportunities for you to grow with us into the future. All Portland employees are invited to apply for positions as a Clarion-at-Home Agent. To learn more about Clarion-at-Home please go to http://www.clarionathome.com. Available Clarion-at-Home positions have also been posted for your convenience at http://apply.clarionathome.com.

We will be continuing to operate business as usual through May 15, 2014, and we expect you to continue to perform your duties and responsibilities and to act in the best interests of Clarion and our clients with absolute loyalty, dedication, and professionalism.

In case you want to determine if you are interested in Clarion-at-Home, or if Clarion-at-Home is not able to immediately offer you a position, please know that you will have the opportunity to take an Administrative Leave of Absence for up to 30 days. This status will allow your employment to continue uninterrupted along with continuation of any fringe benefits you may currently have. You must notify Employee Relations (ER) or Human Resources (HR) if you elect this option no later than three (3) business days after receiving this notice. Your HR representatives have all the forms and information you need to make an informed decision. Those employees who remain with Clarion in good standing until released will receive any accrued unused PTO, as of the last day worked, May 15, 2014.

continued

FIGURE 7–4 Document for Exercise 3

To answer questions and to assist you as necessary during this time of transition, service representatives from management and ER will be available immediately following the distribution of this announcement. You may also contact Sam Givens, Regional Employee Relations and Training Manager at (555) 742-2500 or SGivens@clarion.com.

In addition, please verify that HR or Payroll has your most up-to-date contact information for tax purposes.

We sincerely thank you for your dedication and hard work and wish you the best of luck in the future.

FIGURE 7-4 *continued*

From: **Jordan Williams, President**
To:
Cc:
Subject: Phased Separation Program

I am writing to remind you about Phoenix Engineering Service's Phased Separation Program. The program offers employees who are considering separating from Phoenix the opportunity to scale back their percentage effort to 50% for a designated period, usually three years, while retaining regular employee status, eligibility for raises, and fringe benefits. To view the policy guidelines and complete your application, visit www.phoenix.com/HR/PSP. Human Resources has also prepared a document comparing the benefits of the Phased Separation Program relative to retirement.

I understand that many of the recipients of this message will consider themselves neither ready for nor interested in this program, but I am writing because I think it would be helpful to all employees to learn more about how this program is intended to benefit both those who participate and Phoenix itself.

The Phased Separation Program offers venerable employees the opportunity of a more gradual transition to retirement, which might also take the form of working only half-time each year or full-time for six months of the year, while remaining active members of the Phoenix community.

Funds made available by the program through salary savings to date have been used to invigorate and grow the company. Indeed, this program is in part what has allowed us to hire an unprecedented 16 new college graduates and 24 college interns this year. My hope is that this program will allow us to continue our momentum and do so in a way that primes eager new employees for success here at Phoenix.

This is a rolling program with no deadlines for submitting an application. Please do not hesitate to contact me with any questions or concerns about the Phased Separation Program. I expect you will give it your serious consideration.

FIGURE 7-5 Document for Exercise 4

Eleutherian College Classroom and Chapel Building

Eleutherian College Classroom and Chapel Building
National Park Service

Perched atop "College Hill" in rural Lancaster, Indiana, Eleutherian College was created as one of the first desegregated educational institutions in America. In 1848, Oxford, Ohio, minister Reverend Thomas Cravens started the Eleutherian Institute in Lancaster after being impressed with local abolitionists' enthusiasm during an 1846 visit. The name "Eleutherian" stems from the Greek word "Eleutheros" which means "freedom and equality."

The classes initially met in local meeting halls before construction of the limestone building on College Hill. After moving into its permanent home, the Institute changed its name to Eleutherian College and became the first Indiana school to offer post-secondary instruction to African Americans. Soon Eleutherian College built men's and women's dormitories, and annual enrollment reached between 70 and 150 students from 1855 to 1861. During the 1860s and 1870s, when public schools for free blacks became more common, Eleutherian College's enrollment dipped.

Constructed between 1854 and 1856, the surviving Greek Revival-style college building is a three-story stone building designed on a rectangular plan, with a square bell tower above the entrance. Its gable-front roof creates a triangular wood pediment above the school's third floor windows. Finished limestone windowsills, lintels, and corner quoins offer a contrast to the building's rough-hewn coursed stone walls. The north-facing façade features three bays with two wood doors with transoms around a central window.

The building's interior reflects its original function as a school and chapel. A small entryway situated inside the exterior doors leads to a two-story high chapel with double-height windows running the remainder of the building's length. Two classrooms are located above the entryway on the second story, but the bulk of instruction space, five more rooms, is on the third floor.

Lancaster Township purchased the building in 1888 and used it as a public school until 1938. Restoration projects in the 1960s and during the 2000s have returned the building to its historic appearance.

Historic Eleutherian College, Inc now owns the Eleutherian College Classroom and Chapel Building. It was designated a National Historic Landmark on February 18, 1997.

Plan your visit

Eleutherian College, a National Historic Landmark and part of the National Park Service National Underground Railroad Network to Freedom is located at 6927 W. State Rd. 250, Lancaster, IN, 10 miles northwest of downtown Madison. Click here for the National Historic Landmark file: text and photos. To reach Eleutherian College, take IN-7 north to SR 250. Head east on SR 250 2.5 miles. The college is situated off the road in the middle of a field. It is open to the public by appointment only. To set up a tour, call 812-866-7291.

FIGURE 7-6 Document for Exercise 5

Source: US Department of the Interior, National Park Service. "Madison, Indiana: Eleutherian College Classroom and Chapel Building." *Discover Our Shared Heritage Travel Itinerary.* http://www.nps.gov/history/nr/travel/madison/Eleutherian_College_Classroom_and_Chapel_Building.html.

Technical Reports

Reports, like correspondence—e-mails, texts, memos, and letters, discussed in the previous chapter—remain the most commonly written workplace documents. Reports serve a variety of functions: they provide information, instructions, analysis, conclusions, and recommendations based on analysis.

Given the quantity of information generated and shared in most work environments, getting your report read has become an increasingly critical consideration. In this chapter, we will explain the basics of designing clear and accessible reports that enhance the likelihood of readers' willingness to find out what you have to say.

Quick Tips

None of your readers will read your entire report, unless it's a single page. Different readers will read different sections of your report according to their needs. Almost all of your readers will read the summary: make sure this section exemplifies clarity and conciseness. The busier your reader, particularly those who receive dozens and hundreds of documents daily, the more important directness, clarity, and conciseness become if you want this reader to read your document.

If a report doesn't have a concise, readable summary, readers will flip to the end to find the conclusions. If they can't find those easily, they likely will not read any portion of the report.

KINDS OF REPORTS

Many routine reports provide information, but others go beyond simple reporting. They categorize and then analyze information or data. From the analysis, the writer may evaluate the information, draw conclusions, and perhaps recommend action based on those conclusions. Analytical reports often defy rigid classification, but for the purpose of learning to write analytical reports, we can generally identify the following types:

- If the analysis focuses on a recommendation, the report may be called a *recommendation report.*
- If the analysis emphasizes evaluation of personnel, data, financial options, or possible solutions to problems or avenues for exploration, the report may be called an *evaluation report.*
- A *feasibility report* analyzes a problem, presents possible solutions to the problem, determines criteria for assessing the solutions, assesses the solutions against the criteria, and then shows the best solution(s) based on the reported analysis of the solutions. The analysis underpins the conclusions and any recommendations. Based on available resources, the writer determines criteria against which possible solutions may be measured and methods of applying criteria and analyzing solutions against criteria selected. Many feasibility reports are long because they represent the way a problem has been defined, studied, and resolved. Results of feasibility studies can lead to major financial expenditures.
- Many reports both inform and analyze: a *progress report* or a *status report*, which we will discuss in Chapter 9. Status reports describe and evaluate the work done on a project, the money expended, and problems encountered.
- A *trip report* documents information gathered on a trip, evaluates this information, and may suggest actions based on findings.
- A *personnel report* describes an employee's performance, analyzes effectiveness, suggests methods of improvement, and estimates the employee's potential for promotion.
- An *economic justification report* explains the cost of a project or action and then argues for the cost-effectiveness of the project.

REPORT CATEGORIES—INFORMAL AND FORMAL

Reports can generally be divided into two categories: informal reports, used for daily internal communications in business organizations, and formal reports, generated for readers outside the organization or for internal use. The two kinds of reports share similarities in structure and major components. They differ principally in length and in format.

INFORMAL REPORT HEADING

Internal reports use a memorandum heading that may include many of the following items. Organizations usually have their own specific requirements for internal reports that use a memo heading. Many organizations have a memo template available on the intranet. Common elements of memo headings usually include the following:

Date:
To:
From:
Subject:
Reference:
Action Required:
Distribution List:

Subject line. The subject line should state clearly and concisely the content of the report. The clearer the subject line, the better the odds that your readers will read your report. Your subject line should make your readers want to read your report.

Reference. Many reports respond to previously written reports, to company policies, or to items such as engineering specifications and legal issues. Be sure your readers have what they need and understand how your report links to previous reports and documents.

Action required. If you need your reader(s) to respond to your report by a certain date, then say so here. For example, "Your approval needed by 5/10/2012." (See Case 7–1, which shows a memo with an action-required statement.)

Distribution list. This report segment, which can occur at either the beginning or the end of the report, indicates who will receive copies as well as the report's file name for later reference. Be sure to include on this list everyone who should receive a copy. Failing to include all readers can create problems for employees as omitted readers may resent their being eliminated from the communication chain.

PARTS OF AN INFORMAL TECHNICAL REPORT

Introduction. Reports should begin with an introduction, a section that introduces readers to the content covered. For routine memo reports, keep your introduction concise but long enough to ensure that readers understand what they are reading and why they are included on the distribution list. Also remember that a report will be kept in a file (virtual or paper) and will be accessible to readers many months and years after you have written it. The introduction should ensure that all readers understand what follows.

Thus, a memo report introduction should contain the following elements. How much you develop each item will depend on the particular report situation—what you think readers will know.

Subject or report topic.

Report purpose. What do you want your report to accomplish? Often, you can combine the report topic and purpose into one sentence. If possible, begin the report with a sentence or two that states the subject and purpose. For many routine reports, a subject/purpose statement will give readers what they need.

> As you requested, I have prepared a brief report on the best options in alternative fuels. Current research over the past 5 years suggests that biofuels, compressed natural gas, and alcohol fuels may present the best alternatives to conventional fossil fuels. This report presents the advantages and disadvantages of each and should offer interesting discussion points for the bimonthly brown-bag lunch meetings.

Background or rationale. What issues led to the need for this report? Be concise. If you have to have more than a full paragraph, place the background or rationale in a separate paragraph. If your main readers don't need this information, then they can skip this paragraph. Readers on the distribution list or future readers will likely appreciate the information.

> This report recommends that the university begin developing online versions of our high-demand courses, particularly those most requested by students who leave campus for the summer.
>
> The university loses substantial funding when students do not enroll for our summer classes. Many would enroll for online versions of these courses, if we offered them. The loss in revenue has increased over the past 3 years. Determining the courses we could develop for online/hybrid delivery, the cost required for redesign, and the time required for effective conversion would provide us information on which to make a decision. Discussion of each point follows.

Report development. What topics will you cover? Simply state how you will develop the report, mentioning each topic you will cover in the order you will cover it. This quick overview lets readers know what to expect and helps prevent surprise or disappointment.

Summary. The summary determines the success of many reports. Many readers who receive copies, simply as information, will read only the introduction and the summary. The summary should tell your readers the essential points and findings in the report. Informal reports may combine the introduction and summary into one paragraph, consisting of the key points and the purpose.

Discussion. This section develops each point you state you will discuss in the introduction.

Conclusion. The conclusion ties together what you have presented and states the main ideas you want readers to remember.

Recommendations. If you need to recommend action, then include that here. To improve readability, list and number your recommendations. If you need to discuss them, you can do so following your list.

Attachments. Attachments include any material, provided on hard copy or electronically, needed to support your discussion, conclusions, or recommendations. Attachments may include additional data, calculations, or previously written documents on the same subject.

Figure 8–1 exemplifies an effective informal report prepared by the chairperson of a small-college marketing committee. The committee's purpose is to inform the president, who dislikes long reports, of their initial findings on how to market this college.

DEVELOPING REPORTS

How reports develop varies, and here we can provide only guidelines and a few examples. In the introduction, you state how you will present your report and the topics that your reader will find. Thus, you can have a main heading for each topic, and readers who have interest in one or two particular issues can go immediately to those issues.

> **Note:** You want to develop a reputation for writing readable, informative, and accurate reports. Don't make readers hunt for information—they won't. If your reports don't fit these criteria, readers will lose trust in you as a writer and perhaps as a colleague.

How you arrange the main components should depend on what you believe your readers will want. Note that in Figure 8–1 the writer begins with the main marketing recommendations. The introduction is brief, as the president of CSU expects the report.

Richard Harrington

TO: Justin Perry, President   Date: June 4, 2012
           Cambridge State University

FROM: Richard Harrington, Professor of Finance
           Chair, CSU Marketing Committee

SUBJECT: Marketing CSU—Status Report

Our Backgrounds

The CSU Marketing Committee, formed by Provost Davis in June of 2011, emerged from concerns that many students and their parents who visited our campus knew little about CSU. The committee, composed of parents of CSU graduates and faculty, is grateful for the opportunity to share its views of the university. We solicited input from nearly 300 parents. Based on our combined experiences with CSU, we offer our recommendations, compiled as of now, based on focus groups, surveys, and our own experiences as former students from other universities. Much work remains, and we solicit your suggestions.

Marketing Recommendations

1. Improve CSU's visibility. CSU is the best-kept secret in quality small colleges. In "getting out the word" about CSU, the university needs to know the audiences it wants to reach and their concerns in responding to college advertising.

2. Leverage on the recent *Princeton Review* (*PR*) recognition of CSU. Highlight the features that the *PR* noted.

3. Don't allow what a marketing firm thinks to control marketing decisions. Stakeholders often have valid views based on their experiences, backgrounds, and education.

4. In attempting to build name recognition and prestige, do so on the basis of qualities that CSU offers that other state schools do not offer.

Marketing CSU   Page 1

FIGURE 8-1 Effective Informal Report

5. Include, in developing a marketing plan, issues that concern parents, grand-parents (who often help with college costs), as well as high school counselors. Businesses should also be a consideration as they will hire CSU graduates and be approached for financial support.

Cost

1. Show how CSU's cost makes it attractive. Discuss the Wallace Scholarships and the amount of money available. Provide comparisons between CSU and your main competitors.

2. Point out that car insurance and the cost-of-living index in our town and county are much lower than those costs in urban areas.

3. Emphasize how teaching, rather than research, provides the foundation of a CSU degree.

4. Discuss how CSU has improved the quality of life for students with the new facilities available to all students: the new student center, the new residence halls, the refurbished dorms and on-campus apartments, and the recreation/exercise facilities. We plan a survey of graduating seniors at the end of every term.

Safety

Note that parents expressed extreme concerns about on-campus safety. They read/listen to news and know that universities, particularly large ones or those in urban areas, do not offer safe havens for students. Point out the low on-campus crime rate at CSU and the low crime rate of Carrington in general. Compare this rate with those in nearby urban areas. These statistics are easy to find and provide a powerful argument for the benefits of a small college community.

Values

We list the following recommendations in no particular order. Different target audiences value each differently, but all seem to us important based on interviews with parents.

1. Parents also want a university that has a culture that supports traditional values and has rules for conduct in its housing and on campus.

Marketing CSU   Page 2

continued

FIGURE 8-1 *continued*

2. They also want an environment of zero drug tolerance, where the athletic program has a zero-tolerance policy on drugs and tests athletes regularly.

3. Parents also like a campus that is clean and inviting. On this point, CSU excels. Marketing material for CSU should have effective photos of the campus, particularly areas on either side of the mall. Show trees, flowers, some of the great classrooms, and the integrated campus concept.

4. Parents want a university focused on education as well as career preparation. Discuss student service projects that have required critical thinking.

5. Parents want to know that a degree from CSU will "mean something." Having testimonies from former students who have succeeded will be extremely helpful.

6. Parents want faculty who care about students and demonstrate that concern. Parents also value research that "makes a difference," but they want students placed first. Parents want a college that offers personal development opportunities, such as on-campus leadership development programs and internships. They also want undergraduate research programs for students who seek those opportunities. Discuss faculty whose creative work and research have been recognized. Teachers whose research supports their teaching is also a useful point to highlight.

We look forward to meeting with you and your team during Parents Week in September.

Marketing CSU   Page 3

FIGURE 8-1 *continued*

Case 8-1

Matt Lunsford, who has worked in Bradshaw Engineering's fuel analysis and development division for 5 years, recently received an assignment from Jerry Bradshaw, one of the three principals of Bradshaw Engineering, during an office discussion: "Matt, are we moving in the correct direction? This 'sustainability' issue puzzles me: What issues do you see on the horizon? What questions do we need to be asking? What are the main issues BE faces? Write a report and a summary. I think I want to post the summary on the BE website for everyone to read, but I also want a longer report for our management team. Where we go from there I don't know."

Based on the conversation, Matt sees a plan: a well-developed, readable summary for the BE website (Figure 8-2) and then an informal report (Figure 8-3) addressed to Bradshaw with copies to the principal engineers. Both can stand alone. Matt also attaches both documents to a memo of transmittal, as shown in Figure 8-2. Or, as you will see in the example formal report in Appendix C, a report and its summary can be combined. Both documents contain the following items:

> short introduction—subject, purpose, and brief rationale
> recommendation
> conclusion

The report itself contains the

> discussion—included in the report
> references—Matt cites all sources he uses in his report.

Reports may inform, analyze, recommend, or do some combination. Figure 8-1 does both. Figures 8-2 and 8-3, the response to Case 8-1, informs, analyzes, and recommends. Note that Matt uses mostly nontechnical language as he knows that many employees in the company could read both the summary and the report. When you plan to cover several topics, as Matt does, consider listing the topics you will cover to alert your readers. Examine Matt's summary and the introduction to his report.

TO: Jerry Bradshaw   DATE: November 12, 2012

FROM: Matthew Lunsford

SUBJECT: **Report on Current Issues in Chemical Engineering: Sustainability**

As we discussed in our last staff meeting, employees at BE need to understand the current issues facing chemical engineering, particularly those that affect the definition of our field. In response to your request, I have devoted the past two months to research on the direction that chemical engineering seems to be moving. My goal: to help BE in our decisions about hiring, continuing education, and company development.

In light of the changes in the field of chemical engineering, I recommend (1) that engineers at BE participate in continuing education in the areas of alternative fuels and biotechnology and (2) that BE look to recruit college graduates and engineers who have had education or experience in these areas. Following these recommendations would help BE adapt to changes in the field of chemical engineering and continue offering excellent engineering services.

As you suggested, I have prepared a summary of my report for our company website and a report for you and the BE management team. I look forward to your comments and recommendations.

FIGURE 8-2 Transmittal Memo and Summary

Current Issues in Chemical Engineering: Sustainability

by
Matthew Lunsford

Summary

Report Purpose

The purpose of this report is to inform BE of changes occurring in the field of chemical engineering. This report focuses on sustainability and the technologies that will help move us toward a sustainable future as well as the challenges that chemical engineers still face. Based on the report findings, I close with several recommendations for BE that will help us stay up to date with these emerging technologies.

Research Methods

To gather information and obtain direction for this report, I interviewed Dr. Chen, a professor of chemical engineering at Texas A&M University, and researched the topic of sustainability using databases such as ScienceDirect. These databases were available to me through the Texas A&M library.

The Importance of Sustainability

Sustainability, meeting our needs presently without compromising the ability of future generations to meet their own needs, has become a major issue in chemical engineering. Chemical engineers have shifted from focusing solely on profits to focusing on being profitable within the context of sustainability. Three main realizations have led to this change.

1. The fossil fuels that we currently rely upon for energy and chemical production will eventually run out, leaving future generations without oil and gas.
2. Burning fossil fuels destroys the environment by causing acid rain and possibly contributing to global warming.
3. As the world's population increases and becomes more affluent, the demand for energy and chemicals will increase dramatically.

Possible Solutions and the Chemical Engineer's Role

Renewable energy is a growing source of electricity that will help us move toward a sustainable energy future. Wind, geothermal, hydro, and solar power are

continued

FIGURE 8-2 *continued*

all examples of renewable energy. Renewable energy is attractive because it uses limitless natural resources such as wind, water, sunlight, and heat from the earth to produce energy. The high price of electricity from renewable energy sources is the main obstacle keeping renewable energy from being more widely produced. Engineers can overcome this challenge by developing more efficient equipment for renewable energy production.

Alternative fuels, especially bioethanol, also present a solution to the problem of meeting our energy needs sustainably. Bioethanol producers use a fermentation process to make ethanol from renewable resources such as crops and other plants. Finding a sustainable feedstock for making bioethanol is one challenge involved in ethanol production. Cellulosic biomass, which includes prairie grasses and some trees as well as sewage sludge and municipal waste, appears to be the most sustainable feedstock available. However, the process of producing ethanol from cellulosic biomass is too expensive. Engineers must decrease the cost of production by finding low-cost pretreatments and by optimizing the process.

Another valuable tool that we can use to reach a sustainable future, especially in the chemical industry, is biotechnology. Biotechnology refers to the use of living organisms to accomplish tasks. Some companies are already using microorganisms to create products sustainably and efficiently. Understanding and manipulating complex metabolic pathways is a major challenge that researchers must overcome in this area. Engineers must also receive the education necessary to create, run, and optimize processes involving biotechnology.

Conclusion

Though energy and chemical producers must utilize each of the technologies discussed in order to achieve sustainability, cellulosic ethanol and biotechnology are particularly important and should be the focus of chemical engineering research.

Recommendations

To keep up with the changes in the field of chemical engineering, I recommend that

1. chemical engineers currently employed at BE participate in continuing education in the areas of alternative fuels and biotechnology
2. BE recruit college graduates and experienced professionals that have education or experience with alternative fuels and/or biotechnology

FIGURE 8-2 *continued*

TO: **Jerry Bradshaw**   DATE: **October 21, 2013**

FROM: **Matthew Lunsford**

SUBJECT: **Issues in Chemical Engineering: Sustainability**

DIST: **M. Malkin, J. White, K. Gwynne, E. E. Garza, R. Cantu, A. Hairen**

Introduction

Over the past 20 years, a growing awareness of our world's limited resources, our environmental impact, and our increasing energy needs has sparked a change in thinking among chemical engineers. Formerly, they focused solely on profits. Now, however, they are placing greater emphasis on preserving our world's resources for future generations. Researchers have coined the term "sustainability" to describe this idea. Sustainability is one of the largest challenges currently facing chemical engineers, and it will continue to be just that for years to come.

This report covers sustainability in the chemical and energy industries and how chemical engineers play a vital role in achieving a sustainable future. I have written this report to keep BE up to date on important developments in the field of chemical engineering. The remainder of this report will (1) discuss the importance of sustainability, (2) highlight solutions that could help move us toward a sustainable future, and (3) close with several conclusions and recommendations that apply to BE.

The Importance of Sustainability

Companies and media often toss the word "sustainability" around, and several valid definitions exist. The most common definition that I encountered during my research is that sustainability is "development that meets the needs of the present without compromising the ability of future generations to meet their own needs" [1]. This statement carries the idea that we must be looking ahead to the good of those coming after us as we currently seek to meet our needs, whether they are financial or energy related. Otherwise, our future children and grandchildren may be left without a way to meet their own needs.

A major issue that we must overcome if we want to reach sustainability is meeting our chemical and energy needs with resources that are fully sustainable, ones that will not run out or destroy the earth that we live on [1]. Even major oil

1|

continued

FIGURE 8-3 Matt's Report

companies such as Exxon, which has invested $600 million in alternative fuels research, are taking notice of this issue [12]. Universities have also changed their chemical engineering curricula to include courses in biotechnology, which can be used in the production of sustainable fuels. For the following reasons, the focus of the chemical engineering field has shifted away from solely profits and toward being profitable within the context of sustainability.

1. One reason is that the resources that we rely heavily upon will run out. Currently, the world meets 80% of its energy needs with fossil fuels such as oil, natural gas, and coal. The chemical industry also relies on these same resources as raw materials. Analysis shows that remaining fuel reserves will last about 40, 60, and 150 years, respectively [2]. So, if current consumption continues without the development of sustainable energy resources, 80% of society's energy needs will be left unmet 150 years down the road. Researchers and engineers must focus their knowledge and skills on developing energy resources and raw materials that will last indefinitely, or future generations will suffer the consequences of energy shortages and a lack of natural resources.

2. Destruction of the environment is a second reason that striving for sustainable energy demands the attention of the chemical engineering field. The fossil fuels that we currently burn for energy release are chemicals that harm the environment. Chemicals such as sulfur oxide (SOx) and nitrogen oxide (NOx) are particularly harmful because they react with water vapor in the atmosphere to form acids. These acids then fall to the earth as acid rain and can harm ecosystems that are sensitive to changes in acidity [1]. These ecosystems include oceans, lakes, rivers, and forests [1]. Electricity generation, residential heating, and industrial energy use contribute about 80% of the SOx emissions, and motor vehicle exhausts contribute about 48% of the NOx emissions [1].

   The burning of gasoline and other fossil fuels also releases carbon dioxide (CO_2) into the atmosphere, which, many scientists suspect, contributes to global warming through a greenhouse gas effect. CO_2 helps to trap heat within the earth's atmosphere, which may be causing temperatures on earth to rise. If this trend continues, temperatures on the earth could rise between 2 and 4 degrees Celsius in the next century, which would cause ocean levels to rise [1]. Rises in ocean levels could flood populated coastal areas and displace agriculture, threatening whole populations [1]. Engineers must find energy resources that do not continue to degrade the environment so that society in the future still has an earth to live on.

2|

FIGURE 8-3 *continued*

3. Finally, engineers must develop sustainable resources that can keep up with increasing demand. By the year 2050, researchers expect the population of the world to double [1]. Along with that, they expect energy demands to grow by 1.5 to 3 times [1]. Large countries such as China and India continue to develop and require more energy for industry and motor vehicles as their populations become wealthier. This ever-increasing demand for energy will put our limited fossil fuel resources under more stress as time goes on. Therefore, the need exists for sustainable energy resources and raw materials for chemicals that will be able to meet this large demand indefinitely.

Therefore, the challenge for chemical engineers is this: develop energy resources and raw materials for chemicals that (1) will not run out, (2) will not destroy the environment, and (3) will meet the demand of an increasingly large and affluent population.

Possible Solutions and the Chemical Engineer's Role

Chemical engineers lead the effort to develop sustainable energy resources and raw materials. One reason for this reality is that researchers have already amassed a large amount of base knowledge on the subject of alternative fuels and renewable resources. Scientists have already discovered the basic principles behind each of the possible solutions that I will discuss [1]. The challenge now is to overcome the economic barriers that keep these solutions from being competitive with the nonrenewable fuels that we currently use [1, 3]. Because researchers cannot easily quantify all of the benefits of using sustainable resources, high costs still keep sustainable resources from being used more widely [2]. Further research is also necessary in some areas, such as in biotechnology. The following paragraphs discuss a few possible solutions to the problem of developing sustainable energy and chemical resources and mention some challenges that engineers must still overcome.

Renewable Energy

Renewable energy is one solution that has and will continue to help move us toward a sustainable energy future. Wind, geothermal, hydro, and solar power all fall into this category of energy production. Renewable energy attracts attention as a solution for a sustainable future because it uses limitless natural resources such as wind, water, sunlight, and heat from the earth to produce energy. Renewable energy is already a growing source of electricity with growth of 60%, 28%, and 13%

3|

continued

FIGURE 8–3 *continued*

for solar power, wind power, and geothermal power, respectively, between 2000 and 2004 [5]. However, renewable energy still comprised a small portion of the world's total energy production in 2004, as shown in Figure 1.

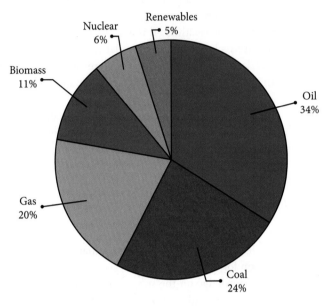

Figure 1 World Energy Supply, 2004
Information taken from Goldemberg [2].

Some sources of renewable energy are more sustainable than others. Wind energy, especially, shows great promise as a major energy contributor for the near future [4]. Though they can be loud and disturb some wildlife, wind turbines have a great upside. They do not pollute the environment, they are relatively efficient compared to some other renewable energy sources, and they produce electricity at a reasonable price [4]. Research shows that wind power is currently the most sustainable form of renewable energy [4].

Some other sources of renewable energy, such as hydropower and geothermal power, come at a higher price to people and to the environment. Damming rivers to produce hydropower often displaces people, animals, and agriculture; and geothermal power plants must carefully process waste water in order to avoid releasing toxic chemicals such as hydrogen sulfide, ammonia, mercury, and arsenic

4|

FIGURE 8-3 *continued*

into the environment [4]. Solar power is the least efficient and most costly of the renewable energy sources listed [4].

Challenges

The main challenge associated with renewable energy is the cost compared to fossil fuels. Currently, electricity from burning coal and gas still costs less than any of the sources of renewable energy, which prevents companies and countries from using renewable energy more widely [4].

Possible Solutions

As engineers make more efficient designs for solar cells and wind turbines and as the renewable energy industry matures and grows, costs of the electricity produced will drop and renewable energy will become more competitive with conventional energy sources. Then, renewable energy will be a more significant contributor to a sustainable energy future.

Alternative Fuels

Though renewable energy will play a part in working toward a sustainable energy future, we cannot rely on it completely. Renewable energy resources such as rivers are not distributed evenly throughout the earth, and places that do not have large amounts of renewable energy resources need other forms of sustainable energy that can be transported to them. Also, the majority of motor vehicles currently use combustible fuels, not electricity, for their energy supply. Therefore, if we want to move toward a sustainable energy future, we must develop sustainable alternative fuels that we can transport and use in motor vehicles. I will focus here on bioethanol because this is the most promising alternative fuel currently being produced.

Bioethanol is a fuel made by fermentation of sugars found in renewable sources such as plants and wastes. Bioethanol producers most commonly use either corn or sugarcane to make ethanol, but sorghum and cellulosic biomass also show promise as ethanol feedstocks. For environmental reasons, researchers see fuels made from plants as a very sustainable option. As shown in Figure 2, biofuels release CO_2 that plants such as corn take in during growth [3]. So, besides fossil fuels burned for feedstock growth and ethanol production, bioethanol does not pollute the environment. Also, because farmers can plant crops year after year, ethanol feedstocks are available indefinitely as opposed to oil, which will eventually run out.

5|

continued

FIGURE 8-3 *continued*

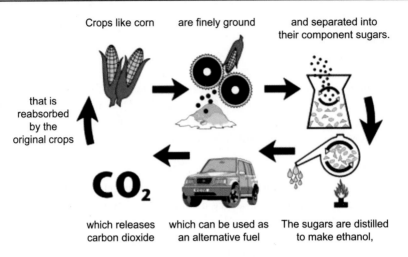

Crops like corn   are finely ground   and separated into
their component sugars.

that is
reabsorbed
by the
original crops

CO_2

which releases   which can be used as   The sugars are distilled
carbon dioxide   an alternative fuel   to make ethanol,

Figure 2 The Carbon Cycle for Bioethanol
Taken from http://www.globalization101.org/biofuels-and-ethanol/.

Challenges

One of the major challenges associated with alternative fuels is finding a feedstock that is sustainable, one that will not destroy the agricultural environment and can keep up with demand. Countries such as the United States use corn as an ethanol feedstock. Corn, however, is not a sustainable feedstock because it degrades soil, cannot keep up with increasing demand, and causes food prices to increase [6, 7]. On the other hand, sugarcane seems to be an attractive and sustainable feedstock. Brazil, which has established a successful ethanol industry that is now independent of government support, uses sugarcane to produce ethanol [2]. The use of sugarcane avoids negative consequences, such as rising food prices and soil degradation, that are inherent in other feedstocks such as corn [2, 6]. However, not all countries have the climate necessary to grow sugarcane.

Another feedstock that will not compete with food or damage soil is cellulosic biomass. Cellulosic biomass includes a wide variety of plants from tree species such as willow, maple, and black locust to grasses like switchgrass and reed canary grass [8]. Wastes such as woodchips, sawdust, municipal residues, and sewage sludge can also be used as cellulosic biomass and made into ethanol [8]. Many researchers view cellulosic biomass as an attractive ethanol feedstock because of its low price and sustainability. But the cost of producing ethanol from cellulosic biomass currently keeps this feedstock from being used on an industrial scale.

6|

FIGURE 8-3 *continued*

For each feedstock the main step in the ethanol production process is very similar. During this step yeast or bacteria take the sugars obtained from the feedstock and produce ethanol in a fermentation process [9]. The key difference between the different raw materials is what occurs before the fermentation step. For grain feedstocks such as corn, ethanol producers must use enzymes, which are biological catalysts, to break down the starches into simple sugars before fermenting [8, 9]. This extra step increases the cost of the process. Cellulosic ethanol requires even one more step. Before breaking down the feedstock into simple sugars, ethanol producers must expose the carbohydrates, which are enclosed by a rigid shell called "lignin," in a step typically called "pretreatment" [9]. This additional step causes cellulosic ethanol to be more expensive to produce than ethanol from other feedstocks.

Possible Solutions

For cellulosic ethanol to be produced and used on an industrial scale, chemical engineering researchers must determine ways to make it competitive economically with petroleum fuels. Decreasing production costs is a way that they can overcome this obstacle [9]. Researchers are currently giving a lot of attention to the pretreatment step as this step comprises almost 40% of the total cost of production [10, 11]. Current pretreatment methods involve chemicals such as sodium or potassium hydroxide, which are too expensive [11]. Other pretreatment methods, such as physically grinding or using steam, either are too expensive or produce yields of carbohydrates that are too low [11]. Possibly viable pretreatment methods include the use of the dilute acids such as sulfuric acid or fungi which degrade the lignin [9]. However, using fungi for pretreatment is typically slow compared to other methods [9].

Chemical engineers have several other options available for decreasing production costs of cellulosic biomass. These options mainly consist of taking techniques currently used in the production of petroleum fuels and applying them to cellulosic ethanol production. Using simulation software to model and optimize the ethanol production process would help ethanol producers make the most ethanol at the lowest cost [10]. This technique is used pervasively in refineries. Another technique that chemical engineers can use to optimize production is heat integration [10]. Engineers are able to conserve energy by taking waste heat from one part of the process and using it to heat up another part of the process through heat exchangers.

Biotechnology

Biotechnology is another powerful tool that chemical engineers can use to work toward a sustainable future. Biotechnology refers to the use of living organisms,

7|

continued

FIGURE 8-3 *continued*

such as fungi and bacteria, to accomplish tasks. Researchers can genetically modify these microorganisms to produce enzymes which are much more effective than traditional catalysts [3]. Through altering the natural metabolic pathways, researchers can also engineer bacteria that will generate desired products efficiently and from renewable resources [3].

This technology is especially useful in the chemical industry. For example, DuPont uses a genetically altered strain of *E. coli* to produce polymers from corn [3]. Normally this same polymer is made from fossil fuels. Hoffman La Roche also uses a genetically altered strain of bacteria to produce vitamin B_2 [3]. This one-step process replaces an older process that involved six steps, resulting in savings on energy and decreased burning of fossil fuels [3]. Biotechnology has the potential to allow chemical companies to make products from renewable resources and burn less fossil fuel in the process.

As previously stated, ethanol producers also use biotechnology in the form of enzymes and bacteria.

Challenges

The main challenge associated with the use of biotechnology is the complexity of the microorganisms and the need for more knowledge and research. To create a desired product using a microorganism, researchers often have to modify the metabolic pathways of that organism [13]. This process is carried out by modifying the organism's DNA [13]. Therefore, researchers must have in-depth knowledge of the extremely complex metabolic pathways as well as the DNA that governs that pathway. Then, the researcher must know exactly how to alter the DNA so that the desired change in the metabolic pathway takes place.

For the engineer, the main challenge is to gain the knowledge necessary to efficiently run processes that involve biotechnology. Though some of the basic principles are the same for a petrochemical reaction and a reaction involving enzymes, there are also many key differences. For example, while using biotechnology, an engineer must be concerned with not only keeping the microorganisms alive but also with what stage of growth they are in as this can affect what they produce [13]. In a petrochemical plant, chemical engineers have to worry about much different things such as the high temperatures and pressures often used.

Possible Solutions

The need for more research seems to be the limiting factor in the use of biotechnology in the chemical industry. Genetically engineering microorganisms

8|

FIGURE 8-3 *continued*

is a relatively new art. So, if chemical producers want to use biotechnology to make more products efficiently and sustainably, researchers must continue to study and gain knowledge on the metabolic pathways of microorganisms. Then engineers and researchers must develop processes using these microorganisms that can compete with petrochemicals. As the knowledge base grows, researchers will continue to develop strains of bacteria and fungi that can yield products sustainably. This developing technology could be the future face of a sustainable chemical industry.

For engineers to create, run, and optimize processes involving biotechnology, universities must prepare them to face the challenges of this relatively new field. Some universities have already begun to change their degree programs to reflect the move toward biotechnology. For example, the University of California, Berkeley, has changed the name of its chemical engineering department to the Department of Chemical and Biomolecular Engineering [14]. Also, MIT now offers degrees in both chemical engineering and chemical–biological engineering [15]. However, some universities, such as Texas A&M University, require very few biology-type courses in their undergraduate chemical engineering programs. Universities must continue to expand undergraduate education in bioengineering if engineers are going to be prepared to work in the changing chemical industry.

Conclusion

Despite posing a large challenge, sustainability in the energy and chemical industries seems to be a realistic future. No one technology, however, will bring us there. Energy and chemical producers must make use of a wide variety of options including renewable energy, alternative fuels, and biotechnology. As these technologies mature and develop, industry will call on chemical engineers to develop more efficient designs, processes, and microorganisms that will make sustainable energy and chemicals more competitive with those produced from fossil fuels.

Cellulosic ethanol and biotechnology are especially promising. Ethanol producers could potentially supplant much of the fossil fuel–derived energy resources with sustainable fuels. And biotechnology gives the chemical industry the opportunity to produce chemicals that are also made from renewable resources. These technologies are very important and should be the focus as we move toward a sustainable future, one where the needs of future generations will not be threatened.

9|

continued

FIGURE 8–3 *continued*

Recommendations

In light of the changes that are occurring in the field of chemical engineering, I recommend that (1) chemical engineers currently employed at BE participate in continuing education in the areas of alternative fuels and biotechnology and (2) BE recruit college graduates and professionals who have education or experience with alternative fuels and/or biotechnology. These measures will ensure that BE can offer the highest-quality services to our clients who have needs in these developing areas.

10

FIGURE 8-3 *continued*

References

1. Dincer, Ibrahim. "Renewable Energy and Sustainable Development: A Crucial Review." *Renewable and Sustainable Energy Reviews* 4, no. 2 (2000): 157–175.

2. Goldemberg, José. "Ethanol for a Sustainable Energy Future." *Science* 315, no. 5813 (2007): 808–810.

3. Rogers, P. L., Y. J. Jeon, and C. J. Svenson. "Application of Biotechnology to Industrial Sustainability." *Process Safety and Environmental Protection* 83, no. 6 (2005): 499–503.

4. Evans, Annette, Vladimir Strezov, and Tim J. Evans. "Assessment of Sustainability Indicators for Renewable Energy Technologies." *Renewable and Sustainable Energy Reviews* 13, no. 5 (2009): 1082–1088.

5. Armaroli, Nicola, and Vincenzo Balzani. "The Future of Energy Supply: Challenges and Opportunities." *Angewandte Chemie International Edition* 46, no. 1–2 (2006): 52–66.

6. Pimentel, David. "Ethanol Fuels: Energy Balance, Economics, and Environmental Impacts Are Negative." *Natural Resources Research* 12, no. 2 (2003): 127–134.

7. Iowa State University. "Corn for Biofuel Production." *American Cattlemen.* Accessed October 6, 2012. http://www.americancattlemen.com/articles/corn-biofuel-production.

8. Solomon, Barry D., Justin R. Barnes, and Kathleen E. Halvorsen. "Grain and Cellulosic Ethanol: History, Economics, and Energy Policy." *Biomass and Bioenergy* 31, no. 6 (2007): 416–425.

9. Sun, Ye, and Jiayang Cheng. "Hydrolysis of Lignocellulosic Materials for Ethanol Production: A Review." *Bioresource Technology* 83, no. 1 (2002): 1–11.

10. Cardona, Carlos A., and Óscar J. Sánchez. "Fuel Ethanol Production: Process Design Trends and Integration Opportunities." *Bioresource Technology* 98, no. 12 (2007): 2415–2457.

11

continued

FIGURE 8-3 *continued*

11. Yang, Bin, and Charles E. Wyman. "Pretreatment: The Key to Unlocking Low Cost Cellulosic Ethanol." *Biofuels, Bioproducts and Biorefining* 2, no. 1 (2007): 26–40.

12. Mouawad, Jad. "Exxon to Invest Millions to Make Fuel from Algae." *New York Times*, July 13, 2009.

13. Shuler, Michael L., and Fikret Kargi. *Bioprocess Engineering: Basic Concepts*, 2nd ed. Upper Saddle River, NJ: Prentice Hall, 2002.

14. Department of Chemical and Biomolecular Engineering, University of California, Berkeley. Accessed November 8, 2012. http://cheme.Berkeley.Edu/.

15. MIT Course Catalog 2013–2014, Department of Chemical Engineering. Accessed November 8, 2012. http://web.Mit.Edu/catalog/degre.Engin.Chemi .Html.

12|

FIGURE 8–3 *continued*

Case 8-2

Remember that reports differ, depending on topic, audience(s), and purpose. For example, Case 8-2 shows how the chair of the Academics Committee of a State University writes a report to the president of the University Senate. The goal is to explain to the president the activities of the committee during the previous year. The University Senate president has asked the head of each Senate committee for a report on the previous year's activities. Katheryn attempts to make the report readable by using clearly worded headings and concise descriptive information under each heading.

Note that she gives the main point in the first sentence and provides a succinct opening introduction/summary. She chunks the report itself into segments introduced by clearly worded headings. Coffman can read those that interest him.

This represents perhaps the most common type of routine report. The writer designs it to inform quickly and concisely.

TO: Michael Coffman DATE: June 9, 2011

FROM: Katheryn Rios

SUBJECT: **Academic Committee and List of Concerns**

The AC has not had significant business this year because of the ongoing financial stresses on the university. However, among the list of topics presented here, several need consideration by the full Senate. I call AC meetings when a meeting seems needed. Some of our business can be conducted by e-mail, and I have used that method, which works better for members who are out of town.

Should the University Switch to Plus/Minus Grading?

The AC prepared a memo for members to deliver to their college deans. We received little response and certainly no consensus for implementing this system. Liberal Arts expressed interest. In short, $+/-$ grading would work well in some colleges but not in all. Our computing system could handle this feature, but our CIO did not want to assess the cost for programming unless more than one college wanted this change.

Student Evaluations of Teaching

• **Online or Paper?**

No consensus emerged for moving to one system, even though the university Teaching Committee would prefer having to maintain only one system.

Faculty who want to have their students use the online system evaluation have trouble getting their students to respond. Paper evaluations seem to provide the most complete number of responses.

• **What Will Happen to Student Teaching Evaluations?**

Two years ago, the University Senate approved a streamlined group of questions that contained five questions required for all college teaching evaluations. Each college could choose an additional 15 questions. This method would allow comparison among colleges.

The AC then recommended that all evaluations require TWO questions: (1) was this an effective teacher? and (2) was this a good course? Still no action from the Council of Deans.

1|

FIGURE 8-4 Example Internal Report

Extension of the Withdrawal Deadline from 50 to 60 Days

Request for this change came from the Student Senate. James Cornyn and I prepared an e-mail that went to advisors. Few responded. Thus, we phoned as many departments as we could. Results: Only departments which have few majors thought the deadline "might" be extended. The undergraduate dean opposed the idea, along with advisors from several large departments. Thus, this item was rejected.

Containing Textbook Costs

Helena Stone and I worked with Charlene Bangor on the bookstore contract. We never saw the final contract. Our concern: the 30% markup on books is too high, and that markup is added to the publisher's suggested price.

The AC approved an e-mail that was sent to the colleges for distribution to departments. It asked that faculty consider the cost of books required and the rationale for books required. The Student Senate reported that many required books are never used in classes. The library also has a website for posting class material.

The E-Book, Book Rental option continues to grow. The AC just voted to send this e-mail every term to notify faculty of this option.

University Rule on Shortened Courses

The AC approved the rule I devised with help from the registrar's staff. Karan offered an amendment (see attachment, p. 4), which clarifies the approval process. I will send this to the AC for their input/approval.

Students Connecting Through Service

This plan, which would be launched before our reaccreditation, was approved by the AC this spring. We do not know what it means, other than that it should broaden students' learning experiences. I sent an e-mail to the major service organizations, and four have responded positively thus far. The premed society and the dean want to work with local physicians on the importance of inoculations. The AC will have a report from the premed society at its next meeting.

Assessing Academic Advisors

The Student Senate has requested that academic advisors be evaluated. I responded to that request by stating that several questions would have to be clarified that deal with logistics: cost, number of advisors and which ones, procedure, method, and reports.

2|

continued

FIGURE 8-4 *continued*

Recommendations for the Next Faculty Senate Discussion

1. Review of Status of Preparation for Reaccreditation

    What do faculty need to know about the coming reaccreditation visit?
    What is the status of recommendations from the Stephens report?
    What will happen with the core curriculum after courses we approve are
        sent to the state?

2. Discuss Ways to Control/Lessen Textbook Costs

    Do faculty consider costs when they select books?

3|

FIGURE 8-4 *continued*

Elements of Formal Reports

Long reports should be developed as formal reports. The word *long* generally means 10 pages or more; but as a writer, you have to decide based on the report context. Writers should select elements to help readers find their way through the report. The longer the report and the more diverse the readership, the more elements will be required to help readers find what they need. Each element has a different purpose and often targets the needs of different readers. Reports are not novels: most reports will not be read cover to cover but selectively.

In general, formal reports may include any or all of the following elements. How these elements appear in practice differs widely and depends on the organization preparing the report. We will describe the general characteristics and content of each element and provide several examples. Most organizations develop report templates to fit the content most often generated.

> Prefatory Elements
>> Letter of Transmittal or Memo of Transmittal
>> Title Page
>> Submission Page
>> Table of Contents
>> List of Illustrations
>> Glossary and List of Symbols
> Abstracts and Summaries
>> Informative Abstract
>> Descriptive Abstract
>> Executive Summary
> Discussion, or Body of the Report
>> Conclusion(s)
>> Recommendation(s)
>> References
> Appendices

Prefatory elements. Readers initially access a report from the prefatory elements. These show readers what your report will discuss and how it will approach the topic. Long, complex reports that will likely be read by many readers require prefatory elements. The letter/memo of transmittal, title page, table of contents, abstract, and/or summary are often your reader's first—and only—experience with the report. Each of these elements should help your readers grasp and accept the rhetorical purpose of the report. Never consider prefatory elements routine "paperwork" needing only perfunctory writing. No matter how well written and researched the report content, the effectiveness of the report begins with the effectiveness of the elements that state the content. Many readers may decide to read the body of the report if they find the prefatory elements compelling.

Letter/memo of transmittal. The letter/memo of transmittal is addressed to the individual who will initially receive the report. This person may not be the primary reader but the individual responsible for routing the report to appropriate readers who will digest and use the content. In consulting reports, the CEO of an organization that has solicited the consulting work is usually the person addressed in the letter of transmittal. Or the transmittal letter can address the person who has authorized and requested the information, analysis, work, or recommendations covered in the report.

Reports prepared for internal distribution will have a memo of transmittal, and reports prepared for external distribution will use a letter of transmittal. Both can be organized the same way.

The letter/memo of transmittal should include the following information at minimum:

- Statement of transmittal—subject and purpose of the report
- Reason for the report

In addition, transmittal documents may include the following items:

- Background material—the larger issue or problem addressed by the report
- Mention of earlier reports (or additional reports that may be needed)
- Information that may be of special interest or significance to the readers
- Specific conclusions/recommendations that might be of special interest to the person to whom the report is addressed
- Financial implications
- Acknowledgments—list of those who provided help in the project

Both Case 8–1 and 1–1 include a transmittal memo. The sample formal report in Appendix C includes a transmittal letter.

Title page.   Title pages perform several functions. Basically, they provide critical identifying matter and may contain a number of identifying items to distinguish the report from others on a similar subject or from reports received in response to specific projects. Many organizations have a standard format for title pages. Whatever items you need—or your organization requires you to include—be sure to make the title page attractive. The following information often appears on title pages:

- Name of the company or individual(s) preparing the report, perhaps with his or her signature
- Name of the company or course for which the report was prepared
- Title and subtitle of the report
- Date of submission
- Code number of the report
- Contract numbers under which the work was performed
- Company or agency logo
- Proprietary and security notices
- Names of contact/responsible individuals
- Descriptive abstract

Submission page.   Reports may use a submission page, which includes the list of contributors to the report and/or the names and signatures of the authorizing officer or project leader. Submission pages emphasize the point we make throughout this book: reports require accountability. Signatures on a submission page indicate that the authors stand behind the content. The submission page usually either precedes the title page or follows it. Or the title page and submission page may be combined.

Table of contents. The table of contents performs at least three major functions. First, it indicates the page on which each major topic begins and thus serves as a locating device for readers who may be searching for specific information. Second, it forecasts the extent and nature of the topical coverage and suggests the logic of the arrangement and the relationship among the report parts. The table of contents contains all major headings used in the report. For that reason, major headings, like the title of the report, should reflect the content of the material that follows. Skimming a table of contents should give readers a clear idea of the topics covered, the content presented under each heading, the amount of coverage devoted to that topic, the development of the report, and the progression of information. (See the table of contents in the report in Appendix C.) While you may wish to use "introduction," "conclusion," and "recommendations," avoid the term "discussion": it tells your reader nothing about how you will present the report content. Third, the table of contents should reflect the rhetorical purpose of your report—what you want your readers to know and why your report is important to them.

List of illustrations. If a report contains tables, graphics, drawings, photos— any type of visual—it's customary to use the page heading "Illustrations" or "Exhibits" to list all the visuals. These are usually divided into "tables" and "figures," but specific types may also be listed if you have an array of illustrations: for example, maps, financial statements, photographs, charts, and computer programs.

Glossary and list of symbols. Reports dealing with specialized subject matter often include abbreviations, acronyms, symbols, and terms not known to readers outside a specialized group. Readership should determine if a glossary or list of symbols is needed. Many reports prepared by government agencies for the general public provide glossaries to ensure that any reader can assess the report.

Glossaries sometimes appear as prefatory elements, but they can also appear in an appendix at the end of the report. When you first use a symbol or a term that you will include in the glossary, tell your reader where to find the list or the glossary. You may wish to place an asterisk (*) by a word or symbol that will be covered in the glossary. If you have only a few terms that you will need to define, you can define these as footnotes at the bottom of the page and use asterisks (* or **) to alert your readers.

Abstracts and summaries. Abstracts and summaries are the most important prefatory items in a report. The title page, table of contents, abstract, and summary may be the only parts of the report many recipients read. Thus, carefully plan these elements as each provides a slightly different perspective. While the title page and table of contents outline the report's content and direction, abstracts and summaries provide the essence of the report: topic, purpose, results, conclusions, and recommendations. Each item can stand alone or be designed in conjunction with other prefatory elements, but the wording of the table of contents should echo throughout an abstract or summary. These elements, when well designed, enable readers to quickly find a specific segment in the body of the report.

Often, the abstract follows or appears on the title page. Summaries may also follow the title page. While abstracts and summaries contain similar information, summaries usually provide more extensive information than abstracts and may be written for decision makers whose needs differ from those of readers who want only the essence of a report. (See the table of contents and the summary in Appendix C.)

Abstracts fall into two categories, informative or descriptive. Abstracts, often accompanied by keywords, are prepared for use in online indexes and databases. Keywords allow your report, if it is stored in a database, to be retrieved. Thus, you should think carefully about the keywords that characterize the content of your report. What words will your readers most likely use in a database search? Abstracts with keywords may be separated from the report but linked to the full report. Based on the information in the abstract, readers can decide if they need to move to the full report.

Differences in abstracts have tended to disappear, and some abstracts have the characteristics of both traditional informative and descriptive styles. Organizations and journals usually have specific requirements for abstracts and may require, for example, a short descriptive abstract on the title page and an informative abstract after the table of contents.

Because the quantity of information that bombards readers increases, well-written abstracts have become critical to report access. The abstract should explain the purpose of the report, findings, conclusions, and recommendations—anything of significance. Because many readers now use abstract services, they may read only your abstract, or they may decide to retrieve/order your complete report if the abstract clearly shows the relevance of the content and key points.

A good way to plan an abstract is to create a file that contains the names of its main parts:

Subject/Purpose
Methods
Findings or Results
Conclusion(s)
Recommendation(s)

Next, insert information under each heading. When you complete your abstract, you can remove the headings. Be sure to check the length of the abstract; publications rarely want abstracts longer than 100 to 150 words. This method allows you to track length easily. However, some publications want the headings included in the abstract. The example abstract (Figure 8–5) contains these items. Color helps you see each item.

Informative abstract. This type of abstract includes the research objectives, research methods used, and findings, including principal results and conclusions. Recommendations may also be included. Informative abstracts usually range from 50 to 500 words, depending on the length of the report and on the requirements of the disseminating organization and of the abstracting service. Informative abstracts help readers decide if they want or need to access the entire report for more thorough examination. They begin with a statement of the report's purpose, and the remaining sentences give major highlights and conclusions.

Chemical (Chlorpyrifos and Permethrin) Treatments Around Stacked Bales of Hay to Prevent Fire Ant Infestations

Keywords: chlorpyrifos, permethrin, Lorsban 4E, Astro Insecticide, fire ant, hay bale

Abstract: This research evaluated the efficacy of using a chemical barrier applied to the soil area under stacked bales of hay to prevent the red imported fire ant, *Solenopsis invicta* Buren (Hymenoptera: Formicidae), from infesting stacked hay. Specifically, we were interested in determining if we could protect "clean" hay bales stored in fire ant infested fields for up to several weeks. Chemicals selected as barrier treatments were Lorsban® 4E, active ingredient chlorpyrifos, which kills ants on contact, and Astro™ Insecticide, active ingredient the pyrethroid permethrin, which can also act as a repellent to ants. We established a series of 12ft × 12ft plots, with a 10ft buffer between plots along a fence row in a fire ant infested field. Plots were grouped into four blocks of three stacks each. Plots within blocks were randomly assigned to each treatment (four plots treated with Lorsban® 4E and four treated with Astro™ Insecticide, and four control plots). Treatments included spraying a 12ft × 12ft soil area with a 1-gal solution of each chemical and water formulation. After soil treatments, we placed four square-bales of hay, stacked two a side and interlocking in two layers, in the center of each plot. Stacked bales were sampled for fire ant infestation using 2.5 × 2.5cm olive oil–soaked index cards; one bait card was placed on each side of the top layer of hay in each stack. Results from ANOVA show a significant difference in mean infestation levels among treatments. Stacks of hay sitting in the chlorpyrifos plots had fewer ant infestations compared to the permethrin and control plots. Results after one week showed that only one stack in the permethrin and two in the control plots were infested with ants, while none in the chlorpyrifos plots were infested. Results show that after three weeks all four control stacks, three stacks in the permethrin treatment, and two stacks in the chlorpyrifos plots were infested. These results indicate that on a short-term basis, such as 1 to 7 days, chlorpyrifos may be an effective short-term treatment option for protecting stacked hay from fire ant infestations.

Ronald D. Weeks, Jr., Michael E. Heimer, and Bastiaan M. Drees, Chemical (Chlorpyrifos and Permethrin) Treatments Around Stacked Bales of Hay to Prevent Fire Ant Infestations, Texas Imported Fire Ant Research & Management Project, Red Imported Fire Ant Management Applied Research and Demonstration Reports, 2000–2002, Texas Cooperative Extension Service.

http://fireant.tamu.edu/research/arr/year/00-02/2000-2002ResDemHbk.htm #stackedbales.

FIGURE 8-5 Informative Abstract

Descriptive abstract. This type of abstract states what topics the full report contains. Unlike informative abstracts, descriptive abstracts cannot serve as a substitute for the report itself. They begin with the report purpose and then explain content areas or topics covered in the report. Figure 8–6, a descriptive abstract, comes from a transportation report on rail accidents. See also the descriptive abstract on the title page of the sample report in Appendix C.

Executive summaries. Major reports often include executive summaries, which contain all of the items listed at the beginning of this chapter. However, unlike summaries for routine reports, executive summaries provide extensive development of each information segment to provide decision makers the information they need without having to read or search the full report. An example of an executive summary can be seen in the report for a state study on shrub control on rangeland (Figure 8–7). Even though you know nothing about this project, the executive summary will provide you with extensive information. In many proposals, the summary often determines whether the proposal itself will be read.

Discussion, or body of the report. The main part of the report, the discussion, takes most of the writer's development time. The information to be reported constitutes the discussion. The discussion will explain in detail—information

---

Section 403 of the Rail Safety Improvement Act of 2008 (RSIA) requires the Secretary of Transportation to conduct a study of the track inspection process. This report describes a study conducted by the Federal Railroad Administration (FRA), on behalf of the Secretary, in response to the requirement in Section 403 of the RSIA. To successfully fulfill this congressional mandate, FRA believed it was necessary to obtain a "snapshot" of the current track inspection process. A survey of a random sample of actively working track inspectors, as well as interviews with labor union officials and various levels of railroad management, provided the necessary information. FRA also compiled data related to track defects identified by FRA track inspectors over a 4-year period (2006–2009). This data provides an indication of the most prevalent defects as well as those that industry track inspectors fail to identify. In addition to the collection of survey data, interview data, and FRA track defect data, an ideal observer model was constructed for visual track inspection. The ideal observer model provides a theoretical upper limit, based on previously compiled empirical data, on how well the best possible observer can perform.

Source: Time Inspection Track Study, US Department of Transportation, Federal Railroad Administration, Office of Railroad Policy and Development, Washington, DC, July 2011.

FIGURE 8–6 Descriptive Abstract

EXECUTIVE SUMMARY

The disparity between the supply and demand for water in Texas has focused attention on brush control as a method for increasing the water supply for Texans. This review (1) summarizes how shrubs affect the water cycle at the tree, hillslope, small catchment, and landscape scales and (2) examines whether broadscale reductions in woody plant cover might increase the water supply. The greatest information need is at the landscape scale, but research at this scale is costly and difficult. Hence, most field investigations have focused on the tree or hillslope scales. However, extrapolating these data to larger scales is problematic because hydrological processes may change as the scale increases.

A review of the scientific literature has allowed the authors to make the following broad generalizations:

1. The relationship between shrub removal and increased water yields becomes stronger as annual rainfall increases.

2. The linkage between shrub removal and increased water yields is weaker in upland areas and stronger in areas adjacent to stream channels, where shrubs may be using shallow groundwater that is hydrologically connected to stream flow.

3. In upland areas, the linkage between removal of woody plants and increased water yield is stronger where water can move rapidly through the soil or parent material to recharge springs or shallow aquifers.

4. In areas where there is little subsurface water movement and woody plants are not accessing groundwater, shrub control is unlikely to significantly affect either groundwater recharge or stream flow.

The data suggest that water can be salvaged by controlling dense stands of saltcedar in riparian zones, dense stands of Ashe juniper in areas with rapid subsurface flow (e.g., where springs are present), and mesquite on soils that develop deep cracks when dry. However, in each case there is considerable uncertainty with respect to the amount of water that can be salvaged and the fate of the salvaged water.

Hydrologic models have predicted substantial increases in water yields following shrub control, but these results must be used with caution. There is a threefold variation in the predicted water yield increases from different models, and in some cases the predicted increases are at least three times larger than the measured values.

More research is needed to fully understand the extent to which shrub control can increase water yields. Key gaps include community-level evapotranspiration studies, isotope studies to determine the sources of water used by shrubs, groundwater–surface water connections, nested watershed studies, and landscape scale studies on saltcedar in various habitat types.

The hydrological condition of Texas watersheds and the partitioning of water within the hydrological cycle are determined by complex interactions between soils, vegetation, and climate. Brush encroachment is only one change that has affected the health of Texas watersheds—droughts, excessive grazing, water impoundments, urbanization, and water transfers also have altered the hydrological cycle. It would be valuable to shift the debate away from a focus on brush control and water yields to a broader assessment of best management practices for improving watershed health and sustainability.

FIGURE 8-7 Executive Summary

appropriate to the context, the readers, and the purpose of the report—why the report was done, its objectives, methods, findings, results, analysis of results, conclusions emerging from results, and recommendations for dealing with the results. The discussion is the heart of the report: without it, effective summaries and abstracts could not be written. All conclusions and recommendations derive from the discussion. Stated another way, the presentation of information in the discussion allows the report writer to draw conclusions and perhaps recommendations. In short, the discussion must support all conclusions and recommendations.

The discussion can also stand alone, as Matt's does in Figure 8–3. Because the report discussion begins with an introduction, including an explanation of the approach used in developing the report, the discussion is complete in itself. Many discussions end with a factual summary, a concise narrative of the report's main findings.

Ironically, however, the main discussion is the report segment read least. While most readers will look at the summary, abstracts, and table of contents, few will actually delve into the discussion. Nevertheless, the discussion becomes the source, foundation, and documentation for every statement written in the abstract and summary. Your conclusions need to evolve from the discussion. Recommendations need to evolve from the discussion and the conclusions. In a sense, then, the discussion develops as a narrative that comes to a conclusion.

Many readers may need to read parts of the discussion carefully. They will read the introduction, check the table of contents, skip to relevant sections that interest them, and then move to the conclusion or factual summary that pulls together the main results or ideas.

Parts of the discussion. The main body of the report generally begins with an introduction (or introduction + summary) that forecasts what is to follow in the report. It directs the reader's mind to the subject and purpose. It tells the reader how to approach the content by explaining the scope of the report, the plan of development, and any additional information the reader may need. Examine the introduction from a report for the US Department of Transportation (Figure 8–8). Note that it presents the topic, the rationale, and the development plan for the report. Because of the length of the introduction, the writer partitions the parts of the introduction with headings. Also examine the introduction to the sample report in Appendix C.

Note that the introduction should always include the report subject, purpose, and plan of development. Some reports place the background and the scope in separate sections that follow the introduction, if these two items are extensive. What you include in the introduction depends on your readers. If your readers will expect your report, you can write a short introduction. However, if your report will be archived and read later by people who know little about the context, you will need to provide a longer, more informative introduction (see Figure 8–9). Avoid long introductions: focus on the subject and purpose in terms that explain the relevance of the report to readers. But do anticipate your readers—who will read this report and when—as you plan the introduction. Without a proper introduction, readers will have difficulty following the main discussion.

Collecting and grouping information.   As you gather information, try grouping your material and notes into specific categories. Label these categories, and then begin your report. You may want to develop your report according to major

1.   Introduction

1.1   Purpose

The Federal Railroad Administration (FRA) continues to develop tools for managing fatigue in railroad operations. A previous report established a statistically reliable relationship between train crew performance effectiveness (inverse of fatigue) and the risk of an HFs accident (Hursh et al., 2006, 2008). The purpose of the current report was to use the data from the FRA database of railroad incidents, augmented by work schedules for the train crew involved, to investigate the relationship between the economic impact (cost) of an incident and the estimated fatigue of the train crew at the time when the incident occurred.

1.2   Background

In 2006, the FRA completed the third phase of a research program to demonstrate a method to validate and calibrate fatigue models for use in predicting and managing fatigue in railroad workers. A fatigue model offers the possibility of objectively assessing and forecasting fatigue so that employees and employers can schedule work and rest to avoid fatigue. A useful fatigue model needs to be calibrated to the demands of a particular job so that the measures from the model can be related to the risk of meaningful failures of human performance. One important part of calibration of a fatigue model for use as a fatigue management tool is an assessment of whether the tool can predict an increased risk of an HFs error or risk of having an HF-caused accident. As part of this assessment, FRA sponsored a project, in partnership with the five Class I freight rail carriers, to examine 2.5 years of data on accidents.

1.3   Scope

This report describes an extension of analyses performed for the FRA that investigated the relationship between accident cost and estimated crew impaired effectiveness from fatigue. The original effort developed estimates of train crew performance effectiveness using the Sleep, Activity, Fatigue, and Task Effectiveness (SAFTE) fatigue model implemented in the Fatigue Avoidance Scheduling Tool (FAST).[1] These estimates were then used to estimate accident risk based on crew member estimated effectiveness at the time of accidents, and to compare risk between accidents with HFs-related causes and those whose causes were not HFs-related. The work described herein extends that analysis to investigate the relationship of HFs-related accident (HF-related accidents) cost to estimated train crew effectiveness.

Source: Analysis of the Relationship Between Operator Effectiveness Measures and Economic Impacts of Rail Accidents, US Department of Transportation, Federal Railroad Administration, Office of Railroad Policy and Development, Office of Safety, Washington, DC, May 2011.

FIGURE 8-8 Report Introduction

sections—introduction, followed by the topics or categories. Open and name the file and begin. For example:

- Introduction: state the purpose of your report—what you expect to accomplish, what you want your reader(s) to know. You can add information later. Stating your purpose at the beginning of the draft helps you stay focused.
- Category/topic 1: phrase describing the issue you want to present
- Category/topic 2: etc.
- Category/topic 3: etc.

Next, begin inserting information under each topic or subject category. Focus initially on inserting information that pertains to the topic heading.

This method helps employees who may be trying to work on a report, answer phone calls, answer/send e-mail, and deal with other routine business events throughout the workday. In short, you can arrange, insert material, save what you have, and add other material as time permits. You may also write notes to yourself or use different colors and fonts for text you may want to move, delete, or revise. How you design your report will depend on (1) the kind of report you are writing, (2) your readers' informational needs, and (3) the purpose of your report. Introductions should be as short as possible while providing readers with the information they need to understand the report.

Strategy for presenting the discussion. The discussion should be planned around each topic mentioned in the plan of development. Note that the logic of the discussion should be evident from the headings as these are repeated in the table of contents.

Each paragraph in the discussion should begin with a topic sentence, followed by supporting sentences, data, and visuals.

Reports with standard arrangement patterns. Some kinds of reports have fairly standard arrangement patterns. Many government funding agencies require that all reports submitted follow a specific plan, and content may be inserted into a template and submitted online. Many organizations also have a standard plan

---

Wind Power in the United States: Technology, Economic, and Policy Issues

Introduction

Rising energy prices and concern over greenhouse gas emissions have focused congressional attention on energy alternatives, including wind power. Although wind power currently provides only a small fraction of US energy needs, it has grown more rapidly than any other electricity source. Wind energy already plays a significant role in several European nations, and countries like China and India continue to rapidly expand their capacity both to manufacture wind turbines and to integrate wind power into their electricity grids.

This report describes utility-scale wind power issues in the United States. We partition the report into the following sections:

- Background on wind energy
- Wind resources and technology
- Industry composition and trends
- Wind power economics
- Policy issues

FIGURE 8-9 Standard (and Readable) Report Introduction

for their policies—what sections to include and the order in which they appear. Policy and procedure manuals, detailing the rules that apply to the employees of a business entity, usually follow a standard pattern.

Often, however, as in Case 8–2, you will need to write a report in which you, as the writer, must decide how to present your research or the information you need to convey. Generally, you have two basic choices: topical arrangement and chronological arrangement.

Topical arrangement. In topical arrangement, the order in which you present your ideas should be logical and inclusive. For example, in a report on disease management of citrus fruit, the writer arranges the report by grouping information about specific citrus diseases. He describes, in parallel arrangement, the main diseases:

> Introduction—Description of treatments for citrus diseases
> Disease #1—Melanose
> > Description
> > Factors to be considered before the application of fungicides for melanose control
> > > Table I. Chemical controls for melanose
> Disease #2—Greasy Spot
> > Description
> > Factors to be considered in the management of greasy spot
> > > Table II. Chemical controls for greasy spot
> Disease #3—Foot Rot
> > Description
> > Factors to consider in managing the disease
> > > Table III. Chemical controls for foot rot
> Disease #4—Citrus Nematode
> > Description
> > Sampling instructions to determine presence of citrus nematode
> > > Table IV. Citrus nematode counts considered low, medium, or high at specific times during the growing season
> Conclusion: Factors to consider before applying nematicides
> > Table V. Chemical controls for nematodes (summary)

Once the citrus researcher has arranged his information, he can begin inserting information beneath each topic heading.

Chronological arrangement. Some topics can be presented by time. You explain or present information sequentially, in the order in which it occurred. The following outline of a literature review, *Cultural Control of the Boll Weevil—A Four Season Approach—Texas Rolling Plains*, illustrates chronological arrangement. This technical report surveys and reviews existing research on how boll weevils can be controlled throughout the agricultural year. Note, too, that the segments use parallel development, and each segment ends with a summary of that segment. This approach allows the reader to choose where to begin and how much to read. For example, the reader may wish to read only the Summary and Introduction and the factual summary for each season:

Summary
Introduction
Spring Cultural Control
    Prepare the land for planting
    Utilize delayed planting
    Use uniform planting
    Summary
Summer Cultural Control
    Shorten the growing season
    Change the microclimate
        Row direction
        Bed shape
        Row spacing
    Summary
Fall Cultural Control
    Utilize harvest-aid chemicals
    Role of planting date
    Terminate irrigations in August
    Summary
Winter Cultural Control
    Eliminate the overwintering habitat
    Modify the overwintering habitat
    Avoid the overwintering habitat
    Summary
Conclusions
Acknowledgments
Supporting Research Studies

Persuasive arrangement and development. Many times reports argue for a specific point or position. When you argue, you need to persuade your readers. Understanding what objections you will need to overcome will be critical to planning your report and the presentation of your arguments. Reports that must produce conclusions and recommendations may require writers to develop documents in which the conclusions and recommendations are not what readers will welcome. Or these reports may be prepared for readers who have no preconceived ideas. In each situation, the report can be designed to anticipate the perspective of readers.

Case 8-3

Allan Harper was asked to visit a research facility that his firm has discovered is likely going to be for sale soon. Dick Crandall, the vice president for operations, asks Allan to provide a brief overview on the usability of the building as a corporate training facility. Allan visits the building, takes notes of major usability features that he knows will interest Dick, and writes a brief report for the vice president, known for wanting only important facts and recommendations. Allan begins with a simple report purpose statement, followed by a bulleted list of the facility's features of interest to Dick. Allan uses notes to himself (in color) as he drafts. He concludes with a brief recommendation of what he believes should be done if Dick thinks the building should be purchased. The report responds to the information needs and reading (skimming) style of a specific reader who wants "the facts."

CASE DOCUMENT 8-3

TO: Dick Crandall   DATE: October 9, 2014

FROM: Allan Harper

SUBJECT: Space Assessment of RAMP, 6004 Highway 7

As you requested, I toured the RAMP research facility during my trip to Atlanta October 4–8, 2014. Based on your questions about using the facility for corporate training, I have concluded the following:

Conclusions

- RAMP offers us the space we need and appears to be in excellent condition. Modifications will be necessary, but those required can be done in stages. We can focus on Building 1, then renovate Buildings 2 and 3 as we want to. The RAMP is composed of two main buildings and one small building (Buildings 1, 2, and 2A on the attached site map).
- Each small office has two Internet connections. The main lecture hall has one Internet connection and no projection equipment. [Check on price of wireless for the entire building. DC will probably want to know ASAP.]
- The current RAMP buildings are divided into offices for start-up projects. Some of the RAMP income is derived from the space leased by small start-up companies. Cost: $22/sq. foot. [Is this price low/high?] [Call Dave Redding about laboratory contracts. Cheryl Kempe has his phone number.]

continued

- The debt on RAMP is slightly over $12M with approximately $5M due in April of 2012. (See p. 4 of the Notes to Consolidated Financial Statements.) [Get current balance owed from Fiscal. Perhaps delete for now? Call Debbie in Fiscal.]
- For us to use RAMP for training, instead of research, the entire first floor of Building 1 would need to be redesigned. We could relocate some of the start-up companies to Buildings 2 and 2A, which would preserve the income.
- An attached sketch shows how renovation of the first floor of Building 1 might look. Note:
  Eight office sites for trainers
  Two computer labs
  Five training rooms—two with breakout rooms off the main area
  One tiered lecture room

Conversion could be made gradually, depending on how many training rooms are needed ASAP. The attached sketch, included along with the original plan, shows one redesign plan.

Assessment of the RAMP Building

- The facility seems to be in good repair. Maintenance seems to be satisfactory. I found no indication of leaks.
- HVAC system needs to be inspected.
- Wireless capability could be easily installed in the main lecture room. ~~The proposed computer laboratories, carved from 12 offices, offer 24 Internet connections.~~
- New projection equipment needs to be purchased. [Cost?]
- Main lecture hall could be converted into a tiered lecture room. The two rooms on either side of the lecture hall could be eliminated to further expand the main lecture hall. The kitchen would have to be moved or eliminated—perhaps moved to Building 2.
- Renovation costs will be high, close to $1M (according to rough estimates by Gavin Newberry—see attached e-mail answer to my query) even if only the first floor of Building 1 is refurbished. Current office furniture is usable, but new furniture, IAV equipment, and computers would need to be purchased for training and meeting rooms. Closets/storage would have to be added to all training rooms. Floor coverings would need to be replaced.
- Location—RAMP is easily accessible from three main arteries.
- Estimated remodeling time: 4 months.

Attachments: 3

Let's assume, for example, that Harper believes the RAMP building should be seriously considered, while Crandall has decided to purchase another site. Crandall told Harper to visit RAMP, but he never expected that RAMP would be another site consideration. To respond to this situation, Harper might begin his report as follows:

> As you requested, I toured the RAMP research facility during my trip to Atlanta October 4–8, 2014. Based on my findings, I believe that you should seriously consider the RAMP facility as the site for our CE operation. RAMP offers us what we need in terms of space and location. Renovation costs can be done in stages to reduce cost, and the facility offers us the space and arrangement we need.

In short, the reader's perspective often determines how persuasive you need to be if you want your ideas considered.

Conclusion(s). Reports end with a statement of the primary issues covered in the discussion. In long reports, a factual summary comes after the discussion and before the conclusions and recommendations. The factual summary does what its name implies: it covers the essential facts but without interpretation. (Interpretation of the factual summary is done in the executive summary.) The longer the discussion, the more useful a factual summary will be to the reader who actually reads the body of the report. The conclusions themselves provide judgments about the subject based on the evidence explained in the discussion.

Recommendations. Recommendations, if required or needed, emerge from conclusions. However, many reports end with only a conclusion. The type of ending depends on the type of report. Recommendations are proposed actions to be taken based on the conclusions.

Appendices. Appendices include documents that support or add to information you include in the discussion. While the appendix may not be read, it serves an important role. An appendix may contain tables of supporting data, statistical studies, spreadsheets, or any material that supports the points or arguments you make in your discussion. If your report recommends an action, having letters or e-mails requesting that action in the appendix provides convincing support for your recommendation.

LETTER REPORTS

Reports can be prepared as letters. See Figure 8–10, prepared by an engineer at Bradshaw Engineering, who has been asked to explain the differences in corn and sugar ethanol to undergraduate engineering students at Texas A&M. Note the purpose statement and the plan of the report, in addition to the short summary statement. The effective document design allows her report to be read quickly.

BRADSHAW ENGINEERING, LLC
2104 RIVERSHIRE PLACE
HOUSTON, TX 77036

February 18, 2013

Dr. Jim Freeman
P.O. Box 3998
MSC Station
College Station, TX 77834

SUBJECT: Advantages and Disadvantages of Ethanol from Brazilian Sugar

Dear Jim:

Introduction

I was interested in your questions about ethanol and Brazilian sugar, which you introduced during your class last week. Mr. Bradshaw suggested that you might like to share the following information with your engineering interns.

With a growing demand for ethanol in the United States, producers are finding new, cheaper crops to convert to ethanol. Corn remains the major source of ethanol production in the United States. Recent discoveries find ethanol from sorghum in China and sugarcane in Brazil to decrease the overall demand for corn-based ethanol because of their advantages. This letter examines the advantages and disadvantages of Brazilian sugarcane–based ethanol from an economic, social, and environmental perspective.

Economic Impact

Brazilian Sugar Is Cheaper and More Effective

For the past two years, prices for sugar have steadily decreased because of a global surplus in production.[1] Because of a large demand for ethanol, both in Brazil and globally, ethanol prices are much higher than raw sugar prices. This pushes Brazilian millers to focus more on producing sugar-based ethanol, helping them make a profit faster. A larger supply of ethanol from sugar creates an economic competition with corn ethanol.

Ethanol from Brazilian Sugar
Tasnim Mohamed

FIGURE 8–10 Letter Report

If prices of corn ethanol remain higher, more consumers will prefer to use sugar ethanol. McKenna also argues that the use of corn ethanol is inefficient. In 2007, 27% of all the corn grown in the United States was dedicated to ethanol production.[2] Even if all the corn was dedicated to ethanol production, he states, gasoline usage would only reduce by 15%.[2] The same reduction of gasoline usage could also be achieved with increasing fuel efficiency by 3.5 miles per gallon.

Increase in Food Prices

The increase of corn prices has greatly affected the food industry as well. A chain reaction began when prices of corn increased because many consumers depend on corn to make a variety of foods. Corn is often used to make animal feedstock, so if these prices increase, so will the price of meat. The prices of other common corn-based foods can increase as well, such as cornmeal, tortillas, corn bread, and many more. A recent drought at the end of 2012 kept the price of corn high.[3] High corn prices cause many farmers to lose their businesses.

Using sugarcane-based ethanol can also have adverse effects on the price of raw sugar products, too. The graph in Figure 1 shows the increase of food prices in fats and oils, grains, and other foods (sugar, citrus, bananas, shrimp, and meats).

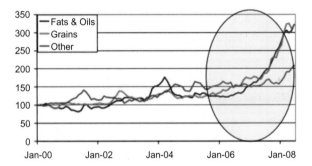

Figure 1 Trend in Food Prices from 2000 to 2008[4]
(Nominal $ Index, Jan 2000 = 100, world export value weights)

Social Impact

Tension with the Indigenous Brazilian People

The mass production of sugarcane in Brazil currently presents a problem to its society. Most sugarcane plantations in Brazil are located on land reserved for the indigenous people, the Guarani. These people hold little power with the Brazilian

FIGURE 8-10 *continued*

government. Because of Brazil's increasing profit from exporting sugar-based ethanol, sugarcane millers are free to extend their plantations as they please, forcing
the Guarani out of their homes and off their land.[5]

The Guarani have no other choice but to work on the plantations under horrible
conditions and low wages. To decrease the negative social aspect of sugarcane production in Brazil, the landowners "need to be held accountable" and respect the
rights of the local people. [4]

Jobs Created for the Uneducated
Despite tension with indigenous peoples in Brazil, sugarcane production has created a large workforce for the uneducated population. Table 1 lists a few demographics of people with various jobs in Brazil. Based on the table, the people
working with sugarcane crops have the lowest education. Clearly, many people
work in this industry relative to the number of people in other job positions.

Table 1 List of Demographics for Various Jobs in Brazil[6]

Statistic	Sugar Cane Crops	Sugar	Ethanol	Food and Beverages	Fuels
People (× 1,000)	789.4	126.0	67.0	1,507.0	104.7
Mean Age (years)	35.1	36.6	35.6	34.4	37.1
Mean Education (years)	2.9	6.5	7.3	7.1	8.9
Mean Income	446.6	821.3	849.9	575.0	1,281.1

Environmental Impact

Ethanol in the Carbon Cycle
Both sources of ethanol have an equal environmental benefit. By converting any
kind of biomass to fuel, the carbon dioxide (CO_2) absorbed by plants (such as corn
and sugarcane) is released again into the atmosphere. Because CO_2 is constantly
removed from and replaced to the atmosphere, burning bioethanol resembles the

FIGURE 8-10 *continued*

carbon cycle. Unlike bioethanol, burning gasoline simply adds CO_2 to the atmosphere without initially removing it.

Reduction of Greenhouse Gases

The production, distribution, and consumption of ethanol create nearly 20% less greenhouse gases than the same process used for gasoline.[7] However, if an increased demand for cropland leads to the clearing of large forests or grasslands, ethanol would be useless. Forests and grasslands naturally absorb most of the atmospheric CO_2, and depleting this would greatly increase the amount of atmospheric oxygen even with the use of ethanol rather than gasoline.

Sincerely,

Tasmin Mohamed

Tasmin Mohamed, P.E.

Attach: Sources

FIGURE 8-10 *continued*

ATTACHMENT

Sources

1. Almeida, Isis. "Brazil Ethanol Above Sugar Shows Mills May Favor Biofuel." *Bloomberg BusinessWeek* 2013. http://www.businessweek.com/news/2013-02 -11/brazil-ethanol-topping-sugar-signals-millers-may-favor-biofuel.

2. McKenna, Phil. "Corn Biofuel 'Dangerously Oversold' as Green Energy." *NewScientist* 2007. http://www.newscientist.com/article/dn12283.

3. McDonald, Kay. "Paying More for Food? Blame the Ethanol Mandate." CNN.com, last modified August 20, 2012. http://www.cnn.com/2012/08/20/ opinion/mcdonald-corn-ethanol.

4. Mitchell, Donald. "A Note on Rising Food Prices." Policy Research Working Paper 4682. The World Bank Development Prospects Group. July 2008. http:// www-wds.worldbank.org/external/default/WDSContentServer/IW3P/IB/2008/ 07/28/000020439_20080728103002/Rendered/PDF/WP4682.pdf.

5. Villarreal, Ryan. "Deadly Ethanol: Brazil's Sugarcane Farms Take Toll on Indigenous People." *International Business Times* December 8, 2012. http:// www.ibtimes.com/deadly-ethanol-brazils-sugarcane-farms-take-toll -indigenous-people-912116.

6. Smeets, Edward, Martin Junginger, Andre Faaij, Arnaldo Walter, and Paulo Dolzan. "Sustainability of Brazilian Bio-ethanol." Report NWS-E-2006-110. Copernicus Institute. August 2006. http://web.archive.org/ web/20080528051851/http://www.bioenergytrade.org/downloads/ sustainabilityofbrazilianbioethanol.pdf.

7. Elmendorf, Douglas W. "The Impact of Ethanol Use on Food Prices and Greenhouse-Gas Emissions." Report. US Congress Congressional Budget Office. April 2009. http://www.cbo.gov/sites/default/files/cbofiles/ftpdocs/100xx/ doc10057/04-08-ethanol.pdf.

FIGURE 8–10 *continued*

EXAMPLE REPORT FOR STUDY

An example report appears in Appendix C. Two additional annotated reports can be found on the book's website: **www.oup.com/us/tebeaux**.

WRITING COLLABORATIVELY

In the workplace, you may need to write reports, instructions, and letters on your own and as part of a team. Collaborative writing occurs in numerous ways that continue to evolve as technology evolves: you may draft a document or part of a document and then submit it, as an e-mail attachment, to several additional employees who will contribute information to your draft. Each writer may suggest changes to others' work by using a comment tool, such as that found in most word processing programs. Team members establish a website on the company intranet and post material that can then be viewed, revised by all team members, and then discussed via the company intranet.

A team member may establish a wiki on the organization's server—never the Internet because of security issues. On the intranet wiki, everyone on the team can contribute information and help "build" the document. The group may then use one or more interactive audio/video calls to discuss the material, particularly if contributors cannot meet face to face.

Team members often cannot collaborate in "real time" because of other work responsibilities. Wikis and other forms of virtual collaboration allow teams to work more flexibly but could also raise questions about information security.

> **Warning:** As we discussed in Chapter 1, the need for security of an organization's intellectual property cannot be underestimated. Hackers can destroy an organization by stealing intellectual property, trade secrets, contracts, and personnel information. The networking software students use in school cannot offer the security needed by business and research organizations. If used carelessly, social media can ruin lives, businesses, and reputations. Think carefully and cautiously before you use social media for any message.

The secret to effective collaboration is planning:

- Deciding who will do what
- Deciding how to meet the completion time and planning the work schedule for each document segment
- Learning to work together. When people from different countries and cultures collaborate, the need for sensitivity to cultural differences is critical.

• Ensuring one member will assume the task of editing the document to make sure the style shows consistency throughout

The team leader. Collaborative projects must have a team leader. This person may be the project manager or someone appointed. This person will convene meetings, virtual or on-site, and serve as facilitator to keep the project moving. In addition, the team leader

• ensures that all team members know their tasks
• launches the effort and provides everyone with the document requirements (purpose, audiences, rationale), format, timetable for completion, and methods of communicating about the document (see Figure 8–11)
• monitors development so that the report is completed as required
• ensures that the completed document has a single style throughout and contains all information and required segments

Requirements of team leaders. Team leaders, in either face-to-face or virtual settings, need to employ a few fundamental strategies:

• Maintain a positive approach
• Be aware of how you sound; be respectful
• Solicit input
• Encourage timely responses from everyone
• Keep records of all communications

After each draft, the team leader should encourage, as time permits, comments by each team member. The "comment" tool in doc or pdf files allows each team member to recommend changes and assess the effectiveness of the document.

Project name	
Goals	
Team members	
Responsibilities	
Deliverables	
Timeline	
Meetings	
Initial outline	

FIGURE 8–11 Collaborative Team Planning Sheet

Requirements of team members.

- Keep the schedule as determined by the project leader. If you find you have an issue that threatens your keeping the schedule, notify the team leader immediately.
- Be respectful and accept comments, recommendations, and suggestions from other team members and the project leader. Exhibit a good attitude, even when you feel stressed.
- Do your part of the project. Do not assume that if you get behind, another team member can do your work. The quality of the team project stems from the quality of the work performed by the entire team.

REPORT CHECKLIST ✔

Planning

☐ What is the purpose of your report? Have you stated it in one sentence?
☐ What is the scope of your report?
☐ Who are your readers? What are your readers' technical levels?
☐ What will your readers do with the information?
☐ What information will you need to write the report?
☐ How long should the report be?
☐ What format should you use for the report?
☐ What report elements will you need?
☐ What visuals will you need to present information or data?
☐ What elements do you need to include in your introduction?
☐ What arrangement will you use in presenting your report?

Revision

☐ Does your report do the following:
- Introduce the subject and purpose?
- Present enough data in words and visuals to justify any conclusions drawn?
- Discuss and evaluate the data fairly?
- Summarize the data?
- Draw logical conclusions from the data?
- If necessary, present recommendations that are clearly based on the data and the conclusions?

☐ Are your data accurate?
☐ Do your visuals immediately show what they are designed to show?
☐ Is your format suitable for your content, audience, and purpose?
☐ Have you properly documented all information sources?

EXERCISES

1. Visit the website of a major employer in your field. What kinds of reports do you find at this site? What are the elements and characteristics of the reports? How long are the reports? Who are the audiences for the reports? What are the purposes of the reports? Are the reports written by a single author or multiple authors? What do the reports indicate about this employer? What do the reports indicate about your field? How would you prepare to write a report like one of these?

2. Visit the website of the National Transportation Safety Board (www.ntsb.gov) to examine each of its six kinds of accident reports (aviation, hazardous materials, highway, marine, pipeline, and railroad). Describe the elements of the reports. How would you describe the writing style of the reports? What kinds of illustrations do you find integrated into the reports? Do the reports offer satisfactory descriptions of the accidents? Do the reports offer clear and convincing evidence of the causes of accidents? Do the reports make a persuasive case for changes in practices or materials to improve safety?

Proposals and Progress Reports

Proposals and progress reports, two common types of workplace documents, can be written collaboratively or individually depending on the size of the projects. Many projects, such as those funded by the National Science Foundation or the National Institutes of Health, begin with proposals and require regular progress reports during their course. This chapter explains development strategies for both types of documents. However, any work done by an employee may require a status (progress) report submitted to one's supervisor to explain what has been accomplished. A project may begin with a verbal agreement between supervisor and employee, but the employee may be required to submit a progress report either as an e-mail message or as a short memo report. Progress or status reports serve as a major source of documentation to show that employees have performed their work.

To help you understand how to design and write proposals and progress reports, we first discuss the development of proposals and use a student's research project proposal as an example. We also provide additional annotated examples of progress reports to illustrate how these occur in work settings.

The *proposal*, as its name implies, describes proposed work or research, the reasons for the work, the methods proposed to accomplish the work, the estimated time required, and the expected cost.

The *progress report*, as its name implies, describes and evaluates a project as work occurs. Thus, if an individual or an organization decides to begin a work project or research project, particularly one that requires several months or even several years to complete, the individual or organization will report the progress on that project at intervals agreed upon when the proposal is accepted and the resulting agreement or contract is being negotiated.

Employees may also need to report progress on the full range of projects or problems on which they are working. In situations like these, the employee writes a progress report (or "status report," as it may be called) to inform supervisors or other individuals about what has been accomplished in completing a job or solving a problem. The progress or status report thus becomes an official and even a legal record of work performed. Many organizations require employees to prepare annual status reports of their work, their accomplishments, and their objectives for the coming year.

Quick Tips

Keep in mind that proposals and progress reports are persuasive documents. You write to convince your reader of the merit and integrity of your work. In a proposal, you write to persuade your reader that you have a good idea, a good method for implementing the idea, and the experience necessary to manage the implementation—that you can achieve what you propose. In a progress report, you write to persuade the reader either that you are making good progress on your project or that you realize you aren't making good progress but know why and are taking appropriate steps to fix the problem.

PROPOSALS

The starting point of projects is often a proposal. A proposal offers to provide a service or a product to someone in exchange for money. Usually, when the organization—frequently a federal, state, or city government or a business enterprise—decides to have some sort of work or research done, it wants the best job for the best price. To announce its interest, the soliciting organization may advertise the work it wants done and invite interested individuals or organizations to contact it. In a university setting, the research and grants office may notify departments that money is available for research projects in a specific area. Faculty members submit project proposals that explain the research they envision; how much time they will need to complete the project; any financial resources required for equipment, salaries, and release time from regular teaching duties; and the goals and benefits of the research to the individual researcher, the university, and perhaps the public.

When an organization disseminates a description of the work it wants done, this document is usually called a *request for proposals* (RFP) or a *statement of work* (SOW). The soliciting organization may send selected companies an RFP/SOW that includes complete specifications of the work desired.

Alternatively, the soliciting organization may first describe the needed work in general terms and invite interested firms to submit their qualifications (Figure 9–1). This type of request is usually called a *request for qualifications* (RFQ). The organization may also place a notice of intent in newspapers, professional journals, or

-NOTICE OF INTENT-**Texas Department of Transportation** Notice of Intent—
Contract Number 17-8RFP1001. The Bryan District of the Texas Department of
Transportation (TxDOT) intends to enter into two (2) contracts with prime providers pursuant to Texas Government Code, Chapter 2254, Subchapter A, and 43
TAC §§9.30-9.43, to provide the following services. The work to be performed shall
consist of right of way surveys, aerial photography and digital terrain models for
advance planning and PS&E. Work shall include, but is not limited to: on ground
surveys, parcel plats, legal descriptions, right of way maps, establishment of horizontal and vertical control for aerial mapping, etc. for various construction projects
within the Bryan District. A letter of interest notifying TxDOT of the provider's
interest will be accepted by fax at (979) 778-9702, or by hand delivery to TxDOT,
Bryan District, 1300 North Texas Avenue, Attention: Mr. Chad Bohne, P.E.; or by
mail addressed to 1300 North Texas Avenue, Bryan, Texas 77803. **Deadline: Letters
of interest will be received until 5:00 p.m. on Wednesday, September 3, 2012.**
For more information regarding this notice of intent, please visit the TxDOT Internet Home Page located at www.dot.state.tx.us/business/professionalservices.htm or
contact Cecelia McCord at (979) 778-9765. 8-10-08

FIGURE 9-1 Example Request for Qualifications from a Newspaper

newsletters. The US government posts its RFQs online at http://www.grants.gov/.
The responding organizations explain their past accomplishments, giving the
names of companies for which they performed work, describing the work they
did, and giving references who can substantiate the organizations' claims. Based
on the responses it receives, the soliciting organization will send full descriptions
of the work (RFPs/SOWs) to the groups it believes have submitted the best qualifications. The responses submitted to an RFQ or a notice of intent may be called
letters of interest or *intent.*

In short, each aspect of the solicitation process, from the RFQ to the RFP/SOW,
has an appropriate use, but one or more of them is necessary to initiate action on
a project.

Example RFP. The vice president for research at a large university has located
funding opportunities for collaborative research grants. These will be available to
faculty across various disciplines. He issues an RFP via e-mail to the university's
research faculty (see Figure 9–2).

The context of proposal development. Because proposals are time-
consuming to write—most require substantial research and analysis on the part
of the proposing organization—individuals and organizations wishing to respond to an RFP study it carefully. They do not want to submit a proposal that
is unlikely to be accepted. Thus, the proposer—whether a university professor

January 5, 2015

MEMORANDUM

TO: Distribution A

FROM: Dr. Rolfe D. Auston
             Vice President for Research

**SUBJECT: 2015 ABC University-Conacyt: Collaborative Research Grant
             Program**

The Office of the Vice President for Research is pleased to issue a request for propos-als for the ABCU-Conacyt: Collaborative Research Grant Program. This program annually awards one-year grants of up to $24,000 to faculty members to advance inter-institutional cooperation in science, technology, and scholarly activities that have a direct application in industry or government through the complementary efforts of scientists and scholars from ABCU and Mexican institutions.

Two main objectives of the Collaborative Research Grant Program are to provide seed funding to 1) support the completion of a 12-month inter-institutional project, and 2) support the development and submission of proposals for external funding of research from competitive granting agencies both domestic and international (e.g., NSF, NIH, DOE, World Bank, NATO, UNESCO, etc.) and industry.

ABCU and Conacyt have agreed on several research priority areas as noted in the request for proposals. The research proposed must be linked to the private sector and have direct application to solving an industrial or governmental problem. All proposals must include research that directly improves security for the citizens of the region or explores issues relating to security challenges facing both countries.

A principal investigator (PI) is required from both Texas ABCU and a Mexican institution. The PI from ABCU must be a tenured or tenure-track faculty member. The PI from Mexico must be a scientist or scholar from any Mexican institution of higher education and research that is registered with Conacyt. Other investiga-tors may include faculty from branch campuses or Mexican faculty, postdoctoral students, graduate students, or research staff. A letter of intent must be received by 5:00 p.m. on Friday, March 5, 2015, to be eligible to submit a full proposal. Full proposals must be submitted, routed electronically for appropriate signatures, and received by 5:00 p.m. on Friday, April 30, 2015. The request for applications is available on the Web at http://conacyt.abcu.edu.

Additional information may be obtained by contacting Ms. Catharine J. Restivo (916-555-555; cjr@abcu.edu).

FIGURE 9-2 E-mail Request for Proposals

seeking research funds or a highway construction firm seeking to win a contract from a county to upgrade its rural roads—will carefully approach the decision to prepare a proposal.

The individual or the company must first decide whether to respond to the proposal. Can we do the work requested? Can we show that we can do this work, based on what we have already done? Can we do it within the time limit given in the RFP? Businesses responding to RFPs must address economic issues. How much will our proposed approach cost? How much money can we make? Who else will be submitting proposals? What price will competitors likely quote for the same work? Can we compete? What other projects do we have under way? Could problems arise that would make us unable to complete the job on time and at the price we quote? Do we have personnel qualified to work on this project?

The e-mail RFP in Figure 9–2 would elicit other questions. Is my field applicable to the research opportunity described here? Can I develop a collaborative proposal by the deadline? What types of national security topics would be most likely to attract funding? The university research office often has a person who helps answer these questions.

Many business entities requesting proposals will hold a bidders' conference at which companies interested in submitting a proposal can ask questions about the project or seek clarification of the needs described in the RFP. Most RFPs require that proposals be submitted by a deadline and contain specific information. Proposals that miss the deadline or do not contain the information requested are ordinarily omitted from consideration. Therefore, once an organization decides to submit a proposal, staff members carefully study the RFP and identify the information requirements. Each information requirement is given to an individual or a group who will be responsible for furnishing necessary material and data.

Some proposals, such as university research proposals, may be written by one or two persons. In complex proposals, however, different sections may be written by individuals in different areas of the organization. An editor or proposal writer will then compile the final document. This writer/editor may be assisted by readers who help check the developing proposal to be sure that all requested information is included and that the information is correct. Many proposals have to conform to strict length requirements. Proposals that do not conform will be eliminated from consideration.

Once the proposal has been written and submitted, it becomes a legally binding document. The proposing company or individual is legally committed to do what is described in the proposal, at the cost, and within the time limit stated. For that reason, the proposing organization carefully checks all information for accuracy.

When a large number of bidders submit proposals in response to an RFP, the soliciting organization may select several finalists and allow each finalist to give an oral version of the proposal. During oral presentations, the soliciting group asks questions. Representatives of the proposing groups have one more opportunity to argue for the value of what they propose, the merits of their organization, and the justification for the cost attached to the proposed work.

Students may be required to submit proposals for semester research projects or for other university programs, such as research opportunities with faculty. These proposals (two appear in this chapter) provide you with good practice in developing proposals you may write in your career: for example, government grants to fund your research, approvals to launch a work-related project, and actual bid proposals to help your business organization win work projects. Explaining to a potential employer during an interview that you have studied and written at least one proposal and a progress report can help you win a job offer.

Effective argument in proposal development. All writing is persuasive: it must convince the reader(s) that the writer has credibility and that the writer's ideas have merit. However, the success of a proposal rests totally on the effectiveness of the argument—how convincingly the writer argues for a plan, an idea, a product, or a service to be rendered and how well the writer convinces the reader that the proposing organization offers the best qualifications to do the work or research needed. In planning the proposal, the proposer must harmonize the soliciting company's needs with the proposer's capabilities. The writer must be acutely sensitive to what readers will be looking for but not propose action outside the capability of the proposing individual or organization. The proposing individual or organization has an ethical responsibility to explain accurately and specifically what work can be done and what work cannot be done so that no possibility exists of deceiving readers by making promises that cannot be fulfilled.

The following questions can help you analyze the effectiveness of the argument, from the solicitor's perspective, whether in a written or an oral proposal:

- What does the soliciting organization really want?
- What is the problem that needs to be solved?
- What approaches to the solution will be viewed most favorably?
- What approaches will be viewed unfavorably?
- What objections will our plan elicit?
- Can we accomplish the goals we propose?

To answer these questions, the proposer may be required to do research on the organization, its problems, its corporate culture, the perspective and attitudes stemming from its corporate culture, and its current financial status, goals, and problems. As the proposal develops, the writer should examine it from the intended reader's perspective.

- What are the weaknesses of the plan, as we—the writers—perceive them?
- How can we counter any weaknesses and the reader's potential objections?
- How can we make our plan appealing?
- How can we show that we understand their needs?
- How can we best present our capability to do this project?
- What are our/my strengths?
- From our own knowledge of our organization, what are our weaknesses—in personnel and in overall capability to complete this project as proposed?

• Do we need to modify our proposed plan to avoid misleading readers about our ability to perform certain tasks on time, as proposed, and at cost?
• Can we sell our idea without compromising the accuracy of what we can actually do?

As a proposal writer, you should consider each question and determine what evidence you will need to support the merits of your idea and the arguments needed to refute any objections. Every sentence in your proposal should argue for the merits of your plan and your or your organization's ability to complete it. Although the proposal must be designed as a sales document, you have an ethical obligation to present a plan that meets the soliciting organization's needs and requirements. (In considering the ethical issues that confront proposal writers, you will want to review Chapter 3, "Writing Ethically.")

Standard sections of proposals. Proposals generally include three main divisions: a summary, a main body, and attachments. The main body focuses on the three main parts of the proposal: the proposal's objectives (technical proposal), methods for achieving objectives (management proposal), and project cost (cost proposal). Proposals vary, but you will see the following segments embedded in some way:

> Project summary
> Project description (technical proposal)
> > Introduction
> > Rationale and significance
> > Plan of the work
> > > • Scope
> > > • Methods
> > > • Task breakdown
> > > • Problem analysis
> > Facilities and equipment
> Personnel (management proposal)
> Budget (cost proposal)
> Conclusion
> Appendices

Major business proposals are submitted in complete report format, which requires a letter of transmittal, a title page, a submission page (perhaps), a table of contents, and a summary—all items discussed in Chapter 8. Shorter proposals may be written in a memo or letter format. Many government proposals must be submitted online only. Whatever the format, the main elements will be required, although how they appear will vary with each proposal. In most RFPs, the soliciting organization explains what should be included in the proposal (either specific information to be included or major elements). Often, RFPs indicate the maximum number of pages allowed in a proposal. Writers should follow these instructions carefully to ensure that the proposal is not rejected during the initial screening process because it fails to follow the preparation guidelines stipulated by the RFP.

Summary. The summary is by far the most important section of the proposal. Many proposal consultants believe that a project will be accepted or rejected based solely on the effectiveness of the summary, your readers' first introduction to what you are proposing. The summary should concisely describe the project, particularly how your work meets the requirements of the soliciting organization, your plan for doing the work, and your or your company's main qualifications. The summary should be a concise version of the detailed plan, but it should be written to convince readers that you understand what the soliciting firm needs and wants, that what you are proposing can be done as you describe, and that your approach is solid because you have the required knowledge and expertise. After reading the summary, readers should want to read more of your proposal.

Project description (technical proposal). The technical proposal describes what you or your company proposes to do. The description must be as specific as possible. The technical proposal has a number of elements.

Introduction. The proposal introduction should explain what you are proposing, why you are proposing this idea, and what you plan to accomplish. The introduction contains the same elements as any introduction. In short proposals, the summary and introduction can be combined.

Rationale and significance. Much of your success in convincing readers that you should be granted a contract to do the work you propose will rest on your success in convincing them that you understand the project. In the section on rationale and significance, you need to make clear that you understand readers' needs—as stated in the summary or introduction—and that you have designed your goals by analyzing and defining their needs. Although you will clearly be selling your idea, you should recognize and answer any questions your readers may have as you argue the merits of your project. Convincing your readers that you fully understand what they are looking for is critical in establishing your credibility.

In short,

- You may want to define the problem, to show that you understand it.
- You may want to explain the background of the problem and how it evolved, by providing a historical review of the problem.
- If you are proposing a research project, you may want to explain why your research needs to be done and what results can be expected from your research.
- You may want to describe your solution and the benefits of your proposed solution.

Of greatest importance, however, is the *feasibility* of the work you propose. Is your proposed work doable? Is it suitable, appropriate, economical, and practicable? Have you given your readers an accurate view of what you can and will do?

Plan of the work. The section on the work plan is also critical, particularly to expert readers who will attempt to determine whether you understand the breadth of the work you are proposing. In this section, you will describe how you will go about achieving the goals you have stated. You will specify what you will do in what order, explaining and perhaps justifying your approach as you believe necessary. A realistic approach is crucial in that a knowledgeable reader will sense immediately if your plan omits major steps. A flawed work plan can destroy your credibility as well as the merits of the goals or the solution you are proposing.

Scope. The work plan section may need to describe the scope of the proposed work. What will you do and not do? What topics will your study or your work cover and not cover? What are the limits of what you are proposing? What topics will be outside the scope of your project? As the writer of the proposal, you have both an ethical and a legal obligation to make clear to your readers the limits of your responsibility.

Methods. A work plan may also require a statement of the methods you will use. If you are going to do on-site research, how will you do this research? If you plan to collect data, how will you do so? How will you analyze this data? How will you guarantee the validity of the analysis? If your research includes human subjects, how will you make sure that their participation is voluntary and their privacy is protected? If you are going to conduct surveys or interviews, how will you do so and what questions will you ask? If you plan to do historical research or a literature review of a topic, how will you approach such a review to ensure that your findings are representative of what is currently known about a subject area? What precautions will you take to verify that your research is conducted according to applicable ethical and legal standards? A precise, carefully detailed description of your work methods can add to your credibility as one who is competent to perform the proposed work.

Task breakdown. Almost all proposals require you to divide your work into specific tasks and to state the amount of time allotted to each task. This information may be given in a milestone chart, as illustrated in the methods section of the student research proposal (Case 9–1). The task breakdown indicates how much time you plan to devote to each task. A realistic time schedule also becomes an effective argument. It suggests to readers that you understand how much time your project will take and that you are not promising miracles just to win approval of your proposal or business plan.

If a project must be completed by a deadline, the task breakdown and work schedule should indicate exactly how you plan to fit every job into the allotted time. However, do not make time commitments that will be impossible to meet. Readers who sense that your work plan is artificial will immediately question your credibility. Remember, too, that a proposal is a binding commitment. If you cannot do what you propose, what the soliciting organization requires within the required time, you can destroy your professional credibility and leave yourself open to litigation.

Problem analysis. Few projects can be completed without problems. If you have carefully analyzed the problem or work you intend to do, you should anticipate where difficulties could arise. Problems that may be encountered can often be discussed in the rationale section. However, if you discover major obstacles that you believe will occur during the course of the project, you may wish to isolate and discuss these in a separate section. Many organizations that request work or solicit research proposals are aware of problems that may arise. Reviewers in these organizations look carefully at the problem analysis section, wherever it occurs, to see whether the proposer has anticipated these problems and explained the course of action that will be followed in dealing with them. Anticipating and designing solutions to problems can further build your credibility with readers, who will not be impressed if you fail to diagnose points in your work plan that could be troublesome and even hinder your completion of the project as proposed.

Facilities and equipment. The facilities section of the proposal is important if you need to convince the reader that your company has the equipment, plant, and physical capability to do the proposed work. Facilities descriptions are particularly crucial if hardware is to be built at a plant site owned by your organization. Even in research proposals, your readers may want to know what resources you will use. Sometimes existing facilities are not adequate for a particular job and your company must purchase specific equipment. The facilities section enables you to explain this purchase and how it will be included in the cost proposal.

Researchers may need to travel to visit special libraries or research sites. The amount of money needed for this travel will be part of the cost proposal. Thus, the nature of any extra research support, its importance, and its cost to the project should be explained here.

Personnel (management proposal). Any technical proposal or project is only as good as the management strategy that directs it. The management proposal should explain how you plan to manage the project: who will be in charge and what qualifications that person or team has for this kind of work. Management procedures should harmonize with the methods of pursuing the work described in the technical proposal.

Descriptions of your management philosophy and hierarchy should clearly reflect your company's management philosophy and culture. Readers should see the same kind of management applied to the proposed work as to the company and other projects it manages. Any testimony to or evidence of the effectiveness of the management approach will lend credibility to the technical proposal. Proposal reviewers must be convinced that you and your organization have a sound approach supported by good management of that approach.

In research proposals, the researcher who is soliciting funds will want to explain his or her expertise in the subject area proposed. This explanation may focus on educational background, previous projects successfully undertaken, published research on the topic, and general experience.

Budget (cost proposal). The cost proposal is usually the final item in the body of the proposal, even though cost may ultimately be the most crucial factor in industrial proposals. Cost is usually given last and appears as a budget for the length of the proposal period. The technical and management sections of the proposal, with their descriptions of methods, tasks, facilities, required travel, and personnel, should help justify the cost. They should have already explained the rationale for items that will produce the greatest cost. However, any items not previously discussed in the technical and management sections—such as administrative expenses, additional insurance benefit costs, and unexpected legal costs—should be explained. An itemized budget, often submitted as a separate document, includes items such as the proposing organization's liability for not meeting project deadlines, for cost overruns, and for unforeseen labor strikes and work stoppages. Many budget sections include standard statements such as descriptions of union contracts with labor costs, insurance benefit costs, nonstrike costs, and statements of existing corporate liability for other projects—any existing arrangements that affect the cost of the proposed contract. Clearly, the budget should explain exactly how much the project will cost and how the cost is determined. The extensiveness of the budget depends on the magnitude of the project.

Conclusion. The proposal includes a final section that repeats what the proposal offers the potential client or the soliciting agency, why you or your company should be selected to perform the work, and the benefits that the project, when completed, will yield for the client. The conclusion presents the final restatement of your central argument.

Appendices. As in any report, the appendix section includes materials to support information you give in the main body of the proposal—in the technical, management, or cost proposals. For example, the appendix might include résumés of principal investigators, managers, or researchers. These résumés should highlight their qualifications as they pertain to the specific project.

Case 9–1: Research Proposal

The undergraduate research office at a university solicits proposals from under-graduate students who wish to pursue research with a faculty member. Case Document 9–1 is a research proposal written by an undergraduate chemical engineering major, Evan Cherry. The proposal includes sections required by the solicitation for these undergraduate research proposals. Readers of this proposal will be faculty from science and engineering. For that reason, the proposal is highly technical and is prepared to meet the strict length requirements given in the RFP.

CASE DOCUMENT 9–1: EVAN CHERRY

Proposal for Undergraduate Research Scholars Program: Effects of Simvastatin and Fluvastatin on Endothelial Invasion

Summary

Millions of people take cholesterol-lowering medications. This proposal investigates whether the cholesterol medications simvastatin (Zocor®) and fluvastatin (Lescol®) can additionally alter a key step of wound healing. We will specifically determine whether these medications affect angiogenesis, the formation of new blood vessels from preexisting structures, using a model of endothelial cell invasion. We hypothesize that simvastatin and fluvastatin may have additional effects unrelated to reducing cholesterol. The specific aims are as follows:

1. Determine the effects of simvastatin and fluvastatin on endothelial invasion.
2. Determine whether the effects on invasion are a direct result of altered cholesterol levels or due to inhibition of 3-hydroxy-3-methylglutaryl coenzyme A (HMG CoA) reductase activity.
3. Determine the signaling mechanisms affected by simvastatin and fluvastatin.

Introduction

I. Atherosclerosis and Statins

Cholesterol regulation has become one of the primary methods for preventing heart disease. HMG-CoA reductase inhibitors (statins), the first line of treatment for inhibiting cholesterol synthesis, lower blood cholesterol levels and significantly reduce both atherosclerosis[1] and the instance of heart disease by 60%.[2]

Atherosclerosis is the buildup of fatty plaques along blood vessel lining. Various stress factors oxidize the cholesterol transport molecule low-density lipoprotein (LDL). Oxidized LDL adheres to the lining of blood vessels and encourages foam cell accumulation and plaque growth, which reduces blood flow and can cause heart attack.[3] Plaque rupture can result in a stroke.

The Agency for Healthcare Research and Quality found that in 2005 29.7 million people purchased statins for an estimated total of 19.7 billion dollars.[4] Two examples of statins are shown below (Figure 1). Simvastatin is derived from lovastatin, a naturally occurring compound in the fungus *Aspergillus terreus*. Fluvastatin, a synthetic statin, has a visibly different structure.

Figure 1 The structures of simvastatin and fluvastatin differ substantially.

II. Angiogenesis

Patients treated with statins exhibit other cardiovascular protective effects, including a lower incidence of cancer.[5] New blood vessel growth, or angiogenesis, is an important part of tumor growth and wound healing. These data have helped form the basis for our hypothesis that statins may affect angiogenesis.

During angiogenesis, endothelial cells lining blood vessels respond to extracellular signals and change into invading structures (Figure 2). The endothelial cells enter the surrounding matrix and form new blood vessels. In wound healing, the process relies on platelets, fibroblasts, and neighboring endothelial cells to close the wound. Endothelial cells propagate across the gap until they make contact with other endothelial cells and cease multiplying. The simultaneous combination of lipid inducers and growth factors is required for endothelial invasion.[6]

For our project, we will investigate if statin therapy changes invasion responses in a model reproducing angiogenesis during wound healing.

continued

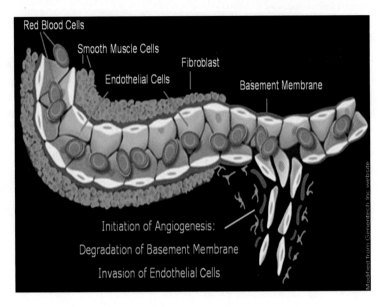

Figure 2 Endothelial invasion begins angiogenesis.

Project Description

We will evaluate the effects of statin therapy on endothelial invasion and determine whether the effects stem from cholesterol mediation or other processes. My project consists of three aims:

1. Determine the effects of statin therapy on endothelial invasion.

We will conduct invasion experiments using statin-pretreated endothelial cells and quantify both the number of cells invading and the invasion depth.

2. Determine whether the effects on invasion are a direct result of altered cholesterol levels or HMG CoA reductase.

We will analyze cholesterol levels in the cells using a commercially available kit and decrease cholesterol levels.

3. Determine how simvastatin and fluvastatin alter signaling.

Preliminary data indicate that several proteins mediate invasion responses. We will analyze levels of activated proteins at various time points of invasion.

Methodology and Timeline

Dr. Bayless's lab implements an artificial angiogenesis model using three-dimensional collagen that mimics the extracellular matrix. We place collagen solution containing nutritive media and sphingosine-1-phosphate (S1P).

S1P is a lipid that signals the endothelial cells to invade the collagen matrix.[7] We add human endothelial cells to the top surface of the collagen. The cells form a single layer and attach to the gel. We then add nutritive medium and growth factors.

After the cells incubate for 18–24 hours, we preserve the cells with formaldehyde before staining with 4′,6-diamidino-2-phenylindole, a molecule that causes nuclei to fluoresce under ultraviolet light. This allows us to count the number of invading cells before we stain them with toluidine blue and cut cross-sectional slices. Photographs show invading cells, and photography software measures the depth of the invading cells (Figure 3).

0 hr   100 μm 6 hr   12 hr   18 hr

Figure 3 Endothelial cells invade the three-dimensional matrix.

To complete *Aim 1* we will place cells pretreated with simvastatin or fluvastatin in assays as shown in Figure 3. The invasion density and depth will be counted. All experiments will be confirmed three times.

To complete *Aim 2* we will measure cellular cholesterol levels using a commercially available kit and measure any differences in cholesterol synthesis between fluvastatin and simvastatin. The cholesterol profile will tell us whether the two compounds comparably inhibit cholesterol.

If we observe any differences in cholesterol levels or invasion, we will assume that these effects are due to decreased HMG CoA reductase activity. To confirm this, we will knockdown HMG CoA reductase by delivering a small hairpin RNA sequence to the endothelial cells using lentiviruses produced in the Bayless lab. We will check gene knockdown by Western blot and analyze the behavior of these cells in our invasion model.

If fluvastatin or simvastatin affect invasion responses, we will assume they are altering signaling events that control invasion. Two of these events are phosphorylation of Akt and p42/44 MAP kinase (Figure 4). To complete *Aim 3* we will treat cells with simvastatin or fluvastatin and extract proteins over time. For the Western blot, we will line up the treated and control extracts on an acrylamide gel and separate them by electrophoresis.

continued

Figure 4 Western blots show Akt and p42/44 levels vary with invasion time.

We will specifically look for differences in activation of Akt and ERK p42/p44. While we do not know all the proteins involved in the invasion response, we do know that these proteins occur early in the process.

The possibility exists that simvastatin and fluvastatin have no effect on these designated proteins; we can then conclude that effects may occur further down the pathway and continue from there.

TABLE 1 The proposed process spans 24 weeks and finishes early April 2009

						Fall 2008						
Week #	1	2	3	4	5	6	7	8	9	10	11	12
Week of	9/21-9/27	9/28-10/4	10/5-10/11	10/12-10/18	10/19-10/25	10/26-11/1	11/2-11/8	11/9-11/15	11/16-11/22	11/23-11/29	11/30-12/6	12/7-12/13
Aim 1	Demonstrating the effects of statins on endothelial invasion											
Aim 2							Measure cholesterol levels					
Report											Compile Data	

						Spring 2009						
Week #	13	14	15	16	17	18	19	20	21	22	23	24
Week of	1/11-1/17	1/18-1/24	1/25-1/31	2/1-2/7	2/8-2/14	2/15-2/21	2/22-2/28	3/1-3/7	3/8-3/14	3/15-3/21	3/22-3/28	3/29-4/4
Aim 3	Transfect cells with HMG-CoA shRNA											
Aim 4						Western blot analysis of pathway proteins						
Report											Draft, review, and finalize thesis	

Qualifications

I am a junior chemical engineering major with an interest in medicine and healthcare. I have been employed as a pharmacy clerk and am a phlebotomist. Following graduation, I plan to enter an MD/PhD program. This past summer I worked with Dr. Bayless on the statin project during the Summer Undergraduate Research Program at the Health Science Center and gained experience performing invasion experiments.

I have completed the following elective courses in addition to my major while maintaining a GPA of 3.9.

- Cell & Molecular Biology (BIOL 213)
- Biochemistry (BICH 303)
- Genetics (GENE 301)

References

1. Nissen, S., S. Nicholls, et al. "Effect of Very High-Intensity Statin Therapy on Regression of Coronary Atherosclerosis." *Journal of the American Medical Association* 295, no. 13 (2006): 1556–1565.
2. Law, M., N. Wald, A. Rudnicka. "Quantifying Effect of Statins on Low Density Lipoprotein Cholesterol, Ischaemic Heart Disease, and Stroke: Systematic Review and Meta-Analysis." *British Medical Journal* 326 (2003): 1423.
3. Libby, P., M. Aikawa, U. Schonbeck. "Cholesterol and Atherosclerosis." *Biochimica et Biophysica* 1529, no. 1–3 (2000): 299–309.
4. Stagnitti, Marie N. *Statistical Brief #205: Trends in Statins Utilization and Expenditures for the U.S. Civilian Noninstitutionalized Population, 2000 and 2005.* Medical Expenditure Panel Survey (2008).
5. Demierre, M., P. Higgins, S. Gruber, E. Hawk, S. Lippman. "Statins and Cancer Prevention." *Nature Reviews Cancer* 5 (2005): 930–942.
6. Davis, G., K. Bayless, A. Mavila. "Molecular Basis of Endothelial Cell Morphogenesis in Three-Dimensional Extracellular Matrices." *Anatomical Record* 268, no. 3, (2002): 252–275.
7. Bayless, K., G. Davis. "Sphingosine-1-Phosphate Markedly Induces Matrix Metalloproteinase and Integrin-Dependent Human Endothelial Cell Invasion and Lumen Formation in Three-Dimensional Collagen and Fibrin Matrices." *Biochemical and Biophysical Research Communications* 312, no. 4 (2003): 903–913.

Case 9-2: **Project Proposal**

The preveterinary honor society at ABC University has invited students to propose and then write, if the proposal is accepted, a discussion of issues important to future veterinarians. Sarah Irving proposes a report that discusses the increasing role of animal rehabilitation in veterinary practice. This proposal includes a summary, a list of topics, methods to be used in developing the topic, a tentative report outline, and an initial list of sources. Since it deals with a nontechnical topic, the level of language needs no technical detail. The audience for the report will be other preveterinary students and perhaps faculty who work with the preveterinary society.

CASE DOCUMENT 9-2

MEMORANDUM

TO: Dr. Elizabeth Tebeaux   **DATE:** February 22, 2014

FROM: Sarah Irving

SUBJECT: Proposal for research report—Is animal rehabilitation the future of veterinary medicine?

Subject and Purpose of the Proposal

Animal rehabilitation is not currently accepted as a veterinary specialty and is not widely developed. The growing willingness of owners to pay for rehabilitation treatments is creating a demand for more research into this field and a way to explain animal rehabilitation to the public. More widespread knowledge of the benefits of this care would encourage more veterinarians to practice or write referrals for animal rehabilitation.

My objective is to create a report that will explain to veterinarians, preveterinary students, and the public where the field of animal rehabilitation is now and how animal rehabilitation will affect the future of veterinary medicine.

Rationale for the Topic and Approach—Its Importance as an Issue

Veterinary patient care is moving toward the level of in-depth planning and quality of care required for human medical care. Owners expect their pets to receive the same level of care they would receive. Options such as physical therapy, prosthetics, and pain management treatments like acupuncture and medication are often used in treating humans. Though these techniques are being adapted for animal patients, the practice of animal rehabilitation is not widely developed or recommended by veterinarians.

In the last decade many advances in procedures have been made that make animal rehabilitation a feasible part of veterinary medicine. Few veterinarians practice rehabilitation because of the lack of research into rehabilitation methods, results of rehabilitation, and patient benefits.

Growing public awareness of animal rehabilitation is leading to increased demand. The newness of the field leaves few resources for veterinarians, pre-veterinary students, and the public to learn about this growing field. I will research what advances have been made in the field of animal rehabilitation and how animal rehabilitation affects patient care.

I will use articles from peer-reviewed journals and answers written by specialists involved in the field to gather data about animal rehabilitation. To obtain firsthand knowledge from specialists in the animal rehabilitation field, I will e-mail surveys to the specialists. To gather articles, I will use databases and Google Scholar.

Issues Addressed in the Proposal

- Definition of animal rehabilitation
- Unresolved issues in the field of animal rehabilitation
- Direction the field of animal rehabilitation is headed
- Why veterinarians and future veterinarians should practice animal rehabilitation
- How animal rehabilitation relates to the future of veterinary medicine

Methods to Research the Topic

To guarantee the validity of my data, I will use articles from peer-reviewed journals and answers written by specialists involved in the field. These specialists will include veterinary rehabilitation practitioners and researchers in the animal rehabilitation field. To gather articles for my project, I will use databases owned by Texas A&M libraries and Google Scholar. These databases include the following:

- Academic Search Complete
- OmniFile Full Text Mega
- Medline (OVID)

To obtain firsthand knowledge from specialists in the animal rehabilitation field, I will e-mail surveys to the specialists. I have collected names of veterinarians who have written articles about veterinary rehabilitation. I will e-mail

continued

them a short three- to five-question survey about their experiences in animal rehabilitation.

Proposed questions

To veterinarians:

- What kinds of treatment do you offer?
- What would you like to tell veterinary students considering rehabilitation practices?
- What would you say the public perception of animal rehabilitation is?
- What is your most common procedure?
- What is the perception of other veterinarians about animal rehabilitation?

To other rehabilitation specialists:

- What would you say the public perception of animal rehabilitation is?
- What is the perception of other veterinarians about animal rehabilitation?
- What would you like to tell veterinary students considering rehabilitation practices?

I will have these e-mails checked for readability by Dr. Tebeaux before I send them. I will make the e-mails as easy to answer as possible, to get the most responses. I will request that all e-mail replies be sent within 2 weeks of receiving the e-mail survey so that I may use the information in my report.

To make sure my data are the most current available, I will continue researching for the next 4 weeks. This will give me time to find all sources created up to 1 month before my report is due.

Work Schedule

To complete my report on time using the most data available, I will follow the schedule below.

Week 1 February 21–27	Draft e-mails to animal rehabilitation specialists, search for more sources, read sources, and gather names and e-mail addresses of specialists
Week 2 February 28–March 6	Have e-mails checked for readability, send e-mails, read sources
Week 3 March 7–13	Continue searching for and reading sources
Week 4 March 14–20	E-mail replies due by this week, start reading replies, continue searching for and reading sources
Week 5 March 21–27	Turn notes on sources and surveys into rough draft
Week 6 March 28–April 3	Edit rough draft, print, read aloud, edit again

Week 7 April 4–10	Have others edit rough draft
Week 8 April 11–14	Final editing
April 15	Turn in project

Tentative Outline of the Report

I. Letter of transmittal

II. Title page

III. Definition of animal rehabilitation

    a. History of animal rehabilitation

    b. Types of animal rehabilitation

        i. Physical therapy

        ii. Hydrotherapy

        iii. Prosthetics

    c. The developing demand for animal rehabilitation

IV. Unresolved issues in the field of animal rehabilitation

    a. The animal pain management debate

    b. Who is qualified to practice animal rehabilitation

        i. The American Veterinary Medical Association position

        ii. The American Physical Therapy Association position

    c. What certification programs exist?

        i. Canine Rehabilitation Institute

        ii. University of Tennessee College of Veterinary Medicine Certificate Program in Canine Rehabilitation

        iii. Canine and Equine Physiotherapy Training in England

    d. Organizations formed to promote animal rehabilitation

        i. Institute of Registered Veterinary and Animal Physiotherapists

    e. State government actions to date

    f. Owner involvement in the rehabilitation process

V. What is the future of animal rehabilitation?

    a. Recent developments in the field

    b. Establishment of a specialty

    c. Animal plastic surgery

    d. Greater public awareness

VI. Why practice animal rehabilitation?

    a. Good for business, good veterinary practice, good for the clinic's reputation

    b. Advice for preveterinary students considering a future as an animal rehabilitation practitioner

VII. Is animal rehabilitation the future of veterinary medicine?

VIII. Bibliography

continued

Initial List of Resources

Brockett, John E. "Why Give Pain Medication—Patient Care, Image or Profit?" *DVM: The Newsmagazine of Veterinary Medicine* 39.2 (2008): 41. Print.

Downing, Robin. "Putting the New Pain Management Guidelines into Practice." *Veterinary Medicine* 102.11 (2007): 704–705. Print.

Foley, Denise. "Life with a Disabled Pet." *Prevention* 58.2 (2006): 198–201. Print.

Nolen, R. S. "Pet Rehab Becoming Mainstream Practice." *Journal of the American Veterinary Medical Association* 235.7 (2009): 798–799. Print.

Payne, Richard M. "Canine and Equine Physiotherapy." *Veterinary Record: Journal of the British Veterinary Association* 165.4 (2009): 122. Print.

Sayre, Carolyn, and Robert Horn. "Fake Fins, Plastic Paws." *Time* 170.10 (2007): 49–51. Print.

Skernitvitz, Stephanie. "Rehab Gives Animals Swifter Road to Recovery, DVM Says." *DVM: The Newsmagazine of Veterinary Medicine* 39.10 (2008): 1s–12s. Print.

Smith, Sarah D. "Petting with a Purpose." *Natural Health* 36.3 (2006): 99–100. Print.

Stein, Robert M., and Stephanie Ortel. "Postsurgical Pain Management: Take a Pre-Emptive Approach." *DVM: The Newsmagazine of Veterinary Medicine* 39.7 (2008): 44–46. Print.

Van Dyke, Janet B. "Canine Rehabilitation: An Inside Look at a Fast-Growing Market Segment." *DVM: The Newsmagazine of Veterinary Medicine* 40.7 (2009): 14S–15S. Print.

Verdon, Daniel R. "An Animal Planet." *DVM: The Newsmagazine of Veterinary Medicine* 36.7 (2005): 24–28. Print.

Progress Reports

When a soliciting organization requests a proposal, it often states that a specific number of progress reports will be required, particularly if the project covers a long time period. As their name suggests, progress reports, sometimes known as "status reports," tell readers how work is progressing on a project. Their immediate purpose is to inform the authorizing person of the activities completed on a project, but their long-range purpose should show the proposing organization's or the individual's competence in pursuing a task and completing it.

Status reports may also be prepared in paper format and submitted to supervisors. These progress reports help you or your work group provide evidence or documentation of your activities. They generally have three goals:

- To explain to the reader what has been accomplished and by whom, the status of the work performed, and problems that may have arisen that need attention
- To explain to your client how time and money have been spent, what work remains to be done, and how any problems encountered are being handled
- To enable the organization or individual doing the work to assess the work and plan future work

Structure of progress reports. Any work project, such as transportation or construction projects, requires regular progress reports. While these differ, they incorporate the same basic segments: goal of the project, work accomplished (for a given period), work remaining/planned (for the next period), problems encountered, and financial expenditures.

Structure by work performed. This is the standard structure for progress reports.

> *Beginning*
> - Introduction/project description
> - Summary
>
> *Middle*
> - Work completed   or   • Task 1
>   Task 1   Work completed
>   Task 2, etc.   Work remaining
> - Work remaining   • Task 2
>   Task 3   Work completed
>   Task 4   Work remaining
> - Cost   • Cost
>
> *End*
> - Overall appraisal of progress to date
> - Conclusion and recommendations

In this general plan, you emphasize what has been done and what remains to be done and supply enough of an introduction to be sure that the reader knows what project is being discussed.

For progress reports that cover more than one period, the basic design can be expanded as follows:

Beginning
- Introduction
- Project description
- Summary of work to date
- Summary of work in this period

Middle
- Work accomplished by tasks (this period)
- Work remaining on specific tasks
- Work planned for the next reporting period
- Work planned for periods thereafter
- Cost to date
- Cost in this period

End
- Overall appraisal of work to date
- Conclusions and recommendations concerning problems

Structure by chronological order. If your project or research is broken into time periods, your progress report can be structured to emphasize the periods.

Beginning
- Introduction/project description
- Summary of work completed

Middle
- Work completed
  - Period 1 (beginning and ending dates)
    - Description
    - Cost
  - Period 2 (beginning and ending dates)
    - Description
    - Cost
- Work remaining
  - Period 3 (or remaining periods)
    - Description of work to be done
    - Expected cost

End
- Evaluation of work in this period
- Conclusions and recommendations

Structure by main project goals. Many research projects are pursued by grouping specific tasks into major groups. Then the writer describes progress according to work done in each major group and perhaps the amount of time spent on that group of tasks. Alternatively, a researcher may decide to present a project by research goals—what will be accomplished during the project. Thus, progress reports will explain activities performed to achieve those goals. In the middle of the following plan, the left-hand column is organized by work completed and remaining and the right-hand column, by goals.

Beginning
- Introduction/project description
- Summary of progress to date

Middle
- Work completed   or   • Goal 1
    - Goal 1   Work completed
    - Goal 2   Work remaining
    - Goal 3, etc.   Cost
- Work remaining   • Goal 2
    - Goal 1   Work completed
    - Goal 2   Work remaining
    - Goal 3, etc.   Cost
- Cost

End
- Evaluation of work to date
- Conclusions and recommendations

Case 9-3

Recipients of research funds from the preveterinary honor society are expected to submit a progress report. Sarah submits her report several weeks before the research reports will be completed and submitted. Note the detail of her progress, which suggests that her work follows the research methods she included in her proposal.

CASE DOCUMENT 9-3

TO:	Dr. Elizabeth Tebeaux	**DATE:**	4/6/14
FROM:	Sarah Irving		
SUBJECT:	Progress on Semester Project		

Summary

- Animal rehabilitation is a developing field many veterinarians do not utilize and many pet owners do not know is available. My objective is to create a report that will explain to veterinarians, preveterinary students, and veterinary students where the field of animal rehabilitation is now and how animal rehabilitation will affect the future of veterinary medicine.
  - I have narrowed the focus of my report to be a report for students interested in veterinary medicine, veterinary students, and veterinarians. My original audience also included pet owners. I believe these changes will enhance the readability and clarity of my report by preventing the report from being cluttered with information the primary readers would not find useful. I will format the report for pet owners to be able to find information that interests them, while focusing on the audience most in need of this information.

- **My report will address the following issues:**

  - Definition and description of animal rehabilitation
  - Unresolved issues in the field of animal rehabilitation
  - Future of animal rehabilitation
  - Rationale for including animal rehabilitation in veterinary practice

- I have read and taken notes on 11 articles that have information that I can use for my paper. I have four more articles with good information to reread and take notes on. I have also e-mailed a survey to 50 veterinarians, veterinary technicians, and physical therapists who practice animal rehabilitation. I have received 13 responses to my survey. I must read the responses and make a citation for each response. I also must send individual thank you notes to the respondents. The sources I have used and am planning to use are listed below.

- I have made good progress on drafting my report. The articles and surveys left to read will fill the gaps in information in my rough draft. I have also reorganized my report outline. The changes to the outline will enhance the logical flow of the report by grouping similar information. The updated outline can be found on page 4 of this progress report.

Research Consulted:

Articles:
1. **Why Give Pain Medication—Patient Care, Image or Profit?**
    a. **Citation:** Brockett, John E. "Why Give Pain Medication—Patient Care, Image or Profit?" *DVM: The Newsmagazine of Veterinary Medicine* 39.2 (2008): 41. Print.
2. **Putting the New Pain Management Guidelines into Practice**
    a. **Citation:** Downing, Robin. "Putting the New Pain Management Guidelines into Practice." *Veterinary Medicine* 102.11 (2007): 704–5. Print.
3. **Life with a Disabled Pet**
    a. **Citation:** Foley, Denise. "Life with a Disabled Pet." *Prevention* 58.2 (2006): 198–201. Print.
4. **Underwater Treadmill Therapy in Veterinary Practice**
    a. **Citation:** Jurek, Christine, and Laurie McCauley. "Underwater Treadmill Therapy in Veterinary Practice." *Veterinary Medicine* 104.4 (2009): 182–90. Print.
5. **Pet Rehab Becoming Mainstream Practice**
    a. **Citation:** Nolen, R. S. "Pet Rehab Becoming Mainstream Practice." *Journal of the American Veterinary Medical Association* 235.7 (2009): 798–9. Print.
6. **Canine and Equine Physical Therapy**
    a. **Citation:** Payne, Richard M. "Canine and Equine Physiotherapy." *Veterinary Record: Journal of the British Veterinary Association* 165.4 (2009): 122. Print.

continued

7. Rehab Gives Animals Swifter Road to Recovery, DVM Says

    a. Citation: Skernitvitz, Stephanie. "Rehab Gives Animals Swifter Road to Recovery, DVM Says." *DVM: The Newsmagazine of Veterinary Medicine* 39.10 (2008): 1s–12s. Print.

8. Petting with a Purpose

    a. Citation: Smith, Sarah D. "Petting with a Purpose." *Natural Health* 36.3 (2006): 99–100. Print.

9. Postsurgical Pain Management: Take a Pre-Emptive Approach

    a. Citation: Stein, Robert M., and Stephanie Ortel. "Postsurgical Pain Management: Take a Pre-Emptive Approach." *DVM: The Newsmagazine of Veterinary Medicine* 39.7 (2008): 44–6. Print.

10. Canine Rehabilitation: An Inside Look at a Fast-Growing Market Segment

    a. Citation: van Dyke, Janet B. "Canine Rehabilitation: An Inside Look at a Fast-Growing Market Segment." *DVM: The Newsmagazine of Veterinary Medicine* 40.7 (2009): 14S–5S. Print.

11. An Animal Planet

    a. Citation: Verdon, Daniel R. "An Animal Planet." *DVM: The Newsmagazine of Veterinary Medicine* 36.7 (2005): 24–8. Print.

Survey

- I e-mailed 50 surveys to veterinarians, veterinary technicians, and physical therapists. The e-mail addresses were found on the University of Tennessee Certificate in Canine Rehabilitation website using the "Find a Practitioner" option. As this is one of the two certificate programs in the United States, I felt it would provide a good example of the opinions of certified animal rehabilitation practitioners on the field of animal rehabilitation. I received 13 responses. Five responses were from veterinarians, seven were from veterinary technicians, and one was from a physical therapist.
  - The surveys I sent to veterinarians contained the following questions:
    1. How would you describe the public perception of animal rehabilitation?
    2. What is the perception of other veterinarians you know about animal rehabilitation?
    3. What kinds of treatment do you offer?
    4. What is your most common procedure?
    5. What would you like to tell veterinary students considering rehabilitation practices?

○ I modified the first and second questions for the surveys I sent to veterinary technicians and physical therapists:
  1. What is the public's perception of animal rehabilitation?
  2. What is the perception of veterinarians about animal rehabilitation?

Work Remaining

Articles to Read

1. **Fake Fins, Plastic Paws**
   a. **Citation:** Sayre, Carolyn, and Robert Horn. "Fake Fins, Plastic Paws." *Time* 170.10 (2007): 49–51. Print.
2. **Changing Times**
   a. **Citation:** Shearer, Tami. "Changing Times." *DVM: The Newsmagazine of Veterinary Medicine* 39.3 (2008): 34–. Print.
3. **State Board Lacks Consistency on Standard of Practice**
   a. **Citation:** Marshak, Doug. "State Board Lacks Consistency on Standard of Practice." *DVM: The Newsmagazine of Veterinary Medicine* 38.11 (2007): 38–40. Print.
4. **The Price of Pain**
   a. **Citation:** Fiala, Jennifer. "The Price of Pain." *DVM: The Newsmagazine of Veterinary Medicine* 38.10 (2007): 30–2. Print.

Survey

- The 13 survey responses must also be read, and I must make the citations for the surveys. I also must draft individual thank you notes to each respondent, as some respondents asked questions at the end of their survey responses. These thank you notes will be sent the week of April 5, 2014.

Changes to the Initial Proposal

- I have narrowed the focus of my report to be a report for students interested in veterinary medicine, veterinary students, and veterinarians. My original audience also included pet owners. This change reflects the fact that many sections of my report would not interest most pet owners. Though I have narrowed my audience, it is possible that pet owners could also be a part of my audience, and the report will be formatted for pet owners to be able to find useful information.

continued

- I have added a section defining animal rehabilitation at the beginning of the report and sections that are required in the major report template. I have also reorganized the report by moving subsections into the section "What Is Animal Rehabilitation." These sections include owner involvement in the rehabilitation process, what certification programs exist, and organizations formed to promote animal rehabilitation. These subsections were moved from the "Unresolved Issues in the Field of Animal Rehabilitation" section after reading sources that suggested the debate on these issues has been largely completed.

- To describe the field of animal rehabilitation more accurately I have changed the section on prosthetics to a section on assistive devices. This change reflects the realization that most veterinarians consider rehabilitation to be work done postsurgically and with minimal invasiveness. I believe these changes will enhance the readability and clarity of my report.

Updated Outline

Letter of transmittal
Title page with descriptive abstract
Table of contents
Summary
Introduction
What is animal rehabilitation
    a. Definition of animal rehabilitation
    b. History of animal rehabilitation
    c. Types of animal rehabilitation
        i. Physical therapy
        ii. Hydrotherapy
            1. Benefits of hydrotherapy
            2. What to consider before choosing hydrotherapy
        iii. Assistive devices
    d. The developing demand for animal rehabilitation
    e. Owner involvement in the rehabilitation process
    f. What certification programs exist?
        i. Canine Rehabilitation Institute
        ii. University of Tennessee College of Veterinary Medicine Certificate Program in Canine Rehabilitation
    g. Organizations formed to promote animal rehabilitation
        i. Institute of Registered Veterinary and Animal Physiotherapists

Unresolved issues in the field of animal rehabilitation
    h. The animal pain management debate
    i. Who is qualified to practice animal rehabilitation
        i. The American Veterinary Medical Association position
        ii. The American Physical Therapy Association position
        iii. State government actions to date
Where animal rehabilitation is headed
    j. Recent developments in the field
    k. Establishment of a specialty
    l. Animal plastic surgery
    m. Greater public awareness
Why practice animal rehabilitation
    n. Good for business, good veterinary practice, good for the clinic's reputation
    o. Advice for preveterinary students considering a future as an animal rehabilitation practitioner
Conclusions: Is animal rehabilitation the future of veterinary medicine?
Bibliography
Appendix A: Proposal
Appendix B: Progress Report

Working Bibliography

Brockett, John E. "Why Give Pain Medication—Patient Care, Image Or Profit?" *DVM: The Newsmagazine of Veterinary Medicine* 39.2 (2008): 41. Print.

Downing, Robin. "Putting the New Pain Management Guidelines into Practice." *Veterinary Medicine* 102.11 (2007): 704–705. Print.

Fiala, Jennifer. "The Price of Pain." *DVM: The Newsmagazine of Veterinary Medicine* 38.10 (2007): 30–32. Print.

Foley, Denise. "Life with a Disabled Pet." *Prevention* 58.2 (2006): 198–201. Print.

Jurek, Christine, and Laurie McCauley. "Underwater Treadmill Therapy in Veterinary Practice." *Veterinary Medicine* 104.4 (2009): 182–190. Print.

Marshak, Doug. "State Board Lacks Consistency on Standard of Practice." *DVM: The Newsmagazine of Veterinary Medicine* 38.11 (2007): 38–40. Print.

Nolen, R. S. "Pet Rehab Becoming Mainstream Practice." *Journal of the American Veterinary Medical Association* 235.7 (2009): 798–799. Print.

continued

Payne, Richard M. "Canine and Equine Physiotherapy." *Veterinary Record: Journal of the British Veterinary Association* 165.4 (2009): 122. Print.

Sayre, Carolyn, and Robert Horn. "Fake Fins, Plastic Paws." *Time* 170.10 (2007): 49–51. Print.

Shearer, Tami. "Changing Times." *DVM: The Newsmagazine of Veterinary Medicine* 39.3 (2008): 34. Print.

Skernitvitz, Stephanie. "Rehab Gives Animals Swifter Road to Recovery, DVM Says." *DVM: The Newsmagazine of Veterinary Medicine* 39.10 (2008): 1s–12s. Print.

Smith, Sarah D. "Petting with a Purpose." *Natural Health* 36.3 (2006): 99–100. Print.

Stein, Robert M., and Stephanie Ortel. "Postsurgical Pain Management: Take a Pre-Emptive Approach." *DVM: The Newsmagazine of Veterinary Medicine* 39.7 (2008): 44–46. Print.

van Dyke, Janet B. "Canine Rehabilitation: An Inside Look at a Fast-Growing Market Segment." *DVM: The Newsmagazine of Veterinary Medicine* 40.7 (2009): 14S–15S. Print.

Verdon, Daniel R. "An Animal Planet." *DVM: The Newsmagazine of Veterinary Medicine* 36.7 (2005): 24–28. Print.

Case 9-4

In research, progress reports are part of doing business. And progress reports take numerous forms. Ohio State University provides its general research reports online (see Figure 9–3). Readers can click on a date, from July 2011 to October 2013, listed in the left-hand menu. The research done during that month appears, and readers can read as many summary monthly reports as they choose. This is a good example of an interactive online report. However, reports to funding agencies will follow the standard progress report guidelines. Again, the form of progress reports, like any reports, depends on audience and purpose.

In business, progress reports are standard. COSMIX (Colorado State Metro Interstate Expansion) posts general progress reports on road projects on the Internet so that the public can track progress through text and pictures that can be seen by clicking through each project. This type of online interactive progress report provides transparency and allows the public to see how COSMIX spends highway expansion dollars. Despite the new emphasis on media-based

continued

FIGURE 9-3 Interactive Progress Report

Source: http://medicine.osu.edu/research/news/archive/2013/05.asps

progress reports, the progress report standards remain: what you have done, tasks that remain, evaluation of work.

Online progress reports that combine text and photos that work interactively provide visual perspective on how the project is being pursued. Readers can toggle between segments of a project and see a slideshow of the project as it develops. See Figure 9–4. Figure 9–5 exemplifies another method of reporting progress, a pdf of a general COSMIX progress report also available online. Nevertheless, when you write progress reports for grants or projects, you will need to provide a detailed project report to the entity that approved the project, its funding, and its timeline.

FIGURE 9–4 Interactive Progress Report

Source: http://www.wilsonco.com/projects/transportation/cosmix

Design-Build

Client
Colorado Department of
Transportation

Location
Colorado Springs, CO, USA

Colorado Springs Metro Interstate Expansion (COSMIX™)

Rockrimmon Constructors, a CH2M HILL and SEMA Construction joint venture, won this design-build contract in January 2005. The project extends and expands Colorado Springs' highway capacity from Bijou Street to the North Academy Boulevard interchange by widening I-25 to six through-lanes, while accommodating the future expansion to eight lanes. The $129 million scope includes 18 new bridges, two culverts, and numerous retaining walls. Other agency costs, including right-of-way acquisition, bring the total value of the project to $150 million.

Segment 1: Bijou Street Interchange

The Bijou Street Interchange (Segment 1), located at the southern portion of the project, will be the last phase of work performed. This complex reconstruction of a 40-year old interchange and 70-year old bridge over Monument Creek will also reduce I-25 curvature and provide a new mainline crossing over Colorado Avenue, and improved storm drainage. The new Bijou Bridge will improve downtown access with two additional lanes and be expandable for two more in the future.

A gravity storm water outfall to Monument Creek for I-25 runoff will be provided under Bijou Street to reduce the flooding currently experienced on I-25 during major storm events.

Heavily traveled and often congested, this elderly "gateway to downtown" will be closed for construction. Our plans to detour traffic from this busy access point include providing shuttle bus services to transport pedestrians and bicyclists across the highway while normal routes are closed. Construction and traffic-related noise at the nearby Monument Valley Park duck ponds, horticultural gardens and baseball fields will be mitigated with sound walls and earthen berms. These features will help preserve the quiet environment and character of this urban open space.

Segment 2: I-25 Expansion at Fillmore to Garden of the Gods Road

The Fillmore segment will widen I-25 from two lanes to three in each direction, widen the median, provide an asphalt overlay south of Garden of the Gods, widen the Ellston Street bridges, and widen and replace the I-25 pavement north of Garden of the Gods to Garden of the Gods Road. The Garden of the Gods Road north ramps will be rebuilt.

Segment 3: Nevada/Rockrimmon Interchange

The interchange will reconfigure the existing southbound left exit ramp to the right side, dramatically improving driver expectations while other additions improve cross traffic flow between neighborhoods on both sides of I-25. A newly configured split-diamond interchange will tie Nevada to Rockrimmon together and add collector-distributor (C D) roads between the two interchanges. Corporate Drive will link into Nevada Avenue, and the

C032005006BDS
COSMIX.JPG
COSMIX_1.JPG

continued

FIGURE 9-5 PDF Progress Report

Source: www.ch2m.com/corporate/markets/transportation/assets/.../COSMIX.pdf

business arterials of Mark Dabling and Corporate Drive will gain improved connectivity with several new means of access.

Ramps will be closed during this construction, which will connect with the recent roadway improvements associated with the south end of the Woodmen Road interchange project.

Segment 4: I-25 from Woodmen to North Academy

This segment adds a third lane to I-25 in each direction, widens concrete in the median from Cottonwood Creek to Pine Creek, and widens asphalt on the outside of the existing roadway pavement and asphalt overlay from Pine Creek to the North Academy Interchange. We will maintain two lanes of traffic in each direction to minimize delays and traffic flow disruptions.

The Finished Product

As the project tasks are completed and segments connected, the miles between Bijou and the North Academy Interchange will provide a seamless network of roadway specifically designed to keep pace with Colorado Springs's continued growth.

Beginning in the summer of 2005, our team's sequenced construction approach accelerates the work dramatically during the life of the project. We expect to complete the COSMIX™ tasks by winter of 2007, one year earlier than required by CDOT.

C032005006BDS
COSMIX.JPG
COSMIX_1.JPG

FIGURE 9–5 *continued*

STYLE AND TONE OF PROPOSALS AND PROGRESS REPORTS

The proposal and its related report documents serve as sales documents, but writers have an ethical commitment to present information about a project in a clear and accurate manner. Proposals, once accepted, become legally binding documents. Because contracts emerge from proposals, organizations must be prepared to stand behind their proposals. Thus, the style should be authoritative, vigorous, and positive, suggesting the competence of the proposer. Generalizations must be bolstered by detailed factual accomplishments. Problems should be discussed honestly, but positive solutions to problems should be stressed. Neither the proposal nor the progress report should resort to vague, obfuscatory language.

CHECKLIST FOR DEVELOPING PROPOSALS AND PROGRESS REPORTS

Proposals

Planning

- ☐ Have you made a list of all requirements given in the RFP?
- ☐ Who are your readers? Do they have technical competence in the field of the proposal? Is it a mixed audience—some technically educated, some not? To whom could the proposal be distributed?
- ☐ What problem is the proposed work designed to remedy? What is the immediate background and history of the problem? Why does the problem need to be solved?
- ☐ What is your proposed solution to the problem? What benefits will come from the solution? Is the solution feasible (both practical and applicable)?
- ☐ How will you carry out the work proposed? Scope? Methods to be used? Task breakdown? Time and work schedule?
- ☐ Do you want to make statements concerning the likelihood of success or failure and the products of the project? Who else has tried this solution? What was their success?
- ☐ What facilities and equipment will you need to carry out the project?
- ☐ Who will do the work? What are their qualifications for doing the work? Can you obtain references for past work accomplished?
- ☐ How much will the work cost (e.g., materials, labor, test equipment, travel, administrative expenses, and fees)? Who will pay for what? What will be the return on the investment?

continued

CHECKLIST FOR DEVELOPING PROPOSALS AND PROGRESS REPORTS *continued*

☐ Will you need to include an appendix? Consider including biographical sketches, descriptions of earlier projects, and employment practices.

☐ Will the proposal be better presented in a report format or in a letter or memo format?

Revising

☐ Does your proposal have a well-planned design and layout? Does its appearance suggest the high quality of the work you propose to do? Do your readings both promote and inform?

☐ Does the project summary succinctly state the objectives and plan of the proposed work? Does it show how the proposed work is relevant to the readers' interest?

☐ Does the introduction make the subject and the purpose of the work clear? Does it briefly point out the importance of the proposed work?

☐ Have you defined the problem thoroughly?

☐ Is your solution well described? Have you made its benefits and feasibility clear?

☐ Will your readers be able to follow your plan of work easily? Have you protected yourself by making clear what you will do and what you will not do? Have you been careful not to promise more results than you can deliver?

☐ Have you carefully considered all the facilities and equipment you will need?

☐ Have you presented the qualifications of project personnel in an attractive but honest way? Have you asked permission from everyone you plan to use as a reference?

☐ Is your budget realistic? Will it be easy for the readers to follow and understand?

☐ Do all the items in the appendix lend credibility to the proposal?

☐ Have you included a few sentences that urge readers to accept the proposal?

☐ Have you satisfied the needs of your readers? Will they be able to comprehend your proposal? Do they have all the information they need to make a decision?

CHECKLIST FOR DEVELOPING PROPOSALS AND PROGRESS REPORTS *continued*

Progress Reports

Planning

☐ Do you have a clear description of your project available, perhaps in your proposal?

☐ Do you have all the project tasks clearly defined? Do all the tasks run in sequence, or do some run concurrently? In general, are the tasks going well or badly?

☐ What items need to be highlighted in your summary and appraisal?

☐ Are there any problems to be discussed?

☐ Can you suggest solutions for the problems?

☐ Is your work ahead of schedule, right on schedule, or behind schedule?

☐ Are costs running as expected?

☐ Do you have some unexpected good news you can report?

Revising

☐ Does your report have an attractive appearance?

☐ Does the plan you have chosen show off your progress to its best advantage?

☐ Is your tone authoritative, with an accent on the positive?

☐ Have you supported your generalizations with facts?

☐ Does your approach seem fresh or tired?

☐ Do you have a good balance between work accomplished and work to be done?

☐ Can your summary and appraisal stand alone? Would they satisfy an executive reader?

EXERCISES

1. Your friend and colleague, Lydia Sasaki, has composed the following brief proposal and asks you to review it. What specific changes would you advise she make so that the proposal is as clear and persuasive as possible?

To: Chief William Ricco
From: Officer Lydia Sasaki
Subject: Surveillance Convention
Date: August 11, 2014

I have received a letter of invitation from the Association of Certified Surveillance Professionals in reference to a convention being held by the association June 8–11, 2015, in Albuquerque. This convention will offer a comprehensive training program designed to thoroughly train police officers in all aspects of acoustic gunfire detection systems. Instructors at the convention consist of surveillance and security specialists from all over the world and the United States.

I am respectfully requesting travel funds to attend the convention in order to represent our department. There are workshops at the convention that deal with the establishment of gunfire detection systems by police departments. There are also two preconvention courses that are being conducted specifically on acoustic surveillance. These are the AS Certification Course and the AS Maintenance Course, both on June 7. The remaining workshops and training sessions will be held June 8–11. The workshops include:

Equipment Selection	Urban Obstacles
Sensor Placement	Acoustic Gun Signatures
Legal Issues	Triangulation and Multilateration
Service and Maintenance	Acoustic-Based Patrolling
Community Relations	Video Surveillance Interaction
Critical Incident Evaluation	Incident Archiving

As soon as we have the ability to acquire a gunfire detection system, it will be necessary for the department to obtain all of the facts regarding acoustic surveillance and to examine what other departments have done to implement and operate their systems. Attached to this memo is a copy of the letter I received from ACSP. In this letter are the details for registration, travel, and accommodations for individuals attending the convention.

If you have any questions regarding the convention, please feel free to contact me at any time.

---

2. Visit www.grants.gov, locate a request for proposals for a project related to your major, and answer the following questions:

- According to the RFP, what is the product or service the agency wants provided? Or what work does it want performed? Or what problem does it want solved?
- Does the RFP specify the categories of information the proposal must include? If so, what are those categories? If not, what would be appropriate categories?
- Does the RFP specify a minimum or maximum length for the proposal? If so, what is that length? If not, what length would you estimate as necessary?
- Does the RFP specify the organization of information? If so, what is that organization? If not, what would be an appropriate organization?
- Does the RFP specify the format for the proposal? If so, what is that format? If not, what format would you consider appropriate?
- Does the RFP identify the criteria by which proposals will be evaluated? If so, what are those criteria? If not, what are likely criteria? How might these criteria guide the writing, organization, and formatting of the proposal?

Write a memo summarizing your findings to distribute to the students in your class.

Instructions, Procedures, and Policies

INSTRUCTIONS VERSUS PROCEDURES

Procedures provide general guidelines for performing a task, while *instructions* provide specific, detailed steps. Procedures will be appropriate for some situations, while instructions may be necessary for readers who need detailed directions to perform a job task or process.

Instructions and procedures can appear in several formats: in letters and memoranda (see Chapter 7 for an example instructional memo), in reports, in technical papers or notes, in stand-alone documents to accompany mechanisms or processes, or in complete manuals. Complex procedures and instructions usually appear as manuals, which use many of the report elements exemplified in the formal report in Appendix C. E-mail is not a suitable method of sending instructions or procedures, unless you send these as an e-mail attachment (in a standard rtf, doc, docx, or pdf file) that can be printed or read on the receiver's computer screen. Differences in e-mail applications can produce awkwardly formatted, unreadable text.

CRITICAL ROLE OF INSTRUCTIONS AND PROCEDURES IN THE WORKPLACE

Instructions and procedures often carry heavy legal liability. Because instructions are written for those who need them, most new products and processes have instructions attached. Instructions may also appear on a website, but many of these do not exhibit the qualities of good instructions. When a product fails because instructions lack clarity, completeness, and/or adequate warnings, the manufacturing company can be held responsible for costs incurred from buyers who, in good faith, attempted to follow the instructions. Even worse, when a customer or an employee dies or suffers major injuries because instructions did not provide complete, readable information and safety warnings, the company can be liable for the injuries or death. Companies that provide incomplete or hard-to-follow instructions on products that must be installed or assembled by the purchasers can lose money on the product when purchasers return the product. Conversely, well-written instructions can improve the image of a company that manufactures products. Many companies will provide on-site, face-to-face instructions for individuals who will have to operate equipment after it is installed. Face-to-face instructions plus written instructions can help purchasers feel confident in the company that sells the equipment. In short, well-designed, clear, complete, accurate instructions can become good sales documents while preventing lawsuits if equipment or products are damaged because users did not or could not follow the instructions.

Quick Tips

Many people resist reading instructions. They try to figure out for themselves how to operate a product or perform a task and will turn to the instructions only if all their efforts fail. When they do read the instructions, they want to understand everything immediately without having to read anything twice. A simple design, plain wording, and clear illustrations will be critical to encouraging readers to pay attention to your instructions or procedures.

1. Use concise headings and subheadings to describe and highlight each section.
2. Leave plenty of white space around headings.
3. Use numbers for every step in a chronological process; use bullets for lists of conditions, materials, and equipment.
4. Use white space to make items in lists easy to find and read.
5. Highlight safety information and warnings. Distinguish between *danger*, *warning*, and *caution*.
6. Keep illustrations as simple as possible.
7. Locate illustrations at the point where the reader needs them. Don't expect readers to locate an illustration that is several pages away from the instruction to which it pertains.
8. Label every illustration, and at the appropriate point in the related text, write "See Figure X."
9. Do not begin an instruction at the bottom of one page and complete it at the top of the next page. Insert a "page break" and move the entire instruction to the next page.

Planning Instructions and Procedures

Instructions and procedures should enable those who need them to perform the tasks covered. Successful instructions and procedures can be easily read, easily understood, and correctly followed by the intended users with minimal difficulty. Understanding your target readers and, in some cases, the context in which the documents will be used is paramount. If readers are confused or misled by statements, they can hold the company liable for injury or financial damage suffered because of poorly written instructions or procedures. Instructions, particularly those that involve processes that could be hazardous, should explain the knowledge level required of the target audience.

Who are the readers? What do your readers know about the subject? These questions will determine how much information you provide and the type of language you use. The audience for instructions should be specified, especially if readers not conforming to that audience should not attempt to follow the instructions. For general instructions, remember that many people resist reading instructions. They tend to avoid reading them, if possible, and attempt to perform the task without them. When they do read the instructions, they want to understand them immediately without having to reread them. Design features (layout), format, choice of language, and use of visuals can be critical to encouraging readers to focus on the instructions or procedures.

What is your purpose? Successful instructions/procedures require that both you and your readers know your purpose. What should your readers know and be able to do after they read these instructions/procedures? Any information that isn't applicable should be omitted.

What is the context in which the instructions/procedures will be used? Knowing how documents will be read can help you design effective documents. Readers may be focusing on them while they are sitting at their desks. They may need to read the entire document before trying to perform the task, or they may perform the task as they go.

Do your readers need to read the instructions/procedures before they begin? Your telling them how to read the instructions may be necessary. For example, you can include detailed instructions for first-time readers and then have a summary of steps for readers to use once they are familiar with the longer instructions. Context-sensitive instructions can be found in numerous settings:

- Some sailboards have instructions for how to right the sailboard—if it turns over in the water—on the bottom of the boat where riders can see the instructions if the boat turns over.
- "Quick Start" instructions for many handheld devices can be found inside the top or lid of the device.
- Some products, such as bookcases, computers, and lamps, come with instructions composed only of pictures, to show the steps in assembling the product. However, if the product will be sold in other countries, "picture-only" instructions should be tested for interpretation.

- Instructions used in machine shops and on manufacturing floors are often printed in large type on eye-saver yellow or green paper and laminated to resist grease. These instructions may be in metal loose-leaf notebooks so strong that pages cannot be torn out. Or they may appear on computer screens projected along the assembly line. Again, the readability, completeness, and accuracy of these screens can be critical to error avoidance and safety of employees.
- Basic operation instructions and troubleshooting information for large clocks can be found inside the clock mechanism access compartment. These are designed to accompany the booklet that contains detailed instructions.

Will the instructions be available online or on paper? If instructions will be available online, focus on readability of the material. Avoid "text-heavy" and lengthy instructions. Partition your instructions into distinct sections hyperlinked in a logical way. Or you may want to tell your readers immediately that they should print a copy of the instructions if they are too long and complex for online reading.

Planning is particularly critical for writing instructions or procedures. Remembering or even using the document analysis checklist at the end of this chapter will help you determine whether you need instructions or procedures, the delivery medium (online or paper), the content you will include, and any design strategies that could help you produce an effective document. Effective instructions require sensitive attention to readers' information needs, their current knowledge of the topic, and what you want them to know as a result of reading the instructions.

STRUCTURE AND ORGANIZATION

First, analyze the audience, purpose, and context; then, begin to structure your procedures/instructions. Examine the following general guide for what elements to include, but remember that what you decide to include will depend on your topic, your audience, your purpose, and the reader's context. Comprehensive instruction/procedure manuals will usually contain most of these elements, while shorter instructions/procedures may require fewer. To illustrate instructions, we will focus on three examples.

Introduction. The introduction should familiarize readers with the task to be performed: specifically, what the instructions will allow readers to do and what skill level the person should have to perform the task successfully.

Theory governing the procedure or instruction. For some types of procedures, readers will perform tasks better if they have a general overview of the process. Knowledge of the process may help them understand when they have made an error and how to avoid errors. If you are going to give instructions for operating a mechanism, some explanation of the value and purpose of the mechanism can prepare readers for the process.

Warnings, cautions, hazards, and notes regarding safety or quality. Given the wide variety of technologically complicated and potentially dangerous products available today, you really cannot "overwarn" readers. And a failure to warn could have costly legal implications for your organization. Any level of hazard needs to be described—what will happen and why—if a particular action is performed. Identify the hazard, followed by the reason or the result, and be sure that all warnings, cautions, and notes are clearly visible. Warnings may also be repeated if they are associated with a particular step or direction. Warnings given at the beginning alert readers to possible problems they may encounter. Be sure that warnings stand out so that readers' eyes are drawn to them. The American National Standards Institute (www.ansi.org) articulates standards for hazard statements. A review of research on hazard messages is available from the US Occupational Safety and Health Administration (www.osha.gov).

Conditions under which the task should be performed. Some instructions, particularly laboratory instructions, may require you to describe the physical conditions required to perform a task: room size, temperature, time required for the entire process and for individual steps, specific safety processes. If time is a constraint, readers need to know how much time will be needed before they begin the task.

Name of each step. For each step, give the following: purpose of the step; warnings, cautions, and notes; any conditions necessary for performing the step; time required to perform the step; and the list of materials needed for the step.

- Place instructions in chronological order.
- Number each step.
- Limit each instruction to one action.
- After you have written the instruction, state in a separate sentence the reason for the instruction but only if you believe that the reader may not follow the instruction without an explanation. Include warnings whenever necessary. And be sure to explain all warnings if you believe that people will not understand the reason for the warning.

Case 10-1: **The Grignard Reaction**

Tasnim Mohamed, a graduate student who supervises a biochemistry lab, decides that the laboratory instructions for the Grignard reaction, presented as linear text in the lab manual, need clarification. She believes that labs should focus on teaching students chemical concepts rather than having them try to read confusing process discussion. She decides to revise these before assigning the lab. A portion of the original instructions appears in Case Document 10-1A. She will expect students to read the original instructions but also to have the revision available when they actually begin the experiment. Her first step in revising is to do a prewriting analysis.

Prewriting Analysis for Revising the Grignard Reaction

1. Identify your reader(s).
Texas A&M students in the second semester of organic chemistry (CHEM238).

2. Describe your reader profile: background, attitude toward following instructions
Most of the students are sophomores, so they may have had lab experience in another class before taking this lab. Generally, all students have already taken the first organic chemistry lab (CHEM237), so they have some idea of what to expect. They already know common lab safety rules such as wearing goggles and appropriate clothing, tying back long hair, no eating or drinking, no horseplay, etc.

Students in organic chemistry lab do not usually do everything correctly to obtain high yield or even the expected results. Most hope to finish the lab as quickly as possible in order to leave early, so they do not take the time to read huge paragraphs of instructions.

3. How much does your reader(s) know about the subject?
The students in the lab are not experts in the subject, so they will need some background information to understand what happens throughout the experiment. The course has a prerequisite of CHEM228, which is the Organic Chemistry II lecture course. Usually, the lecture material aligns with the laboratory experiments, but sometimes the professors teach at a different pace. The lecture gives students some background information, but they will probably still need help assessing what happens when something in the lab goes wrong.

4. Will you have to define any terms that your reader(s) may not understand?
Students should already know most vocabulary words that align with the organic chemistry material. A new technique specific for this lab alone will be the only thing I may have to define.

5. What do you want your reader(s) to be able to do after reading the instructions or procedures?

The reader should be able to fully understand the purpose of the lab, methods that will be used in the lab, and how to conduct the lab in a timely fashion. They will know where to obtain all the material and exactly what equipment to use for a specific setup. They will also know if they need to begin another task while they wait for something else to be completed so that they can safely and efficiently complete their experiment. The reader should know that actually understanding the lab is more important than the result.

6. What is the situation that led to the need for these instructions?

Organic chemistry lab is designed to teach students several lab techniques aligned with material learned through the lecture course. Some techniques include purification, spectroscopy, classification tests, computer modeling, and reaction kinetics. This specific set of instructions guides students through the Grignard reaction and allows them to practice their distillation and infrared spectroscopy techniques.

6a. If you are revising an existing set of instruction, explain why the original needs revision.

The Grignard reaction was the hardest lab of this semester. Students are often not able to perform the complete experiment because they are unable to prepare the Grignard reagent. If the lab instructions are revised, students may stand a better chance of performing the experiment correctly. The format of the procedure and the lab manual as a whole deter the students from actually reading the entire lab instructions, which hinders their success. They also find it difficult to reread bulky paragraphs of instructions while conducting the experiment, causing them to miss some vital steps.

7. How will your reader(s) use these instructions? Will they need to be read every time, or will they be used only for reference?

Readers should be able to read the instructions before lab, understand the lab, and then perform the experiment on their own. However, they can have the instructions available to them while conducting the experiment for a reference guide. Students should not have to read everything over again during the lab.

8. What kinds of pictorials, if needed, will you use?

I will include graphics of some experimental setups that are used during the lab. I will also need to include pictures of the equipment that the setup will need in case the student does not know some of the terms.

9. What kinds of formatting and page design strategies will you use to make the instructions easy to read?

I will make sure to use bold headings to separate main topics and subtopics. I will put each picture or graphic beside the text to which it corresponds. I will also make sure to leave plenty of white space to help the eye find what it needs on the page.

10. What content segments will you include?

I will include the following sections:

- Purpose
- Summary of Methods
- Materials
- Reagents and Solvents
- Methods
- Experimental Procedure
- Safety Precautions
- Discussion for In-Lab Notebook
- Discussion for Post-Lab Report

CASE DOCUMENT 10–1A: ORIGINAL, GRIGNARD REACTION

THE GRIGNARD REACTION

Purpose: To prepare and use the Grignard reagent as a method of studying the utility of organometallic reagents in synthesis and to identify starring materials by using retrosynthetic analysis

Reading: McMurry: See assignment for this lab session on the website.
Review *Separation and Purification Techniques:* "Distillation" (Simple)
Review *Separation and Purification Techniques:* IR, "Extraction"

INTRODUCTION

Organometallic compounds are defined as organic compounds which have a carbon-metal bond. These are valuable tools for use in organic synthesis. The reactivity of the organometallic compound is highly dependent upon the nature, ionic or covalent, of the carbon-metal bond. Organosodium and organopotassium reagents are highly ionic and the carbon acts like a powerful base. These reagents react explosively with water and are pyrophoric (burst imo flame upon exposure to air). Organomagnesium (Grignard reagents) and organolithium reagents are sufficiently ionic in nature to be strong carbon nucleophiles, strong bases, and yet covalent enough to be soluble in many organic solvents. These two reagents are widely used in organic synthesis. In recent years the use of other organometallic compounds, particulacly organocuprates and organomercury compounds, have been widely used in research. However, the very low toxicity of the Grignard reagents makes them a reagent of choice.

continued

Organomagnesium reagents are perhaps the easiest to prepare and use of all the organometallic reagents. These reagents were discovered in 1901 (Victor Grignard won the Nobel Prize in 1912 for this discovery; the reagents are named after him.) Grignard reagents are prepared by the reaction of alkyl or aryl halides with magnesium. The exact mechanism involved in the formation of Grignard reagents is unknown, and despite their wide use, there is some doubt as to their exact structure. Grignard reagents are conventionally written as RMgX, where X is chlorine, bromine, or iodine, although there is some evidence that in solution a di-organomagnesium structure exists in equilibrium with the magnesium halide. Generally speaking, any organic halide can be used in the preparation of a Grignard reagent. There are a few exceptions to this, most notably are aryl chlorides. Aryl chlorides, where the chlorine attaches to the aromatic ring, do not react with magnesium. Methyl iodide (todornethane), a liquid cancer suspect agent, is used instead of either of the poisonous gases methyl bromide or methyl chloride to prepare methyl Grignard. Likewise, ethyl bromide (Qachrymator) is commonly used in place of the gas ethyl chloride to prepare ethyl Grignard. In the preparation of the Grignard reagents, the alkyl halide used can affect the rate of formation with $I > Br > Cl$ (organomagnesium fluorides have not been prepared). Once the reagent is formed, the halide ion has no apparent effect on either the rate or the yield of the subsequent reaction.

The use of an ether solvent is a key step in the synthesis of Grignard reagents. The electron pairs on the oxygen of the ether coordinate with the magnesium of the Grignard reagent and form a soluble complex. Most frequently diethyl ether or tetrahydrofuran (THF) is used as the solvent for both the synthesis and the subsequent reaction of the Grignard reagent. Grignard reagents have been made and used in the absence of ether; however in both situations (synthesis and use) the yields were "disappointingly poor."

$$RX \quad + \quad Mg \quad \xrightarrow[\text{or THF}]{\text{ether}} \quad RMgX$$

$CH_3CH_2-O-CH_2CH_3$
diethyl ether
"ether"

tetrahydrofuran
THF

soluble Grignard complex

Safety! The use of ether is the only serious safety concern associated with the Grignard reaction. Ethers are highly flammable. Exercising a few simple safety precautions dramatically reduces the hazard of fire. Unfortunately that is not the only hazard associated with ethers. Ethers react readily with atmospheric oxygen to form peroxides. Organic peroxides can undergo explosive decomposition and the risk of an explosion is directly proportional to the concentration of the peroxide. This is why it is important *not* to distill to dryness (the distillation pot goes dry). Ethers are usually checked with special peroxide detection strips prior to their use.

Grignard reagents are most often used in reactions with aldehydes, ketones, or esters. The products of these reactions following work up are alcohols. Reaction with formaldehyde produces a primary alcohol that has a carbon chain one carbon longer than the beginning Grignard reagent. Reactions with other aldehydes give secondary alcohols, and reactions with ketones or esters yield tertiary alcohols. Notice that esters react with *two* moles of Grignard. After the first reaction, the intermediate loses alkoxide to form a ketone. The ketone then reacts with a second mole of Grignard to form a tertiary alcohol. Grignard reagents have also been used in reactions with nitriles, R-C/N to yield ketones, with epoxides to yield alcohols, and with anhydrous carbon dioxide, CO_2, to yield carboxylic acids (*Note:* Grignard reagents will also react with oxygen in air. The reaction with air is so slow that it is not normally necessary to run the reaction under an inert atmosphere. Atmospheric moisture is a real problem and will be discussed in detail.)

Tasnim Mohamed
CHEN 301
Spring 2013

Lab 18: The Grignard Reaction

Purpose

You will learn how to prepare and use a Grignard reagent to study the utility of organometallic reagents in synthesis and to identify starting materials by using retrosynthetic analysis.

Introduction

Organometallic compounds are organic compounds with carbon-metal bonds. These compounds are valuable tools for use in organic synthesis. Their reactivity is highly dependent upon the nature, ionic or covalent, of the carbon-metal bond. Organomagnesium (Grignard reagents) are widely used in organic synthesis due to their low toxicity.

Grignard reagents are conveniently written as RMgX, where X is chlorine, bromine, or iodine. When preparing Grignard reagents, the halide used can affect the rate of formation (fastest: I > Br > Cl: slowest). However, once the reagent is formed, the halide ion has no effect on the rate or yield of the reaction.

The use of an ether solvent is the key step of Grignard reagent synthesis. The election pairs on the oxygen of the ether coordinate with the magnesium of the Grignard reagents and form a soluble complex, as shown in the figure below. Most frequently, diethyl ether or tetrahydrofuran (THF) is used as the solvent for the synthesis and subsequent reaction of the Grignard reagent.

soluble Grignard complex

© Beverly A. Clement and Kenn E. Harding

Grignard reagents are most often used in reactions with aldehydes, ketones, and esters to form alcohol products. They can also react with nitriles to yield ketones, epoxides to yield alcohols, and anhydrous carbon dioxide to yield carboxylic acids. Grignard reagents may be viewed as a strong base. They react readily with acids to form alkanes. Even weak acids such as water, alcohols, phenols, and terminal acetylenes will protonate Grignard reagents to give the corresponding alkane. If the synthesis does not call for the formation of an alkane, this side reaction only lowers the yield of the desired product. Also, if the starting material contains an unprotected alcohol, phenol, or terminal acetylene, you must use two equivalents of the Grignard reagent. One equivalent will be destroyed by the proton source.

Warning!

THF is flammable and volatile. Work close to the hood. If flames occur, smother with a watch glass or beaker, and do not use a fire extinguisher.

Diethyl ether is even more flammable and volatile than THF, has a low affinity for water, and reacts readily with oxygen to form peroxides, which can undergo explosive decomposition.

Materials

- ☐ 25mL round-bottom flask
- ☐ Heat gun
- ☐ Magnetic stir bar and stir plate
- ☐ Small beaker
- ☐ Weighing scale
- ☐ Weighing paper
- ☐ Claisen head
- ☐ Air condenser
- ☐ Micro-scale distillation glassware
- ☐ Sonicator
- ☐ Large test tube
- ☐ Plastic pipette
- ☐ Pasteur pipettes

Reagents and Solvents

tetrahydrofuran (THF)
Magnesium turnings
Ammonium chloride
Distilled water
10% sodium bicarbonate
Saturated sodium chloride
Magnesium sulfate

Alkyl Bromides:
  1-bromopropane
  1-bromobutane
  1-bromopentane

Ketones:
  Acetone
  2-butanone
  2-pentanone
  3-pentanone

continued

Methods to Achieve Objectives

Simple Distillation
Extraction
Infrared Spectroscopy

Experimental Procedure

This experiment will take two lab periods. In the first period, you will prepare the Grignard reagent and run the reaction with the ketone. The crude product will be allowed to dry until the second week. In the second week, you will purify and identify the product.

Week 1

Step 1: Obtain your alkyl halide and ketone.

- The alkyl halides and the ketones will be at the front of the lab room. Alkyl halide vials are labeled "#H" and ketone vials are labeled "#K." The # on the vial you obtained should be your workstation number. Record the unknown codes in your lab notebook.
- The alkyl halide vials contain 0.025 moles of one of the 3 alkyl bromides (1-bromopropane, 1-bromobutane, or 1-bromopentane) and the ketone vials contain 0.025 moles of one of the 4 ketones (acetone, 2-butanone, 2-pentanone, or 3-pentanone). Each of the compounds is dissolved in THF.

Step 2: Examine glassware.

- Your 25mL round-bottom flask should contain no cracks. If you find a crack, return the flask to the stockroom and obtain a new one. If the glassware appears dry, you can begin with **Step 3**.
- If the glassware is not dry, dry it with a heat gun. You can speed up the drying process by rinsing glassware with acetone, which evaporates faster than water. All glassware must be *completely* dry before proceeding with the next step.

Experimental Set-up

Step 3: Add a stir bar to the flask and place it in a small beaker.

Tasnim Mohamed   Page **4** of **10**

Step 4: Weigh 0.61g (0.025mol) of magnesium turnings on a piece of weighing paper and add to the flask.

Step 5: Add 4mL of THF to the flask.

> ## Caution!
> THF is flammable. Handle it carefully under the hood at your work station. In case of fire, DO NOT use a fire extinguisher. Smother flames with a watch glass or beaker.

Step 6: Attach Claisen head and air condenser to the flask.

- All ground-glass joints should be greased. If not, notify your TA. Use a Keck clamp to secure all connections.
- Refer to the figure on the next page to set-up the glassware.

Step 7: Use a plastic pipette to add 1mL of your alkyl bromide solution to the flask through the top of the air condenser.

- The diagram below shows an example of the reaction taking place in this step.

© Beverly A. Clement and Kenn E. Harding

continued

Step 8: Swirl the flask.

Your reaction may begin immediately, and the solution will become very warm and cloudy. In this rare case, you may proceed to Step 10.

Step 9: Place the flask with the attached condenser in an ultrasound bath.

- The bath is located underneath one of the main hoods in the back of the lab room.
- Look for agitation of the liquid and magnesium particles. The more agitation, the faster the reaction will initiate.
- After a couple minutes of agitation, the THF solution should appear somewhat cloudy. Warm the flask if it is not already warm.

> **Ultrasound Bath Sonication Tips**
>
> – Make sure the bath is turned on. If it is on, you should hear a constant clicking noise. Turn the time dial if it is off. DO NOT turn on the heat on the ultrasound bath.
> – Sonication is greater near the corners of the bath.
> – Four reaction set-ups can be placed in the bath at the same time.

- Hold the glassware by the condenser to immerse your flask so that the liquid inside is at or below the level of the liquid in the tank.

Step 10: Clamp the glassware so that the flask is centered over a magnetic stir plate underneath the hood at your work station.

- Make sure the mixture is stirring evenly.

Step 11: You can use the same plastic pipette from **Step 7** to add the rest of the alkyl bromide, slowly and in small portions, through the air condenser.

- The reaction should warm the solution to the boiling point of THF (~66°C) but not so much to the point where the solution boils over the top of the condenser.
- If the reaction does become too hot, moderate the temperature with an ice bath, without letting the temperature drop below room temperature (~25°C).
- If the reaction does not become warm and the magnesium turnings disappear, return the flask to the sonication bath.

Step 12: Stir the mixture for 15 minutes after adding all the alkyl bromide.

Tasnim Mohamed   Page **6** of **10**

Addition of Ketone to Grignard Reagent

Step 13: Use the same pipette to add the ketone solution to the flask while continuously stirring.

- Add ketone at a rate to keep a steady, gentle reflux.

Step 14: Stir the mixture for 15 minutes after adding all the ketone.

- The reaction flask now contains the magnesium salt of the alcohol product.
- The diagram below shows an example of the reaction taking place in this step.

Hydrolysis

Step 15: Add 6g of ammonium chloride and 20mL of distilled water in a beaker with a stir bar.

Why ammonium chloride?
The slightly acidic ammonium chloride solution is more effective in hydrolyzing the magnesium alkoxide than water alone, but keeps the acidity of the solution lower than the addition of excess HCl.

Step 16: Stir salt solution vigorously and add contents of the reaction flask.

Step 17: Transfer the solution in the beaker to a large test tube.

- The diagram below shows an example of the reaction taking place in this step.

Step 18: Allow phases to separate and remove the lower *aqueous* phase.

- This phase will later be discarded. However, keep all extractions of the aqueous phase in a waste beaker at your work station. Never discard anything until the end.

continued

Step 19: Wash the organic layer with 5mL of water and remove the lower aqueous phase.

Step 20: Wash the organic layer with 5mL of 10% sodium bicarbonate and remove the lower aqueous phase.

Step 21: Wash the organic layer with 5mL of saturated sodium chloride and remove the lower aqueous phase.

Step 22: Transfer the organic layer to a sample vial.

Step 23: Add magnesium sulfate to the vial to remove any water remaining.

> **Why saturated sodium chloride?**
>
> The saturated salt solution wash helps remove the water from the organic layer. Some water is dissolved in the organic layer because of the presence of THF.

- Cap the vial.
- Label it with your name and section number.
- Store it in your section's storage drawer until next week.

Week 1 Clean-Up Procedures

- Rinse vials that contained your starting alkyl halide and ketone with acetone in the acetone wash bottle in the sink.
- Return the vials to the rack where you obtained them.
- Dispose of plastic pipette by placing it in the labeled plastic beaker in one of the main hoods in the back of the lab.
- Dispose of used Pasteur pipettes in the broken-glass box after rinsing out any organics.
- Dispose of unreacted magnesium metal in your beaker in the solid waste container provided.
- Aqueous washes may be disposed of in the sink.

Week 2

Purification

Step 1: Decant the dry THF solution into a 25mL round-bottom flask with a few boiling chips.

Tasnim Mohamed

Step 2: Set up the micro-scale distillation apparatus.

- All ground-glass joints should be greased. If not, notify your TA. Use a Keck clamp to secure all connections.
- Refer to the diagram on the right to set up the glassware.

Step 3: Distill THF with the steam bath.

- THF should distill around 55°C–65°C.
- Periodically remove the THF from the collection well of the distillation head with a Pasteur pipette.
- Dispose of the collected THF in the appropriate waste bottle in the main hood.
- After most of the THF has distilled (3–4mL remains in the round-bottomed flask), the temperature will begin to fall.

thermocouple

screw cap to adjust height of thermocouple

hole open to the atmosphere

Air Condenser

Side Arm with cork

Distillation Head

insulated

Conicle Reaction Vial

© Beverly A. Clement and Kenn E. Harding

Step 4: Remove distillation set-up from the steam bath and allow product to cool.

- You can speed up this process by placing the flask in a bath of *room-temperature* water, NOT an ice bath.

Step 5: Transfer the liquid in the flask to a conical reaction vial.

Step 6: Attach the vial to the micro-scale distillation head.

Step 7: Distill the tertiary alcohol product using the heating mantle.

- Depending on your starting materials, your product may be expected to boil somewhere between 122°C and 215°C. Refer to **Table 1** on the next page.
- Discard the initial distillate that collects in the collection well.

continued

Tasnim Mohamed

- Periodically transfer the product from the collection well of the distillation head with a Pasteur pipette into a pre-weighed, capped vial.
- At some point when you have already collected some product, and at a constant boiling point, collect a drop of the product from the collection well with a Pasteur pipette into a separate vial to later obtain an IR spectrum.
- Distill until most of the liquid has distilled or until temperature starts to drop.
- DO NOT distill until no liquid remains!

Table 1: Boiling points for possible products

Compound Name	b.p. (°C)	Compound Name	b.p. (°C)
2-methyl-2-pentanol	122	3-methyl-3-heptanol	163
2-methyl-2-hexanol	142	4-methyl-4-octanol	183
3-methyl-3-hexanol	142	3-ethyl-3-heptanol	184
3-ethyl-3-hexanol	160	3-methyl-3-octanol	185
4-methyl-4-heptanol	161	3-ethyl-3-octanol	199
2-methyl-2-heptanol	162	4-methyl-4-nonanol	212

Step 8: Turn off heating mantle and remove it from the distillation set-up.

Step 9: Obtain an IR spectrum of the product while allowing the glassware to cool.

- Identify your product by comparing your spectrum to the spectra of the products in **Table 1**.

> **Caution!**
> DO NOT put the hot conical vial into water. Thermal shock will cause the vial to crack instantly.

- After identifying your product, determine your starting alkyl bromide and ketone.

Step 10: Weigh the vial to determine the yield.

- Calculate the percent yield of your product.

Week 2 Clean-Up Procedures

- Rinse glassware with acetone in the acetone wash bottle in the sink.
- Dispose of THF in the labeled container in one of the main hoods at the back of the lab room.

- Dispose of the magnesium sulfate used to dry the extract in the appropriate waste container hood.
- Aqueous washes may be disposed of in the sink.

Discussion Items for the In-Lab Notebook

- Boiling point of alcohol product
- Identity of alcohol product (attach your IR spectrum)
- Theoretical yield
- Product yield
- Percent yield

Discussion Items for the Post-Lab Report

- Which starting alkyl halide and ketone would give the alcohol product you obtained?
- Write a Grignard synthesis scheme using different starting materials to obtain your same product.

Case 10-2: **Job Instructions**

These procedures were written by an employee who had just completed several years as a runner/floater at an animal hospital. Before she left, she decided to write a set of instructions for the next person who would take her place. Job procedures have become increasingly important in many organizations. These help the organization in writing job descriptions, training new employees, and evaluating employee performance.

CASE DOCUMENT 10-2

How to Be a Runner at Kingsland Animal Clinic

Congratulations! You have the job! You will now need to learn how to perform your job. These instructions will tell you what you will need to do here at the clinic. Keep these instructions close by until you know all the tasks you will need to perform.

Your First Day

You will be training most of your first day. Read these instructions before you arrive, then observe, listen, and practice. Make notes on the instructions to help you remember what to do.

As you learn the job, you will see other ways that you can help. Always ask questions. The better all employees work as a team, the better service and care the clinic can provide. Every person employed at the clinic has an important job. The runner's tasks, as you will see, prevent gaps in service and care and improve efficiency.

What the Runner/Floater Does

As the runner/floater, you will experience all aspects of a working veterinary clinic: reception, technical, kennel cleaning, pharmacy, and any other job that the clinic does. If someone needs your help, respond immediately. The runner/floater position exists to make everyone's job easier and more efficient.

IMPORTANT: Realize that other employees do not always know when they need help, or they do not want to bother you by asking for it. Learn to know when someone needs you. Take initiative. If someone is having trouble, offer to help. Communicate openly with the receptionists, technicians, and vets to find where you are needed at any given time. Don't be afraid to ask questions about where and how you can help.

CAUTION: Since you will work with animals, remember that sick animals are often frightened. Remain alert and aware of your surroundings. Treat all animals with caution. Just because an owner tells you that her chihuahua is a "sweetie" doesn't mean that it won't try to bite you when you take its temperature.

Your Responsibilities—Discussed in These Instructions

- Knowing the Phone Room
- Helping the Receptionists
- Helping the Technicians
- Completing End of Day Duties
- Working on Other Tasks When You Have Time

The Phone Room

If you are assigned to the phone room, when one of the receptionists is at lunch, or if no one is in the phone room, **answer the phone**. A receptionist will teach you to work the phone.

- Greet callers with "This is [insert your name] at Kingsland Animal Clinic. How can I help you?"
- Always have a notepad ready to write down owner's name, animal's name, and any other important information.

Most common calls you will receive:

1. **Make an Appointment.** Use the Cornerstone program on the computer in the phone room to make an appointment. A receptionist will show you how to work Cornerstone. Practice using the program so that you don't make mistakes.
2. **Make a Boarding Reservation.** The boarding book should always remain in the phone room, on one of the desks. Write the owner's last name under the requested dates **in pencil**. If the reservation changes, you can easily erase.
3. **Ask about the Status of a Pet.** If the owner dropped off a pet for any reason, the owner may call to find out how the animal is doing. Put the owner on hold; run to the back; and ask the techs, "Does anyone know if [pet name] is ready to go?" or "[pet name]'s owner wants to know how he or she is doing. Does anyone know?" When you find out, run to the phone and relay the information.
4. **Ask about Prices.** The price book, like the boarding book, is located in the phone room. This book lists prices for common procedures, surgeries, details, etc. You may put the client on hold for a moment as you find the information.

continued

5. **Ask a Pet Health Question.** Transfer the caller to the phone in the tech station. Put the caller on hold and first make sure a tech can take the call, then transfer the caller. If no tech is free, get the caller's information and tell them a tech will call shortly. Pull the file, and put it in the Drop Off box with a note attached. Describe question clearly.

6. **Refill Medication.** Simply transfer the caller to the pharmacy station. Check first to make sure someone is available in pharmacy.

7. **Other.** Many calls do not fit into these categories. If you do not know the answer, put the caller on hold and find someone who does know the answer. Any employee will be glad to answer your questions.

NOTE: Make sure to "roll the phones over" when the clinic closes so that no more calls come in.

Helping the Receptionists

The receptionists should never leave the front desk vacant. Your job: make sure they do not leave unless someone else takes their places. You must constantly run between the reception area and the back of the clinic to relay information from the receptionists to the vets and to retrieve animals. This requirement describes the "runner" part of your job. You will keep the receptionists from leaving by doing the following tasks for them:

- When a client drops off a pet, take the pet into the back.

  1. Make a cage card for the animal. Include name, date, and reason for the visit.
  2. Write the animal's name on one of our temporary collars.
  3. Put these in your pocket.
  4. Walk to the front and gently take the animal from the owner. Leave the dog's collar with the owner. CAUTION: Be gentle in dealing with an unknown animal.
  5. Take dogs to the kennel in the back. Take cats to the cat room. Take exotics to the exotics room.
  6. Put the temporary collar on the animal, and tape the cage card on the cage.
  7. Take the animal's folder, which the receptionist will have given you, and place it in the Drop Off box.

- When a pet is ready to go home, bring it up to its owner.

  1. Ask the receptionist which animal needs to go home.
  2. Ask the owner for leash and collar or carrier.

3. Find the pet.
4. Take it to its owner.

- When a client arrives for an appointment, the receptionist will prepare the file and hand it to you. Take it to the back and place it in the appropriate appointment slot. Tell the techs that Dr. [vet's name]'s appointment has arrived or that a walk-in has arrived.
- If a receptionist needs to leave the front for some reason, ask if you can help so that he or she doesn't have to leave. You can help when bathroom breaks are needed.

Helping the Technicians

If the phone room is quiet and the receptionists seem to have things under control, you can head to the treatment area. Here you will find the techs, who can almost always use a hand. How you can help:

1. Restraining an animal. Usually, two people are needed to deal with an animal. If you see a tech trying to take an animal's temperature or trim nails with no help, offer to hold the animal. Special ways are needed to restrain the animal for a blood draw: the tech will be glad to show you.
2. Taking temperatures. You can help hold the animal, particularly the head if the animal decides to bite.
3. Trimming nails. If you do not know how to trim nails by looking for "the quick," ask a tech to teach you.
4. Weighing an animal.
5. Drawing blood. If you do not know how and would like to learn, the techs will teach you. This process requires two people.
6. Fecal tests. A tech will teach you how to attain a fecal sample and set up the fecal test.
7. Giving vaccines. Once you learn how to use needles, the techs and vets will let you draw and administer some vaccines, such as *Bordetella* and leptospirosis vaccines. You can also learn to give Adequan shots.
8. Pharmacy work. Once you have worked at the clinic for several months, you will be allowed to fill prescriptions. The tech will print a label, which will tell you how many and what kinds of pills to place in the pill bottles. Apply the label to the bottle and initial it.
9. Anything else the tech asks you to do. NOTE: You will not be asked or allowed to perform any task for which you have not been trained. Take every opportunity to learn.

continued

End of the Day Duties

End of day clean-up duties depend on how busy the clinic has been that day. Whenever business begins to slow down toward closing time, you can begin working on the following tasks:

- **Take out the trash.** Try to get someone to help you carry the bag to the dumpster.

    1. Every trash can in the clinic must be emptied at the end of EVERY DAY.
    2. Retrieve one large black trash bag from under the sink in the cat room.
    3. Begin dumping the contents of every small trash can into the large bag. You do not need to change the trash bags in the smaller trash cans unless they are particularly dirty.
    4. Make sure you get every trash can—the ICU, employee lounge, and every treatment room. The cans in the cat room and the kennel will be cleaned by the kennel kids.
    5. Take the now full trash bag out to the dumpster.
    6. Replace any of the small bags that you had to throw away.

- **Vacuum**

    1. The vacuum is located in the freezer room, off the exotics room.
    2. Insert one end of the vacuum into the various outlets in the hallway.
    3. Begin vacuuming the entire clinic (aside from the cat room and kennel).

Other Tasks to Complete When You Have Time

When you have time, such as a lull in activity or no appointments scheduled, other tasks must be completed.

1. Filing. The receptionists will show and help you with filing.
2. Restock the boxes of flea/tick/heartworm prevention on display in the reception area. (New boxes are stored in the stockroom. One of the vets must unlock that room for you.)
3. Restock the shelves of shampoos, ear cleaners, collars, ointments, etc. Most of these items are stored in the pharmacy area.
4. Clean the lobby. If at any point during the day the lobby looks dirty, sweep it. The broom is stored in the filing hall.
5. Clean up any messes the techs have made in the back (blood, nail trimmings, animal hair, etc.) by spraying the treatment tables with Trifectant, found in spray bottles.

6. Clean thermometers with alcohol swabs from the metal tins on the treatment tables.
7. Rinse the fecal loops and cups in the sink and place them in the Trifectant bath along with the urine collection trays.
8. Clean any dirty cages in ICU, isolation, or the cat room. For each cage,
   - Remove the mat from the cage and take it to the nearest bathtub.
   - Rinse off the mat and spray it with Trifectant. Leave it while you complete the next tasks.
   - Remove the newspaper from the cage and throw it away.
   - Spray the cage down with Trifectant.
   - Wipe the Trifectant with paper towels.
   - Line the cage with fresh newspaper found in every room of the clinic.
   - Go back to the mat—rinse it, dry it, and put it back in the cage on top of the newspaper.
9. Make sure all the animals have water, unless they will have surgery that day, and that all cats have litter boxes.
10. Check to see if the kennel kids need help cleaning cages, feeding, or walking dogs in the kennel.

Final Words—If you don't understand, always ask!

Case 10-3: Instructional Letter

Jonathan Varner, president of a professional organization, writes the incoming vice president to explain how elections for new officers should be conducted. The letter can be used as a reference by Dr. Dawson as she plans and executes the elections (see Case Document 10-3).

CASE DOCUMENT 10-3

May 12, 2014

Dr. Gabrielle Dawson
Department of Geography
San Diego State University
San Diego, CA

Dear Gabrielle,

As we discussed at AAGG in April, the VP of AAGG is responsible for conducting elections. From my experience in conducting the last election, I recommend that you begin this process immediately. The more persistent you are, the more efficient the process. Do as much as you can before faculty leave campus for the summer.

Timetable

In general, the election process begins during the late spring–early summer. Ballots are developed and mailed by mid-September. Ballots should be returned and tabulated by mid-January and the results reported to the vice president. As vice president, you will inform me and then inform each person whose name appeared on the ballot about the results of the election.

Election Procedures

The following procedures should help you conduct the elections. Ultimately, when the AAGG procedures manual is complete, these will appear in the manual. Until that time, however, each president will be responsible for informing the vice president of election procedures.

Materials Needed

- Election File—contains ballots of previous elections and names of nominating committees
- AAGG Directory—names, phone numbers, and e-mail addresses of all members

Gabrielle Dawson   -2-   May 12, 2014

- Annual Meeting Agendas—protocols for announcing newly elected officers

Step 1: Creating a Nominating Committee

1. At the annual meeting in April, the president asks the general membership to consider individuals who should be asked to run for the offices of vice president and member-at-large. Members are invited to send the names of individuals to the president.

2. The president also organizes a nominating committee. At AAGG, I asked three individuals—Steve Jones, Pam Souther, and Casey Morgan—to serve as the nominating committee. All agreed to serve. Casey Morgan agreed to chair the committee.

Note: Former officers of the organization are good choices for the nominating committee. They know what is involved in executing the duties of each office. I attempted to select a nominating committee of individuals from different regions in the United States. With that point in mind, you could select four members of the committee, but more than four makes the committee unwieldy.

3. The nominating committee must recommend a minimum of six nominees for three member-at-large positions. Eight nominees will be fine, too. Three vice presidential nominees should be chosen. The chair of the nominating committee should ask each candidate for a one-paragraph (150-word) biographical profile.

Step 2: Selecting Nominees for Offices

1. In selecting candidates, you will want to consider several factors:

    - Interest in AAGG and Its Activities. Look for people who have shown interest in the organization: attendance at regional and national meetings, people who engage in conversation on the listserv, people who make solid suggestions.

    - Interest in Geography and/or Geoscience. AAGG needs individuals who are effective spokespersons for both as fields of teaching and research, i.e., individuals who are effective and prolific researchers; people who are in charge of programs; people who are regular presenters at meetings.

    - Record of Dependability. VERY IMPORTANT! Try to determine how reliable an individual is. Stay in touch with the nominating

continued

Gabrielle Dawson   -3-   May 12, 2014

committee. Once you get a list of individuals you are considering, please post the list on the Executive Committee listserv. If any of the potential nominees has a track record of nonperformance, this is the time to determine that! Many people are "all talk" but slow on performance.

Note: Choose the VP very carefully. This is the person who will succeed you as president of AAGG. Because the president and vice president have so much responsibility and because they will need to work as a team, be sure you choose someone with whom you can work and someone who will continue to strengthen the organization. AAGG has very strong individuals in its thirty-year history. Each group of officers must be committed to developing AAGG further.

Step 3: Developing the Ballot

1. Be sure to let Casey Morgan know when you want the final list of names. I would suggest that you ask for the list by the second week in September.

2. Contact each individual whose name was submitted by the nominating committee. Be sure that each person has agreed to be nominated and understands what is involved.

3. Develop the ballot and a letter that will be sent along with the ballot.

4. Send the ballot and the letter to Dan Jameson. He will make copies. Try to get this material sent to Dan no later than the third week in September.

Step 4: Monitoring the Election Process

1. Once he has the materials, Dan will duplicate and mail the ballots to the membership. He knows the procedure: each ballot has a self-addressed, stamped return envelope, to ensure better response.

2. Dan will want all ballots returned by early January. He will count the ballots and report the results to you. As soon as you have the results, please call—rather than e-mail—each candidate and report the results.

Step 5: Preparing for the Annual Meeting

1. Plan to introduce each of the officers at the meeting. After my farewell speech, I will hand the reins of the organization to you.

2. To help prepare for the meeting, discuss your agenda as president with the incoming VP. I would also suggest that you plan an informal gathering of all the new officers BEFORE the Executive Committee meeting. At the EC, as you are aware, we always have several hours of business to conduct. Meeting with the incoming officers will give you planning time with them.

Gabrielle Dawson   -4-   May 12, 2014

3. Be prepared to tell the general membership at the annual meeting your
goals for the next two years. The newly elected VP and members-at-
large should be allotted time to express their concerns. Usually we allot
fifteen minutes for this segment of the annual meeting.

Final Thoughts

The election process has been developed to attempt to involve a variety of
individuals. The nominating committee should seek members who want to
become more involved in AAGG. In no sense does the election process seek
to exclude individuals other than those who are not interested in geography
and/or geoscience and the development of AAGG as an organization.

As you work with the elections, keep notes on ways that the EC can improve the election process. Then add these to the agenda for the 2010 meeting.

As always, if you have any questions, please call me.

Sincerely,

Jonathan

ONLINE INSTRUCTIONS

Online instructions need to follow the same readability considerations as those
used in letters, reports, and stand-alone instructions. Otherwise, the instructions
will create confusion for many users.

Case 10-4

A university explained how to download class rosters from the faculty e-mail system and insert these into an Excel spreadsheet (see Case Document 10-4A). Note how the application of principles discussed in this chapter improves the instructions (see Case Document 10-4B). The revision supplanted the original, which diminished the number of calls to the computing help desk from faculty who were confused by the original instructions.

CASE DOCUMENT 10-4A

Original Online Instructions

How exactly do I get my roster from NEO, put it into an Excel spreadsheet, and make a comma separated value (.csv) file for uploading?

1. Log into your neo account, go to the Class Roster tool, and go to the appropriate semester. Click on Download Class Roster, and you'll see your class roster as a .csv file. Copy and paste this file into a blank Excel spreadsheet and save. There's no way right now to actually download it.
2. To get rid of the quotes, go to Edit, then Replace. You want to find every " and replace it with a blank, so in the Find what box type " and put nothing into the Replace with box. Click on Replace All, and the quotes disappear.
3. Under File, select Save As option, and save the file as Formatted Text (space delimited)(.prn). Choose OK and Yes to the questions that pop up. Close the file, and answer no to the popup box. You don't want to save the changes.
4. Open the .prn file. The Text Import Wizard will pop up. Leave the file type as delimited and click next. Change the delimiter from tab to comma and click next. Holding the ctrl key down, click on the last four columns, and choose Do not import column (skip), then choose Finish.
5. Under File, select the Save As option. Change the file name to something that does not include the .prn, and save the file as CSV (Comma delimited) (.csv). Always answer yes to the question, "Do you want to keep the workbook in this format?" and when you close the file, answer no to "Do you want to save the changes?" Voila! This is the file the administrator needs.
6. To upload your students into your course, you only need the UIN numbers as a .csv file. So just delete the name columns and any rows above your data and save. Always answer yes to the question, "Do you want to

keep the workbook in this format?" and when you close the file, answer **no** to "Do you want to save the changes?" Upload your file according to the directions in CPR. [**top**]

Revised Online Instructions

How to Download Your Roster

1. Click on **download class roster**. Note that your class roster now appears as a comma separated value (.csv) file.
2. Without closing the roster on Neo, create an Excel spreadsheet.
3. Go back to your roster. Copy the names.
4. Paste the roster into the Excel spreadsheet just as you copied it.

Get rid of the quotes (" ").

1. Go to **Edit** on the toolbar, then to **Replace**. You want to find every " " and replace it with a blank.
2. In the **Find what** box, type " ", and put nothing into the **Replace with** box.
3. Click on **Replace All**, and the quotes disappear.

Prepare to Format the Information in the Excel Spreadsheet.

1. Under **File**, select **Save As** option, and save the file as **Formatted Text (space delimited) (.prn)**.
2. Choose **OK** and then **Yes** to the questions that pop up.
3. Close the file and answer **no** to the popup box.
4. Open the **.prn** file. The **Text Import Wizard** will pop up.

Select the Correct Boxes on the Wizard.

1. Leave the file type as **delimited**. Then click **next**.
2. Change the delimiter from **tab** to **comma**. Then click **next**. You will see the opening rows of your roster appear in neat columns.
3. Under **File**, select the **Save As** option.
4. Save the file as an Excel file. Be sure to save with a name descriptive of the contents.

With your file in an Excel spreadsheet, you can now create a grade or attendance sheet.

CHECKLIST FOR DEVELOPING INSTRUCTIONS/PROCEDURES ✔

Because instructions require careful analysis of readers from many perspectives—the context in which the instructions will be read and used—following a checklist can be helpful in ensuring that you have considered critical issues needed for the instructions.

☐ Who are your readers? Describe them in terms of their knowledge of the subject: educational level, technical level, responsibilities in the organization.

☐ What do you want them to be able to do as a result of these instructions/procedures?

☐ What is the situation that has led to the need for these instructions/procedures to be written?

☐ How will these instructions/procedures be used? Will readers need to read all of them before they begin the task? In what context will readers be using these instructions?

☐ What problems could readers encounter in attempting to use these instructions/procedures?

☐ What types of problems in safety and/or quality control do you need to emphasize? What warnings or notes will you need to include?

☐ What topics do you want to be sure to include/exclude?

☐ What format will you use: online, online to be printed, paper, manual, poster?

☐ Given the context in which the instructions/procedures will be read, what formatting strategies do you need to use to enhance accessibility?

☐ What types of visuals will you need to include?

☐ What is the basic outline of your instructions? Does this outline meet the needs of your readers? Will it achieve your purpose?

EXERCISES

1. You are employed by a small manufacturer of all-organic ready-to-eat breakfast cereals. The company would like to develop its sales overseas, especially in China. Because ready-to-eat cereals would be relatively new to the majority of the population in China, your package will have to include instructions on how to prepare and eat a serving of cereal. Compose step-by-step instructions, including rough illustrations, to fit on the side of the package.

2. Revise a set of instructions, from a course in your field, that you have found difficult to use, incomplete, or confusing. Submit both the original and your revision. Complete the prewriting analysis.

3. Working individually or collaboratively, find a set of instructions you have tried to use with limited success: for example, instructions for posting to a blog or wiki, for checking the security settings on your Internet browser, for operating a new mobile device, or for assembling the several components of a home entertainment system. Develop a slide presentation for your class explaining the deficiencies in the original instructions and the revisions you perceive as necessary.

4. Working individually or collaboratively, revise the document on Laser Toys (Figure 10–1). Before you begin, answer the following questions:

   What problems do you find with the instructions?
   How does the writer perceive the readers?
   What problems might readers have with this document?
   Assuming the document must be one page, what design changes would you make?

Consumer Health Information
www.fda.gov/consumer

Laser Toys: Not Always Child's Play

M any a kid (and parent) who has seen Luke Skywalker battle Darth Vader with a lightsaber thinks lasers are cool.

What they may not know is this: When operated unsafely, or without certain controls, the highly-concentrated light from lasers—even those in toys—can be dangerous, causing serious eye injuries and even blindness. And not just to the person using a laser, but to anyone within range of the laser beam.

The Food and Drug Administration (FDA) is particularly concerned about this potential danger to children and those around them, and has issued a draft guidance document on the safety of toy laser products.

According to Dan Hewett, health promotion officer at FDA's Center for Devices and Radiological Health, "A beam shone directly into a person's eye can injure it in an instant, especially if the laser is a powerful one."

However, laser injuries usually don't hurt, and vision can deteriorate slowly over time. Eye injuries caused by laser light may go unnoticed, for days and even weeks, and could be permanent, he says.

Some examples of laser toys are:
• lasers mounted on toy guns that can be used for "aiming;"

• spinning tops that project laser beams while they spin;
• hand-held lasers used during play as "lightsabers;" and
• lasers intended for entertainment that create optical effects in an open room.

FDA Regulates Lasers

A laser creates a powerful, targeted beam of electromagnetic radiation that is used in many products, from music players and printers to eye-surgery tools. FDA regulates radiation-emitting electronic products, including lasers, and sets radiation-safety standards that manufacturers must meet. Hewett explains that this includes all laser products that are marketed as toys.

Toys with lasers are of particular interest to the FDA because it's often children who are injured by these products, says Hewett. He notes that because advertisers promote them as playthings, parents and kids alike may believe they're safe to use.

"For toys to be considered minimal risk, we recommend that the levels of radiation and light not exceed the limits of Class 1, which is the lowest level in regulated products," Hewett says. Lasers used for industrial and other purposes often require higher radiation levels, he explains. But in toys, those levels are unnecessary and potentially dangerous.

In recent years, Hewett says, lasers have increased markedly in power and have gone way down in price. And while adults may buy a laser pointer for use in work, kids often buy them for amusement.

"Low-cost, compact laser pointers

used to be quite low in power," Hewett says; but, in the last 10 years, many laser pointers have increased in power 10-fold and more. The fact that lasers can be dangerous may not be evident, particularly to the children who use them as toys, or to the adults who supervise them.

Tips to Keep in Mind

• Never aim or shine a laser directly at anyone, including animals. The light energy from a laser aimed into the eye can be hazardous, perhaps even more than staring directly into the sun.
• Do not aim a laser at any reflective surface.
• Remember that the startling effect of a bright beam of light can cause serious accidents when aimed at a driver in a car or otherwise negatively affect someone who is engaged in other activity (such as playing sports).
• Look for a statement that it complies with 21 CFR (the Code of Federal Regulations) Subchapter J on the label.

"If you buy a laser toy or pointer and you don't see this information in the labeling, it's best not to make any assumptions about its safety," Hewett says. FDA

Find this and other Consumer Updates at *www.fda.gov/ForConsumers/ConsumerUpdates*

Sign up for free e-mail subscriptions at *www.fda.gov/consumer/consumerenews.html*

1 / FDA Consumer Health Information / U.S. Food and Drug Administration

AUGUST 2013

FIGURE 10-1 Document for Exercise 4

Oral Reports

Your ability to give oral reports can be as critical to your success as your ability to write reports. In today's work environment, knowing how to speak effectively and use presentation slides has become an increasingly valuable asset, whether you work in an academic or a nonacademic environment. In applying for a job or admission to a graduate program, you may need to give an oral presentation on a specific topic or perhaps on a research project you have conducted. As an employee, you may need to prepare a written document of an oral presentation you have given. Or an employer may ask you to make an oral presentation of a written document. Whatever the venue, you will benefit from knowing and then practicing basic strategies of developing effective oral presentations.

Quick Tips

- In the introduction, interest the audience in the subject, and enumerate the key points you will cover (e.g., "I will cover the following five points.").
- In the discussion section, help your audience pay attention and track your progress in your presentation by numbering each point as you cover it (e.g., "Now let's proceed to my second point.").
- When you reach the conclusion, announce it. You want your audience to pay extra attention. Audiences know that you will now summarize what you've said, and they can catch anything they missed earlier. Once you've said "in conclusion," you have about a minute before you must stop talking. Audiences grow hostile to speakers who promise to stop but then don't. In the remaining minute of your presentation, fix in the audience's mind the one or two ideas you want them to remember or take away from your presentation.
- Keep your presentation as short as possible. Nobody wants to listen to a long presentation.

UNDERSTANDING THE SPEAKING–WRITING RELATIONSHIP

Effective oral and written presentations share the following requirements:

- Analyze your audience. Know what they will expect.
- Articulate your purpose clearly.
- Develop sufficient and appropriate supporting material.
- Understand the context in which your presentation will be received.
- Organize the material so that your audience can follow your points easily.
- Choose a speaking style—level of language, approach to the subject, and tone—suitable to your role as well as your audience and purpose.
- Select the presentation format—text, text and visuals, text and sound and/or animations—that will enhance your audience's understanding of your message.

ANALYZING THE AUDIENCE

Analyzing your listening audience follows the same principles discussed in Chapter 2:

- How much does my audience know about the subject?
- How much do they know about me?
- What do they expect from me?
- How much interest will they have in what I say?
- What is their attitude toward me?
- What is their attitude toward my subject?
- What is their age group?
- What positions do they occupy in the organization?
- What is their educational background?
- What is their cultural/ethnic background?
- What is their economic background?
- What are their political and religious views?
- What kinds of biases will they likely have toward me and my topic?

To be an effective speaker, you must know your audience, establish an effective relationship between you and them by being sincere and knowledgeable about the subject, and then conform to their expectations about dress, demeanor, choice of language, and attitude toward them and the topic. When you speak to people from other countries, you should plan to research the culture of that country. Be aware that hand gestures you use routinely may have different meanings in other cultures. Also, the clothing you choose to wear should be selected with the culture of the audience in mind: you want your audience to pay attention to what you are saying instead of what you are wearing. If your audience includes individuals with limited hearing or vision, you will need to make appropriate accommodations.

Determining the Goal of Your Presentation

Oral presentations, like written presentations, must be designed for a specific purpose. By knowing what they will hear from the beginning of the presentation, audiences can understand the reason for the content. State your goal in one sentence as you begin to plan. Announce your purpose early in the presentation, to prepare your audience for the main ideas to come.

Remember that oral presentations, like written presentations, can enhance your reputation within an organization. Therefore, consider every speaking opportunity as a chance to sell not only your ideas but also your competence and your value to the organization.

Choosing and Shaping Content

Preparing the oral presentation often requires careful research:

- Determine what information you will need.
- Choose information that will appeal to your audience—particularly their attitudes and perspectives toward the topic.
- Consider a variety of information types: statistics, testimony, cases, illustrations, history, and particularly narratives that help convey the goal you have for your presentation. Be sure that every item you include pertains to the goal of your presentation.

Analyzing the Context

Analysis of context and analysis of audience are often inseparable:

- What broader concerns underlie the need for the presentation?
- What immediate issues underlie the presentation?
- How does your presentation relate to these issues?
- What will be happening in the organization when you make your presentation?
- How does your presentation fit into the organizational situation?
- If you are one of several speakers, what kinds of presentations will other speakers make?
- In what surroundings will you make the presentation?

Choosing the Organization

Like written communications, oral presentations must reflect your audience's needs and perspective:

- Does your audience have interest in what you will say?

- What main ideas do you want to convey?
- Based on your purpose and the needs and expectations of the audience, in what order should you present these ideas?

Answers to these questions will help you decide how to go about organizing your presentation. Generally, however, like the written presentation, oral presentations have distinct parts—an introduction, a main body, and a conclusion. The introduction should clearly tell the audience what the presentation will cover so that the audience is prepared for what is to come. The body should develop each point stated in the introduction, and the conclusion should reiterate the ideas presented and reinforce the purpose of the presentation.

As you do your research and collect content for your presentation, begin organizing your information. Divide your content into main categories of ideas:

    I. Idea 1
    II. Idea 2
    III. Idea 3, etc.

Once you have the main content "chunks" or divisions, begin to subdivide each main idea. Order the subdivisions so that the information moves in a logical sequence:

    I. Idea 1
        A, B, C . . .
    II. Idea 2
        A, B, C . . .
    III. Idea 3
        A, B, C . . .

Introduction. Be sure that you state your goal near the beginning. Even if you use some type of opening statement to interest your audience, state the goal of your presentation next. Then, state how you will proceed in your presentation: what main issues you will discuss. The main ideas should be announced and enumerated here.

Main body. Here you must explain each of your key ideas in order. Also try to include at least one example or anecdote for each point to make your ideas more vivid and memorable. If you make your presentation easy for your listeners to follow and understand, the more they (and you) will profit from your presentation.

Conclusion. At a minimum, restate the main issues you want your audience to remember, but be concise. Nothing is worse than a conclusion that drags on and doesn't conclude. Tell your audience exactly what you want them to remember from your presentation. Try to find a concluding narrative or statement that impacts your audience. The conclusion should leave the audience with a positive feeling about you and your ideas.

Choosing an Appropriate Speaking Style

How do you sound when you speak? You may have effective content and ideas and accurate and supporting statistics. However, how you speak your ideas can "make or break" the effectiveness of what you say. Avoid sounding patronizing, rude, overly solicitous, gushing, or insincere. Use a conversational style: short sentences, concrete language, speech that suggests to your audience that you are really talking with them. Remember that your organizational role and your relationship to your audience will dictate the tone and degree of formality of your presentation. Avoid speaking too rapidly.

- Do they know you?
- Is your rank in the organization above or below them?
- Are you speaking to an audience of individuals from all levels within the organization?
- What demeanor, approach, and level of formality does the organization usually expect from those giving oral presentations?
- Is the audience composed of people who understand American English? How well do they understand American English?

If you are speaking before a group composed largely of people from another country, determine beforehand their fluency in American English. If they are not comfortable with American English, speak more slowly, avoid idiomatic expressions, choose concrete words, and speak in relatively short sentences. Limit each sentence to one idea.

Choosing Visuals to Enhance Your Purpose and Your Meaning

Because we live in a time when communication is visual and verbal, illustrations can make your ideas more persuasive and more professional. Many of the guidelines for using visuals in oral presentations mirror those for written documents: they need to exemplify simplicity and clarity and fit the needs of the audience.

Formal presentations should use PowerPoint or similar presentation software to help listeners follow the ideas. PowerPoint allows you to give your listeners the outline of your presentation and insert pictures, graphs, tables, drawings, photographs, diagrams, and flow charts as well as sound and video. But because these will be seen while the audience is listening to you, be sure that all visuals are as simple and as easy to read as possible. The slide presentation in Figure 11–1 exemplifies effective use of the following guidelines:

- Limit the information on any single slide.
- Use a typeface that can be easily read.
- Use a nondistracting background that does not overpower content. Figure 11–4 shows a simple but effective background.

- Limit the typefaces you use to two per visual.
- Avoid using all caps.
- Use a typeface—size and font—that contrasts distinctly with the background.
- Don't crowd the information on your slide: keep it simple.
- Keep your presentation as short as possible; don't use too many slides.

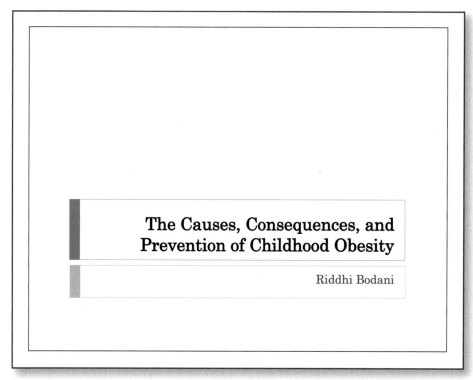

FIGURE 11–1 A PowerPoint Presentation Prepared for a Brown-Bag Luncheon

Purpose

▸ To understand recent developments and research concerning childhood obesity

▸ To outline major causes of obesity in children

▸ To examine consequences of childhood obesity

▸ To identify effective prevention measures

▸ 2

Defining Childhood Obesity

▸ Body Mass Index (BMI)

    ▸ Overweight- BMI in 85th – 95th percentile

    ▸ Obese- BMI > 95th percentile

▸ Defined differently in adults

▸ Controversial as children undergo rapid developmental changes

▸ 3

FIGURE 11-1 *continued*

Prevalence in the United States

▸ Increasing rates of childhood obesity

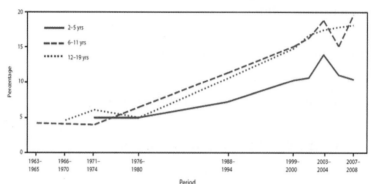

http://www.cdc.gov/mmwr/preview/mmw
rhtml/mm6002a2.htm

▸ 4

Prevalence in Developing Nations

▸ Most overweight or obese children are from
developing nations

▸ Attributed to shift in diet

▸ "Double burden of malnutrition"
  ▸ Infectious diseases and chronic conditions

▸ 5

FIGURE 11-1 *continued*

Genetic Factors

▸ Strong correlation between parental and childhood obesity
▸ Genes associated with obesity:
    ▹ Fat mass and obesity associated *(FTO)* gene
    ▹ Melanocortin-4 receptor *(MC4R)* gene
▸ Influence growth and development at an early age
▸ Twin studies

▸ 6   http://www.zmescience.com/

Environmental Factors

▸ Prenatal and Postnatal Influences
    ▹ Maternal smoking, gestational diabetes and lack of sleep as infant
▸ Poor Nutrition
    ▹ Carbonated beverages
    ▹ Fast food-consumption

▸ Lack of physical activity

▸ Lack of sleep
    ▹ Associated with high ghrelin levels and low leptin levels

▸ 7

FIGURE 11–1 *continued*

Environmental Factors

▸ School Environment
  ▸ Promote junk food consumption
  ▸ Contracts with soda companies pay for school facilities
  ▸ Increases in portion sizes

▸ Television advertising

▸ Low socioeconomic status (SES)
  ▸ Fast food is a cheap alternative
  ▸ Less opportunity for physical activity

▸ Quality of home

▸ 8

Health Concerns

▸ Increased prevalence of chronic conditions among children
  ▸ Cardiovascular problems
  ▸ Type-2 diabetes mellitus
  ▸ Osteoarthritis
  ▸ Depression

http://mathiesenclinic.com/

▸ 9

FIGURE 11–1 *continued*

Other Concerns

▸ Economic Concerns
  ▸ Obesity adds 9.1% to annual health care costs in U.S.
  ▸ Direct Costs: physician visits, medications, etc.
  ▸ Indirect Costs: morbidity and mortality costs

▸ Social Stigmatization
  ▸ Increased since 1961
  ▸ Study by Latner

▸ 10

Role of Parents in Prevention

▸ Mothers should maintain a healthy weight before and during pregnancy

▸ Children model eating behavior of parents

▸ Parents can limit the consumption of sodas and fruit juices

▸ Incorporate more milk, fruits, and vegetables into children's diet

▸ 11

FIGURE 11-1 *continued*

Role of Schools in Prevention

▸ Nutrition education programs

▸ Restricting the sale of soft drinks

▸ Placing nutritional standards on foods sold in vending machines

▸ Promoting physical activity

   ▸ Sports

   ▸ P.E.

   ▸ Recess

http://pickens.clemson.schoolfusion.us/modules/cms/announc eArchive.phtml

▸ 12

Conclusions

▸ Developing nations have high rates of obesity

▸ Increases in soda consumption and fast food intake have increased simultaneously with rates of childhood obesity.

▸ Conditions previously only associated with adults are now prevalent among children.

▸ Depression and obesity in adolescents are linked.

▸ Parents and schools can play the biggest role in preventing childhood obesity.

▸ 13

FIGURE 11-1 *continued*

References

▸ Ogden, C., Carroll, M., Kit, B., Flegal, K. (2012) Prevalence of obesity and trends in body mass index among U.S. children and adolescents, 1999-2010. *J Am Med Assoc* 307:483-490.

▸ Food and Agriculture Association of the United Nations. (2006) Fighting hunger and obesity.
<http://www.fao.org/Ag/magazine/0602sp1.htm>.

▸ Palca, J. (2007) Scientists identify a gene linked to obesity.
<http://www.npr.org/templates/story/story.php?storyId=9545915>.

▸ Child obesity. (2012). *Harvard School of Public Health*
<http://www.hsph.harvard.edu/obesity-prevention-source/obesity-trends/global-obesity-trends-in-children/index.html>.

▸ Latner, J., Stunkard, A. (2003) Getting worse: the stigmatization of obese children. *Obesity Research* 11:452-456.

▸ I4

FIGURE 11-1 *continued*

Other problems that can surface with slide presentations:

- Avoid "busy" or complicated visuals, as in Figure 11–2.
- Avoid making your audience study your slides. Figure 11–3, for example, will require your audience to study the map. If they have to focus to decipher your slides, they will not listen to you. Bar graphs, circle graphs, simple diagrams, and pictures exemplify standard types of visuals. Whatever visual you decide to use, limit content to the concept, data, or point you are trying to make. Figure 11–8 shows a good example of a line graph.
- Do not read your slides to your audience. Let them read each slide as you make it available, then begin to comment. Don't look at your slides as you speak. Look at your audience.
- Be sure that what the visual says is immediately evident, a requirement that Figures 11–2 and 11–3 fail to meet.

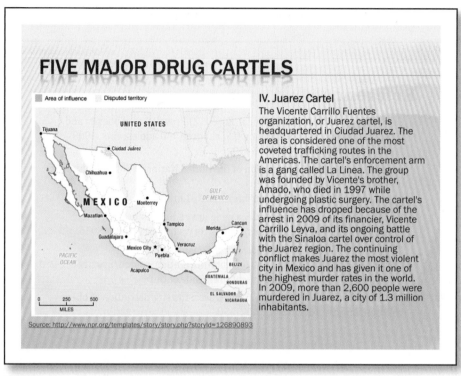

FIGURE 11-2 This slide, from a presentation describing the problems that drug cartels create for Texas, includes too much descriptive text to explain the map.

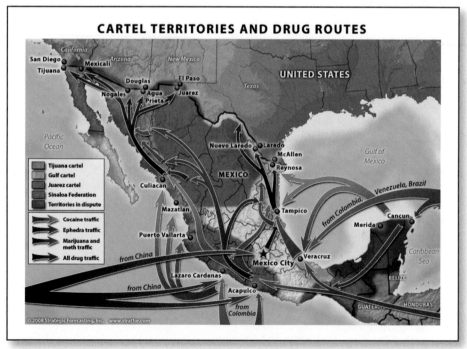

FIGURE 11-3 A Map Selected from an Online Source The quantity of information included could be daunting in a slide but helpful in a report, in which readers have time to study the information.

Avoid using too many slides. Listeners will grow restless if you talk too long and inundate them with too many slides. The nutrition major who prepared Figure 11–1 had been asked to discuss the obesity problem in children during an elementary school workshop. Readers can grasp the ideas easily.

As shown in the next single-slide examples, PowerPoint allows you almost unrestricted use of color and background, but too much color or an ornate slide background can distract your audience. If you need to prepare slides or transparencies for video conferencing, use the plain background and a typeface color that contrasts with the background to enhance readability. The organization for which you work may have a template for their employees to use for all company presentations.

Figure 11–4 exemplifies effective use of a background that does not distract but supports the content of the slides.

Figure 11–5 exemplifies a three-dimensional pie chart, created with Excel and transported into PowerPoint. Viewers have to connect the legend with the segments and read the percentages on each segment. This chart could become more readable if each segment were labeled along with the percentages (see Figure 11–6). As shown in Figure 11–7, the boxed labels improve readability.

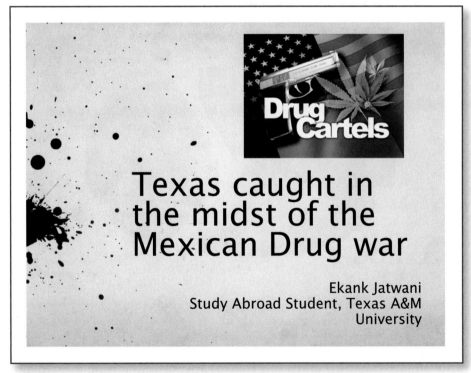

FIGURE 11-4 This slide exemplifies effective use of a background that does not distract but supports the content of the slide. The sound of the pistol firing occurs when this page appears.

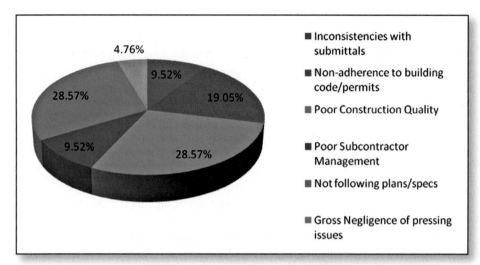

FIGURE 11–5 This slide displays a three-dimensional pie chart created with Excel and transported into PowerPoint. Use of a legend creates difficulties for listeners.

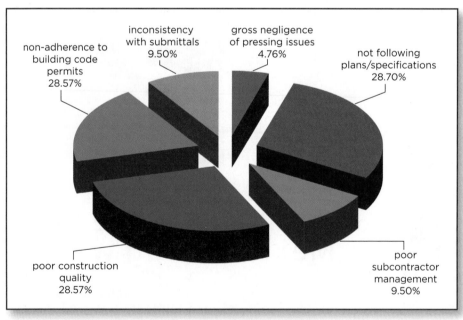

FIGURE 11–6 Figure 11–5 reconfigured to place segment labels outside the graph. The exploded view helps readers see the segment sizes.

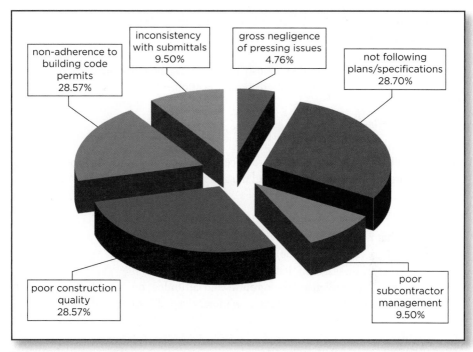

FIGURE 11-7 Use of Boxed Labels

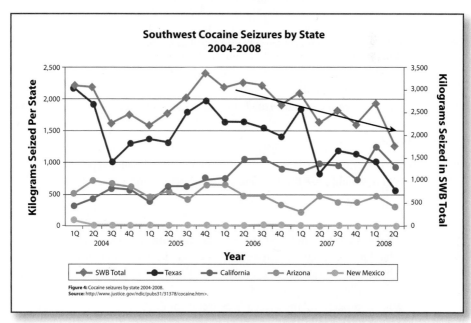

FIGURE 11-8 Effective Slide with a Line Graph

PLANNING YOUR PRESENTATION—QUESTIONS YOU NEED TO ASK

Analyze each point listed previously by answering the following questions, just as you did in planning your written communication.

Audience.

- Who is my audience?
- What do I know about my audience—background, knowledge, position in the organization, attitudes toward me and my subject?

Purpose.

- What is my purpose in giving this oral presentation?
- Is there (should there be) a long-range purpose?
- What is the situation that led to this presentation?
- Given my audience's background and attitudes, do I need to reshape my purpose to make my presentation more acceptable to my audience?

Context.

- At what location will I speak?
- What events have occurred or may transpire in the organization (theirs or mine) that may affect how my audience perceives what I say?

Content.

- What ideas do I want to include and not include?
- Based on the audience and the context, what difficulties do I need to anticipate in choosing content?
- Can any ideas be misconstrued and prove harmful to me or my organization?

Graphics.

- What kinds of visuals will I need to enhance the ideas I will present?
- Where should I use these visuals in my presentation?

Style.

- What kind of tone do I want to use in addressing my audience?
- What kind of image—of myself and my organization—do I want to project?
- What level of language do I need to use, based on my audience's background and knowledge of my subject?
- What approach will my audience expect from me?
- How formal should I be?

Speaking to Multicultural Audiences

As organizations become more international, you may find that you need to give presentations to groups in other countries. Because you will want any audience to respond positively to your presentation, you will need to do research to understand how people from other cultures will likely interpret what you say, how you say it, how you dress, and how you act in your dealings with them. The visuals you use may also have to be changed as symbols in one culture may have an entirely different meaning in another culture.

In your library and online, you will find a list of resources and books that can help you understand the perspectives of international audiences. As you consider your audience and the content you want to present to this audience, remember that your understanding of the cultural profile of your listeners is perhaps as important as your correctly discerning their knowledge of your topic and their interest level.

Designing Each Segment

The oral presentation has to have thoughtful structure for one main reason: once you have said something, the audience cannot "rehear" what you said. In reading, when you do not understand a sentence or paragraph, you can stop and reread the passage as many times as necessary. To help your audience follow what you say, you must design your presentation with your audience's listening limitations in mind.

Audiences generally do not enjoy long presentations. Listening is difficult, and audiences will tire even when a presentation is utterly fascinating. For that reason, look for ways to keep your message as concise as possible. Don't omit information your audience needs, but look for ways to eliminate nonessential material. Again, without carefully analyzing your audience—their attitude toward the subject, their background knowledge of the topic, their perspective toward you—you cannot decide either content or arrangement of content.

Choose an interesting title. Grab your audience's attention by developing a title that, at the very least, reflects the content of your presentation but does so in an interesting way. The title of an oral presentation should prepare your audience for the content you will present.

Develop your presentation around three main divisions. Helping your audience follow your message requires that you build into your structure a certain amount of redundancy. That means that you reiterate main points. In the introduction, you "tell them what you are going to tell them"; in the main body, you "tell them"; and in the conclusion, you "tell them what you told them." This kind of deliberate repetition helps your audience follow and remember the main points you are making. (Again, readers can "reread" text, but listeners cannot "rehear" oral remarks.) To design your presentation with planned repetition, you must clearly know your purpose and what you want your audience to know.

Plan the introduction carefully. During the introduction you focus your audience's attention on your theme and the way you plan to present the theme. Unless the introduction is effective and interests the audience, you will have difficulty keeping your audience's attention. The effective introduction thus tells your audience how to listen, what to expect, and the path you will follow in presenting your message. You may also wish to introduce your topic with an attention-getting device: a startling fact, a relevant anecdote, a rhetorical question, or a statement designed to arouse your audience's interest. Again, the device you choose will depend on the audience, the occasion, and the purpose of the presentation.

Or if your audience is unreceptive or unfamiliar with the subject, you may want to include background material to help them grasp and process your main points:

1. Acknowledge that you perceive the problem that your audience has with you or your topic.
2. Establish a common ground with the audience—your points of agreement.
3. Attempt to refute (if you can do so efficiently) erroneous assumptions that you believe the audience may have about you or the subject.
4. Ask the audience to allow you the opportunity to present your information as objectively as possible.

Design the body. In the introduction you state the main issues or topics you plan to present. Thus, in designing the body of the presentation, you develop what you want to say about each of these main points or ideas. You may want to present your ideas in a chronological sequence, a logical sequence, or a simple topical sequence. This method will help your audience follow your ideas if you are giving an informative speech, an analytical speech, or a persuasive speech. The important point, however, is that you need to demarcate and announce each point in the body as you come to it so that your audience knows when you have completed one point and begun another.

Design the conclusion. The conclusion reinforces the main ideas you wish your audience to retain. How you design the conclusion will depend on your initial purpose. A strong conclusion is nearly as important as a strong introduction because both the beginning and the end will be the parts most likely remembered.

Choosing an Effective Delivery Style

Avoid speaking in a "written" style. Use phrases, and use a variety of sentence lengths. Avoid excessively long, complex sentences as listeners may have difficulty following your ideas. In general, keep your sentences short. If you concentrate on getting your point across by having a conversation with the audience, you will likely use a natural, conversational style. Many suggestions for clarity in writing also apply to clarity in speaking:

- Avoid long, cumbersome sentences. Long sentences can be as hard to hear as they are to read.
- Avoid abstract, polysyllabic words. Instead, use concrete language that your audience can visualize.
- Avoid jargon, unless you are sure that your audience will be readily familiar with all specialized terms.
- Use short, active-voice sentences.

TECHNIQUES TO ENHANCE AUDIENCE COMPREHENSION

Because your audience cannot "rehear" ideas, once you have stated them, look for ways to help your audience easily follow your ideas:

1. Be sure you clearly demarcate the beginning and end of each point and segment of your presentation:

   - Announce each main topic as you come to it so that your audience knows when you have completed one topic and are beginning the next one.
   - Allow a slight pause to occur after you have completed your introduction, and then announce your first topic.
   - After completing your final topic in the main body of your presentation, allow a slight pause before you begin your conclusion.

2. Speak slowly, vigorously, and enthusiastically. Be sure you pronounce your words carefully, particularly if you are addressing a large group or a multicultural group.

3. Use gestures to accentuate points. Move your body deliberately, to aid you in announcing major transition points. In short, avoid standing transfixed before your audience.

4. Maintain eye contact with your audience. Doing so helps you keep your listeners involved in what you are saying. If you look at the ceiling, the floor, or the corners of the room, your audience may sense a lack of self-confidence. Lack of eye contact also tends to lessen your credibility. In contrast, consistent eye contact enhances the importance of the message. By looking at your audience, you can often sense their reaction to what you are saying and make adjustments in your presentation if necessary.

5. Do not memorize your presentation. You don't want it to sound like something you are reciting to your listeners. You want it to seem like your turn in a conversation you are having with your listeners.

6. Do not read your slides to your audience. Allow them to see the information on your slide before you begin discussing the material or points on each slide.

7. Rehearse your presentation until you feel comfortable. Try walking around, speaking each segment, and then speaking aloud the entire presentation. Rephrase ideas that are difficult for you to say—these will likely be hard for your audience to follow. Be sure to time your presentation so that it does not exceed the time limit. Keep your presentation as short as possible.

8. If possible, record or videotape your speech. Listen to what you have said as objectively as possible. As you listen, consider the main issues of audience, purpose, organization, context, content, and style.

9. Listen for tone, attitude, and clarity. Is the tone you project appropriate for your audience and your purpose? Is each sentence easy to understand? Are you speaking too rapidly? Are the major divisions in your presentation easy to hear? Are any sentences difficult to understand?

10. Try not to provide the audience handout material before you begin. To do so encourages your audience to read rather than listen. If you must provide written material, be sure the material is coordinated with your presentation so that you keep your audience's attention on what you are saying as you are saying it.

11. When you use slides, tell the audience what they will see, show them the slide, give them time to digest what they are seeing, and then comment on the slide. Do not begin talking about another topic while a slide depicting a past topic is still showing. Remember: people cannot read and listen at the same time. Avoid using too many slides.

12. When you are planning your presentation, determine how you will handle questions. Prepare for questions your audience may ask and decide how you will answer each one. Again, unless you have analyzed your audience and the reason for your presentation, you will not be able to anticipate questions that will likely arise.

13. Keep the question and answer time moving briskly. Answer each question as concisely as possible, and then move to the next question. If you are faced with a difficult question, reword it or break it into several parts. Then answer each part. Be sure to restate questions if listeners in the back of the room are unable to hear them asked by those near the front.

DESIGNING AND PRESENTING THE WRITTEN PAPER

Papers presented at professional meetings are frequently read from written manuscripts if the material to be delivered is complex. These papers may then be published in the official proceedings of the professional society.

However, presentations may need to be read for other reasons:

1. A presentation that discusses company policy, a sensitive issue, or a topic that must be approved by someone in the organization before the presentation. In

situations like these, the presentation is carefully written and read from the approved, written manuscript to ensure accuracy.

2. A presentation that will be circulated or filed as documentation. In a situation like this one, a spokesperson may read a carefully prepared statement, particularly if a possibility exists that material may be misconstrued by those in the audience.

3. Inexperienced speakers who must deal with a difficult problem may be more comfortable reading from a prepared manuscript. With the manuscript in front of you on the lectern, you don't have to worry about losing your train of thought or forgetting important details.

Written presentations can be effective if the speaker plans and writes the presentation carefully and then utilizes a number of delivery techniques to enhance the effectiveness of the oral reading.

Structuring the written speech. The structure of the written speech is the same as that of the extemporaneous speech. The speech has three main parts: the introduction, the main body, and the conclusion. Each section should be structured like the extemporaneous speech. However, you will need to write each section completely. If you know that the speech will be published, you may wish to write it like an article for publication or a report and use headings and subheadings to reveal the content and organization of the speech.

Writing the speech. After you have designed the content of your paper and made final revisions in your ideas, you will need to give close attention to your sentences and paragraphs since you will be reading these directly from the page.

1. Be sure that you separate each section from the other sections. This means that each section should have an overview that clearly announces that the section is beginning. Each paragraph should also begin with a topic sentence that summarizes the content of the paragraph. In short, in writing a speech to be read, you are making a concerted effort to accentuate every device for revealing organization since your audience cannot stop and re-hear what you have just said.

2. Limit each section and each paragraph within sections to one idea. Watch length so that your audience will not lose track of the main idea you are presenting.

3. Avoid excessive detail.

4. Use enumeration to help your audience follow your main points and to know when one point has ended and the next point is beginning.

5. Avoid long sentences. Long sentences are as difficult (or more difficult) to hear and follow as they are to read.

6. Prune every sentence to make it as clear and concise as possible.

7. Use active voice whenever possible so that your sentences will preserve the natural quality of spoken language.

8. Type your presentation in a large type—12-point type or larger. Triple space and leave wide margins on each side of the page.
9. With a marker, draw a "break" line after the introduction, between each main point in the body, and before the conclusion.
10. Underline or highlight important phrases or sentences throughout the presentation.
11. Consider using visuals, even though these may not be published separately.

Again, choose visuals that will clarify any difficult or important points.

Practicing the presentation.

1. Read each sentence aloud. Rewrite sentences that are difficult for you to say.
2. As you practice reading the presentation, try to look directly at your audience and speak important phrases or sentences to the audience.
3. Use overviews and topic sentences to announce each major topic as you come to it. To further alert your audience to the beginning of a new point, pause briefly, look at your audience, then read your overview statement or topic sentence. If possible, try to speak these to your audience instead of reading them.
4. As you practice reading your presentation, continue to listen for any sentences or words that are difficult to articulate. Recast sentences and paragraphs that do not sound organized, logical, and clear. If possible, replace difficult words with others that are easier to speak.
5. As you read, speak slowly and enunciate clearly and distinctly.
6. Once you can read each sentence with ease and without haste, time your presentation to be sure that it does not exceed a time limit if you have been given one.
7. Read your speech into a recorder. Allow some time between recording and listening so that you can gain some objectivity. As you listen, check for sentences that are hard to follow. Listen for breaks between major sections and major points.

CHECKLIST FOR PREPARING ORAL REPORTS ✔

Audience

☐ Who is your audience?

☐ What do you know about your audience—background, knowledge, position in the organization, attitudes toward you and your subject?

☐ What is the relationship between you and your audience?

☐ What is the attitude of your audience toward you and your presentation likely to be?

☐ Is your audience from a culture markedly different from yours? What adjustments to your presentation will any such differences require?

Purpose

☐ What is your purpose in giving this oral presentation?

☐ Is there (should there be) a long-range purpose?

☐ What is the situation that led to this presentation?

☐ Given your audience's background and attitudes, do you need to adjust your purpose to make your presentation more suitable for your audience?

Context

☐ Where and when will you be speaking?

☐ What events will be transpiring in the organization (theirs or yours) that may affect how your audience perceives what you say?

☐ What equipment, applications, and materials are available to you?
- Slideware?
- Audio?
- Video?
- Internet connection?
- Chalkboard?
- Flipchart and easel?
- Handouts?

continued

CHECKLIST FOR PREPARING ORAL REPORTS ✔ *continued*

Content

☐ What ideas do you want to include and exclude?
☐ Based on the audience and the context, what difficulties do you need to anticipate in choosing content?
☐ Can any ideas be misconstrued and prove harmful to you or your organization?
☐ Do you have a good opening that will interest your audience and create a friendly atmosphere?
☐ Have you limited your major points to fit within your allotted time?
☐ Does your talk contain sufficient examples, analogies, narratives, and data to support your generalizations? Have you repeated key points?
☐ Can you relate your subject matter to some vital interest of your audience?

Graphics

☐ What kinds of visuals will you need to clarify or reinforce the ideas you will discuss?
☐ Where should you use these in your presentation?
☐ Are these visuals immediately readable and understandable?
☐ Do they successfully focus the listeners' attention and augment and clarify your message?
☐ Do they meet the four criteria that govern good graphics?
　　· Visibility
　　· Clarity
　　· Simplicity
　　· Controllability

Style

☐ What kind of tone do you want to use in addressing your audience?
☐ What kind of image—of yourself and your organization—do you want to project?
☐ What level of language do you need to use, based on your audience's background and knowledge of your subject?
☐ What approach will your audience expect from you?
☐ How formal should you be?
☐ Which delivery technique will be more appropriate? Extemporaneous? Manuscript?

**CHECKLIST FOR
PREPARING ORAL REPORTS** ✔ *continued*

☐ If you are speaking extemporaneously, have you prepared a speech outline to guide you?

☐ If you will be reading from a manuscript, have you introduced a conversational tone into your talk? Is your typed manuscript easy to read from?

☐ Do you have a good ending ready, perhaps a summary of key points or an anecdote that supports your purpose? If you began with a story, do you want to go back to it now?

☐ Have you rehearsed your talk several times?

Questions and Answers

☐ What questions might the audience ask? Are you prepared to answer these questions?

☐ How will you keep your answers concise and direct?

☐ Do you have a specific time limit for questions? How will you keep track of your time?

EXERCISES

1. Examine Figure 11–9, a draft of a presentation about ethanol that a college agriculture engineering student will give to a high school FFA class. The speaker will read and glance at the written paper as he does not feel confident in speaking. He will also give the FFA teacher his written speech. Examine the speech in terms of the advice given above for writing a speech that you will read. Can you suggest ways to improve the style?

2. Prepare a five-minute/seven-slide presentation explaining to new majors in your field the job opportunities available to graduates of the program.

3. Prepare a slide presentation of your major semester project, including a copy of your slides to distribute to your audience as a pdf file for their mobile devices.

4. Your supervisor asks you to study the use of Facebook in the workplace and report your findings. He is pleased with your slide presentation (see Figure 11–10) and asks you to prepare a memo to all new hires stipulating the "rules" for using Facebook on the job. Adapt your slide presentation for this memo.

Ethanol in Fuel: Benefits and Problems

Introduction

Good morning! Mr. Bradshaw has asked me to explain to our FFA the benefits and problems of using gasoline that contains ethanol in engines of farm equipment. The abundance of corn and the increasing price of gasoline have made ethanol seem like the obvious solution to the energy crisis. However, when we look deeper into the issue, we find that ethanol has many flaws that possibly lower its value as a fuel alternative. Although ethanol does have benefits, the negative effects, from my perspective, overshadow the benefits.

Ethanol at a Glance—The Main Facts

- Gasoline that contains ethanol is named for the percentage of ethanol in the mixture (i.e., E10 contains 10% ethanol and E15 contains 15% ethanol).
- Congress has approved gasoline to contain up to 15% ethanol.
- Fermented plants such as corn and grain produce ethanol.
- Ethanol oxygenates gasoline, which causes the gasoline to combust cleaner and more completely.
- Those of you who live on farms which have corn as the main crop see ethanol as a financial windfall.

Keep in mind that ethanol has two sides.

Benefits of Ethanol

First, Ethanol burns cleaner than gasoline. As a result, ethanol, when added to gasoline, reduces harmful exhaust emissions. Experts claim that greenhouse gases released by an ethanol–gasoline mixture are 12% to 19% less than pure gasoline.[1] **Second**, ethanol supplements gasoline, which is often imported. Thus, ethanol advocates see it as a way to reduce dependence on foreign oil. Ethanol, on average, is cheaper than its equivalent volume in gasoline. **Third**, advocates see ethanol as a renewable resource, created from plants. Thus, many people see it as a solution to the energy crisis.[2] After ethanol is created, the remains of the corn are used for animal feed. Ethanol can be recycled into the agriculture industry.[1] **Fourth**, agriculture benefits from ethanol production because of the stability that the need for grains brings to the industry. Over the past few years, the price of corn has increased. This increase has a strong correlation to the increased use of ethanol in gasoline. The increase value of corn means more jobs and one way to improve the economy of America.

However, ethanol also has a downside:

FIGURE 11-9 Written Report for Exercise 1

Ethanol's Downside

Reduced Fuel Efficiency

Yes, ethanol can be used as an alternative to gasoline. However, the efficiency of ethanol as compared to gasoline is lower. Studies have shown that E10 and E15 have produced 3–5% less mileage than the equivalent volume of pure gasoline.[3] This trend is magnified in vehicles with low fuel efficiency ratings. Recently, Consumer Reports tested a 2007 Chevy Tahoe with E85 and pure gasoline. The study showed a decrease from 14 mpg to 10 mpg, a 28.5% decrease, when E85 replaced gasoline.[2]

Corrosive Properties of Ethanol

Ethanol, an alcohol, is created through a distillation process.[4] Ethanol is considered a solvent, which means that it can wear away plastics, rubbers, and other materials.[2] This becomes a problem when the worn parts are inside an engine. These parts must be replaced regularly to keep the engine running at its full potential. Repairs can cost hundreds, even thousands, of dollars. The residue from the dissolved parts builds up on filters and valves and further reduces engine efficiency.[3] Seals, hoses, and even fuel tanks can weaken, causing interruptions in the fuel flow and, in worst-case scenarios, leaks of the highly flammable fuel. Older engines and parts are especially susceptible to the corrosive properties of ethanol because engine designers did not make these parts to withstand the damage ethanol could cause.[4]

Tendency to Attract Water

Alcohols have a tendency to attract and retain water, and ethanol is no exception. [3] When you store an ethanol–gasoline mix for long periods of time, the ethanol will attract all of the water in the tank. As a result, water will cause the mixture to separate. If the separated mixture is then used in the engine, the engine will not run properly, and damage could occur.

Conclusion

The information provided should raise concerns about the use of ethanol in mechanized farm equipment. Although the increased use of corn, to distill ethanol, has provided growth and stability in the agriculture industry, the harmful effects of ethanol could outweigh its positive effects. While ethanol reduces the release of greenhouse gases into the air, the loss of fuel efficiency requires more fuel to be burned. The result: release of more greenhouse gases. This factor counteracts any positive effects of ethanol. You need to compare reduction of gasoline price with the cost of engine repair.

continued

FIGURE 11–9 *continued*

Agriculture equipment companies are looking at ways to slow damage of ethanol to engine parts. Using pure unleaded gas offers another solution, but until congressional action occurs, ag engineers must focus on making engine parts resistant to the corrosive effects of ethanol. E10 causes the lowest amount of damage, but many states have already shifted to E20, which has proven extremely corrosive.

One further point: the price of engine repair also has to be considered when you figure the cost of ethanol.

Questions?

Sources

1. Nebraska Ethanol Board, http://www.ne-ethanol.org/facts/facts.htm.
2. "The Problem with Ethanol," *Highlands Today*, http://www2.highlandstoday.com/news/highlands-news/2008/jun/22/problem-ethanol-ar-311037/.
3. Bell Performance, "Bell Performance Fixes Ethanol and Gasoline Problems," http://www.bellperformance.com/fix-your-fuel-problems/ethanol-gasoline/.
4. Energy Consumers Edge, "Problems with Ethanol Fuel," http://www.energy-consumers-edge.com/problems-with-ethanol-fuel.html.

FIGURE 11–9 *continued*

Facebook

Dalton Jones

4

Background

- Facebook is a social networking site where users can interact with friends and family through messaging, posts, pictures, and much more

- Users can choose to have their profile selectively or completely private

- Users can also interact with fan pages that are made by companies or certain groups

FIGURE 11–10 Document for Exercise 4

5

Does Facebook Fit Our Needs?

- Is Facebook a useful tool to review applicants or employees?
  - Is the company following the law when retrieving this information?
- What should the company policy be inside and outside the workplace?
- Should the company have a Facebook page?

6

Part of the Community

- "If you can use social networking sites as ways to connect to real people ... your business will do better"
- The company can respond to the public's complaints or concerns in real time
- Having positive comments is a good way for customers to refer us to their friends
- Do we have a big enough fan base to make a Facebook page?[3]

FIGURE 11–10 *continued*

E-Professionalism

- Employees should also be aware of privacy settings
  - Younger age groups have been shown to be more proactive in changing their privacy settings[4]
- "Companies should adopt a social media policy to address use."[1]
  - "The safest legal path is to have a policy that clearly sets forth prohibited posts"[1]
- Problems with treating Facebook as a casual media site are very similar to the early days of e-mails[1]

Monitoring

- Legal repercussions may occur if discrimination is taking place
- Some states have made it illegal to terminate an employee for "lawful, off-work conduct" [1]
- Policy evolves, and we need to keep updated on it
- The main concern of intellectual property being leaked might be a good enough reason to monitor employees[1]

FIGURE 11–10 *continued*

9

Conclusion

- You must take many issues into account when making these choices
- Legally monitoring employees will help maintain the company's reputation
- Meetings that go over company guidelines will be needed
- Without a public consumer base, the company doesn't need a Facebook page to receive customer feedback

---

10

Small Tip

- The default settings on Facebook don't include secure browsing
- Why?
  - Facebook applications can't assure a secure network
  - Loading times are longer
- If you aren't on a secure connection, your login information could be compromised
- Is your company password the same as your Facebook password?[2]

FIGURE 11-10 *continued*

Works Cited

- [1]Dryer, Randy L. "Advising Your Clients (and You!) in the New World of Social Media: What Every Lawyer Should Know about Twitter, Facebook, YouTube, & Wikis." Utah Bar Journal 23.3 (2010): 16-21. Web.

- [2]"How to Enable Secure Facebook Browsing." Quick Online Tips. N.p., 29 May 2011. Web. 06 Mar. 2013. <http://www.quickonlinetips.com/archives/2011/05/secure-facebook-browsing/>

- [3]Janusz, Ted. "Marketing on Social Networks: Twitter, MySpace and Facebook Demystified." Key Words 17.4 (2009): 124-5. Web.

- [4]Osman, Ahmed, Andrew Wardle, and Richard Caesar. "Online Professionalism and Facebook: Falling through the Generation Gap." Med Teach 34.8 (2012): e549-56. Web.

FIGURE 11-10 *continued*

Résumés and Job Applications

THE CORRESPONDENCE OF THE JOB SEARCH

A persuasive letter of application (sometimes called a *cover letter*) and a clear and organized résumé won't guarantee that you get a job, but ineffective ones will usually guarantee that you don't. Also essential is to make sure that nothing in your online profiles or postings at social networking sites such as Facebook, LinkedIn, and Twitter would raise questions in a potential employer's mind about your suitability for a job. Verify especially that details mentioned in your letter of application or listed on your résumé are identical to the information in your profiles (e.g., names of previous employers, dates of previous employment). Keep in mind that potential employers will almost certainly check your online profiles and postings because it is quick and easy to do so: how you characterize yourself here won't necessarily win you a job but could stop you from getting a job or from keeping a job.

Letter of application. Plan the mechanics of your letter of application carefully. Use a standard font like Times New Roman, and don't use italics or bold: you want the employer to focus on your abilities instead of typographical variations in your letter. Make sure you have no errors in spelling, punctuation, or grammar: you must come across as competent and meticulous. Be brief but not telegraphic. Keep the letter to a single page unless you have extensive pertinent experience to emphasize and explain.

More and more organizations direct job candidates to apply online for positions. If this is the case, you will go online to the organization's job site, enter your name and contact information, identify which job you are applying for, and upload your letter and résumé in the specified format. If no format is specified, submit your materials in one of the widely used formats (e.g., as a doc file, docx file, pdf file, or rtf file) so that it might be easily accessed by the recipient. Make the file read-only so that nothing will be later inserted or deleted.

If you are applying for jobs in the traditional manner, however, make sure to buy high-quality white bond paper and envelopes. This is no time to skimp. Don't send a letter or résumé that has been duplicated. Each copy of your letter must be individually addressed, printed, and signed. Accompany each letter with your résumé.

Pay attention to the style of the letter and the résumé that accompanies it. The tone you want in your letter is one of self-confidence. You must avoid both arrogance and timidity. You must sound interested and eager but not obsequious or desperate. Don't give the impression that you must have the job, but don't seem indifferent about getting it.

When describing your accomplishments in the letter and résumé, use action verbs. They make your writing concise and specific, and they make you seem dynamic and energetic. For example, instead of reporting that you worked as a salesclerk, explain that you maintained inventories, promoted merchandise, prepared displays, implemented new procedures, and supervised and trained new clerks. Here's a sampling of such action verbs:

administer	edit	oversee
analyze	evaluate	plan
conduct	exhibit	produce
create	expand	reduce
cut	improve	reorganize
design	manage	support
develop	operate	promote
direct	organize	write

You cannot avoid the use of *I* in a letter of application, but take the *you* attitude as much as you can. Emphasize what you can do for the prospective employer—how your getting this job will benefit the company. The letter of application is not the place to be worried about salary and benefits. Above all, be mature and dignified. Avoid tricky and flashy approaches. Write a well-organized, informative letter that highlights the skills you have that the company desires in its employees.

The beginning.   Start by explaining that you are applying for a job and by identifying the specific job for which you are applying. Don't be strident or inventive here. A beginning such as "WANTED: An alert, aggressive employer who will recognize an alert, aggressive young programmer" will usually direct your letter to the reject pile. If you can do so legitimately, a bit of name-dropping is a good beginning. Use this beginning only if you have permission and if the name you drop will mean something to the prospective employer. If you qualify on both counts, begin with an opening sentence such as this:

> Dear Mr. García:
>
> Professor Theresa Ricco of Nebraska State University's Department of Biochemistry has suggested that I apply for a research position in the virology division of your company. In June I will receive my bachelor of science degree in biochemistry from NSU. I have also worked as a research assistant in Dr. Ricco's lab for two years.

Remember that you are trying to arouse immediate interest about yourself in the potential employer.

Sometimes the best approach is a simple statement about the job you seek, accompanied by a description of something in your experience that makes you particularly fit for the job, as in this example:

> Your opening for a researcher in the virology division has come to my attention. In June of this year, I will graduate from Nebraska State University with a bachelor of science degree in biochemistry. I have also worked part-time for two years in a cell and molecular research laboratory at NSU. I believe that both my education and my work experience qualify me for this position.

Be specific about the job you want. Quite often, if the job you want is not open, the employer may offer you an alternative one. But employers are not impressed with general statements such as, "I'm willing and able to work at any job you may have open in research, production, or sales." Instead of indicating flexibility, such a claim usually implies that your skills and interests are unfocused—that you would be adequate at several things but truly exceptional at nothing.

In addition, make it clear in this opening that you know something about the company—its history, its achievements, its special projects, its reputation in the industry—anything that will demonstrate that you did your research and picked this company to apply to because something about it was impressive, something about it made you think you could make a key contribution to its success.

The body.   In the body of your letter you highlight selected items from your education and experience that show your qualifications for the job you seek.

Remember always that you are trying to show the employer how nicely you will fit into the job and the organization.

For this section, you need to know what things employers value the most. In evaluating recent college graduates, employers look closely at the major, academic performance, work experience, awards and honors, and extracurricular activities. They also consider recommendations, standardized test scores, military experience, and community service.

Try to include information that employers typically consider important, but emphasize those areas in which you are especially noteworthy. For example, if your grades are good, mention them prominently; otherwise, maintain a discreet silence about your academic record. Speak to the employer's interests, and at the same time highlight your own accomplishments. Show how it would be to the employer's advantage to hire you. The following paragraph does all of these things and is an excellent example of the *you* attitude in action:

> The advertisement for this position indicated that extensive interaction with international colleagues, especially from South America, would be expected. I would like to note that I studied Spanish and Portuguese in college and have a high fluency in both languages. I also visited Argentina and Brazil two years ago for a six-week intensive summer session. I am familiar with the major cities as well as the history and key traditions. I would thus be a confident and articulate representative of the company.

Be specific about your accomplishments; otherwise, it will sound like bragging. It is much better to write "I was president of my senior class" instead of "I am a natural leader." "I was president" is a piece of evidence; "I am a natural leader" is a conclusion. Your aim is to give employers the evidence that will lead them to the right conclusions about you.

One tip about job experience: The best experience relates to the job you seek, but mention any job experience, even if it is unrelated. Employers believe that a student who has worked at a job will probably be more mature and reliable than one who has not: you know what's involved in having a job and what's expected of you. If you have worked at a job, you are more likely to possess effective work habits and a real work ethic. If you have worked at a job, it also indicates that another employer thought you deserved to be hired.

Don't forget hobbies or skills that might relate to the job (e.g., you are training for your pilot's license while applying for engineering positions in the aviation industry). You're trying to establish that you are interested in, as well as qualified for, the position.

Don't mention salary unless you're answering an advertisement that specifically asks you to discuss your expectations. Keep the *you* attitude. Don't worry about insurance benefits, vacations, and holidays at this point in the process. Keep the prospective employer's interests in the foreground. Your self-interest is taken for granted. If you are offered the job, that's the time to inquire about such details if they are not readily provided to you.

If you are already working, you will emphasize your work experience more than your college experience. Identify your responsibilities and achievements on the job. Never complain about your present employer because you will cause a prospective employer to mistrust you and think of you as a complainer.

In the last paragraph of the body, refer the employer to your enclosed résumé. Mention your willingness to supply additional information such as references, letters of recommendation, writing samples, or college transcripts.

The ending. The goal of the letter of application is to get you an interview with the prospective employer. In your final paragraph, you must request this interview. Your request should be neither apologetic nor aggressive. Simply indicate that you are available for an interview at the employer's convenience, and give any special instructions needed for reaching you. If the prospective employer is in a distant city, indicate (if possible) a convenient time and location that you might meet with a representative of the company, such as the upcoming convention of a professional association. If the employer is really interested, you may be invited to visit the company at its expense.

The complete letter. Figure 12–1 shows a complete letter of application. The beginning of the letter shows that the writer has been interested enough in the company to investigate it. The desired job is specifically mentioned. The middle portion highlights the writer's pertinent studies and work experience relating directly to the job she is seeking. The ending makes an interview convenient for the employer to arrange.

Keep in mind that a hiring official skims your letter and résumé in about 30 seconds. If you have not grabbed his or her interest in that time, you are probably finished with that organization.

The résumé. A résumé provides your prospective employer with a convenient summary of your education and experience. As in the letter of application, good grammar, correct spelling, neatness, and brevity—ideally, only one page—are of major importance in your résumé.

The two most widely used types are chronological and functional résumés. Each has advantages and disadvantages.

Chronological résumés. The advantages of a chronological résumé (Figure 12–2) are that it's traditional and acceptable. If your education and experience show a steady progression toward the career you seek, the chronological résumé portrays that progression well. Its major disadvantage is that your special capabilities or accomplishments may sometimes get lost in the chronological detail. Also, if you have holes in your employment or educational history, they show up clearly.

Put your address at the top. Include your e-mail address and your telephone number (including the area code). If you have a fax number, include it.

For most students, educational information should be placed before work experience. People with extensive work experience, however, may choose to put it before their educational information.

List the colleges or universities you have attended in reverse chronological order. Do not list your high school. Give your major and date (or expected date) of graduation. Do not list courses, but list anything that is out of the ordinary, such

Letter of application

300 East 5th Street
Long Beach, CA 90802
January 23, 2014

Mr. Richard Hernandez
Office of Information Technology
Price Insurance Corporation
2163 Artesia Drive
St. Louis, MO 63136

Dear Mr. Hernandez:

Identifies specific job and demonstrates knowledge of company

I am writing to apply for the position of computer security analyst listed in the March issue of *IT Professionals*. Price Insurance has a growing reputation for innovative security solutions, and I believe my education and experience would contribute to your efforts.

Highlights education

I am graduating in May of this year from the University of California at Long Beach with a bachelor's degree in computer science. I have completed courses in software verification and validation, fault-tolerant computer systems, web-based software systems, analysis of algorithms, communication networks, database management systems, and human–computer interaction.

Highlights work experience

In addition, I have a year of experience as a programmer and database manager at a local Internet service provider. I know how to take a project from application of principles through risk analysis, resource management, and adaptation of improvement techniques to delivery and billing. I have also worked in computer support, developing both my technical skills and my ability to interact courteously and effectively with customers.

Refers to résumé

You will find more detailed information about my education and work experience in the résumé enclosed with this letter. I could also supply descriptions of the courses I have taken, copies of the projects completed in those courses, and a complete list of my job duties at each place of employment.

Requests interview

In April, I will attend the regional meeting of the Society for IT Professionals in St. Louis. Would it be possible for me to visit with a representative of Price Insurance at that time?

Sincerely yours,

Gillian Woods

Gillian Woods

Enclosure

FIGURE 12-1 Complete Letter of Application

as honors, special projects, and emphases in addition to the major. Extracurricular activities also go here if they are pertinent to the job you seek.

As you did with your educational experience, put your work experience in reverse chronological order. To save space and to avoid the repetition of *I* throughout the résumé, use phrases rather than complete sentences. As in the letter of application, emphasize the experiences that show you in the best light for the job you seek. Use nouns and active verbs in your descriptions. You may include less important jobs (such as from high school), but summarize these briefly unless they are especially pertinent to the position you seek. You would probably put college internships and work–study programs here, though you might choose to put them under education. If you have military experience, put it here. Give the highest rank you held, list service schools you attended, and describe your duties. Make a special effort to show how your military experience relates to the civilian work you seek.

You may wish to provide personal information. Personal information can be a subtle way to point out your desirable qualities and could build rapport with a prospective employer during interviews if he or she shares your interests. Recent travels indicate a broadening of knowledge and probably a willingness to travel. Hobbies listed may relate to the work sought. Participation in sports, drama, or community activities indicates that you enjoy working with people and are civic-minded. Cultural activities indicate you are a person of wide interests. Avoid mentioning activities that could be considered controversial or dangerous. If you are proficient in a language aside from English, definitely mention this: more and more organizations are looking for employees with the potential to support international business activity.

If you choose to indicate that you are married or have a family, you might also emphasize that you are willing to relocate. Don't discuss your health unless it is to describe it as excellent.

You have a choice with references. You can list several references with addresses and phone numbers or simply put in a line that says "References available upon request." Both methods have an advantage and a disadvantage. If you provide references, a potential employer can contact them immediately, but you use up precious space that might be better used for more information about yourself. Conversely, if you don't provide the reference information, you save the space but put an additional step between potential employers and information they may want. It's a judgment call, but your first goal is to interest the potential employer in you. If that happens, then it will not be difficult to provide the reference information at a later time. If, however, one of your references is a prestigious individual in the discipline or industry, be sure to list your references on the résumé.

In any case, do have at least three references available. Choose from among your college teachers and past employers—people who know you well and will say positive things about you. Get their permission, of course. Also, it's a smart idea to send them a copy of your résumé. If you can't call on them personally, send them

Lawrence T. Morgan
ltmorgan@tamu.edu

Local Address   1221 Locust Street, College Station, TX 77840 (713) 515-8601
Permanent Address   2305 18th Street Lubbock, TX 79401 (806) 543-6039

EDUCATION   **Texas A&M University, College Station, TX**
                         Bachelor of Science Degree
                         Double Major: Biomedical Science and Entomology
                         Graduated: December 2014
                         GPR: 3.3/4.0

EXPERIENCE   **Texas A&M University, College Station, TX**
                         The History of Medicine Study Abroad, Germany,
                         Summer 2013
                         - Shadowed and observed a triple bypass surgery,
                           North Rhine Westfalia Heart and Diabetes Center
                         - Visited German veterinary cloning facilities in Bayer
                           Healthcare Target Research Facilities, Chemical
                           Research Facilities, and fermentation facilities
                         - Participated in research on "Nazi medicine"
                         St. Joseph's Hospital, Bryan, TX
                         Day Surgery Tower Volunteer, August 2011–present
                         - Assisted patients and their families while at the
                           hospital
                         - Aided doctors, nurses, and staff in various duties
                         International Student Volunteers Conservation Project
                         - Traveled to Australia to work with a team on trail
                           building in national forests, planting over 1,000 trees,
                           sages, and grasses and removing invasive species
                           from devastated swamp lands

ACTIVITIES   **Sigma Nu National Fraternity**
                         - Executive Council
                           ○ Candidate Marshall–Fall 2013
                           ○ Lieutenant Commander (Vice President) Spring
                             2012–Spring 2013
                           ○ Sentinel and Risk Reduction Fall 2011–Spring
                             2012

continued

FIGURE 12–2 Chronological Résumé

- Head of Philanthropy Committee 2012–2013, Recruitment Committee, Academic Chair–Spring 2012, Social Chair–Fall 2011, House Manager–Spring 2011, Candidate Class Social Chair–Fall 2010

Biomedical Science Association Fall 2010–present

Texas A&M Pre-Med Society Fall 2012

AWARDS   **Sigma Nu Fraternity**

- Commander's Award Fall 2013–Spring 2014
- Man of the Year Fall 2013–Spring 2014
- Scholar of the Year Fall 2012–Spring 2013
- Big Brother of the Year Fall 2011–Spring 2012
- **Study Abroad Fellowship & International Education Scholarship**
- **National Scholar's Honor Society–inducted Fall 2013**

FIGURE 12-2 *continued*

a letter that requests permission to use them as a reference, reminds them of their association with you, and sets a time for their reply:

Dear Dr. Singh:

In May of this year, I'll graduate from Waterville Polytechnic Institute with a BS in metallurgical engineering. I'm getting ready to look for a full-time job. If you believe that you know enough about my abilities to give me a good recommendation, I would like your permission to use you as a reference.

As a reminder, during the summers of 2012 and 2013, I worked as a laboratory technician in your testing facility at Waterville. They were highly instructive summers for me and reinforced my college studies.

I want to start sending my résumé out to some potential employers by March 1 and would be grateful for your reply by that time. I've enclosed a copy of my résumé so that you can see in detail what I've been doing.

Thanks for all your help in the past.

Sincerely,

At the bottom of the résumé, insert a dateline—the month and year in which you completed the résumé.

Functional résumés. A main advantage of the functional résumé (Figures 12–3 and 12–4) is that it allows you to highlight the experiences and abilities that show you to your greatest advantage. Extracurricular experiences show up particularly well in a functional résumé. The major disadvantage of this format is the difficulty, for the first-time reader, of deciphering the progression of your jobs and education.

The address portion of the functional résumé is the same as that of the chronological résumé. After the address, you may include a job objective line if you like. A job objective entry specifies the kind of work you want to do and sometimes the industry or service area in which you want to do it, like this:

Work in computer security.

or like this:

Work in computer security for a medical insurance corporation.

Place the job objective entry immediately after the address, and align it with the rest of the entries (as shown in Figure 12–3).

For education, simply give the school from which you received your degree, your major, and your date of graduation. The body of the résumé is essentially a classification. You sort your experiences—educational, business, extracurricular—into categories that reveal capabilities related to the jobs you seek. Remember that in addition to professional skills, employers want good communication skills and good interpersonal skills. Possible categories are technical, professional,

Lawrence T. Morgan
ltmorgan@tamu.edu

Local Address	**Permanent Address**
1221 Locust Street	2305 18th Street
College Station, TX 77840	Lubbock, TX 79401
(713) 515-8601	(806) 543-6039

EDUCATION   **Texas A&M University, College Station TX**
Bachelor of Science Degree
Double Major: Biomedical Science and Entomology
Graduated: December 2014   GPA: 3.3/4.0

LABORATORY QUALIFICATIONS

Collegiate Science Coursework

- General Chemistry w/lab I & II
- General Biology w/lab I & II
- Physics w/lab I & II
- Organic Chemistry w/lab I & II
- Biochemistry
- Microbiology w/lab
- Basic knowledge of diseases

Collegiate Entomology

- Identifying Species of Insects
- Collection of Insects
- Forensic Entomology
- Medical Entomology w/lab
- Veterinary Entomology w/lab
- General Entomology w/lab
- Insect Morphology w/lab

Other Skills

- Microsoft Office—Word, Excel, PowerPoint, and Access

EXPERIENCE

The History of Medicine Study Abroad, Germany, Summer 2013

- Shadowed and observed a triple bypass surgery, North Rhine Westfalia Heart and Diabetes Center

FIGURE 12-3 Functional Résumé

Lawrence T. Morgan   2

- Visited German veterinary cloning facilities in Bayer Healthcare Target Research Facilities, Chemical Research Facilities, and fermentation facilities

- Participated in research on "Nazi medicine"

St. Joseph's Hospital, Bryan, TX

Day Surgery Tower Volunteer, August 2011–present

- Assisted patients and their families while at the hospital
- Aided doctors, nurses, and staff in various duties

International Student Volunteers Conservation Project

- Traveled to Australia to work with a team on trail building in national forests, planting over 1,000 trees, sages, and grasses and removing invasive species from devastated swamp lands

ACTIVITIES   Sigma Nu National Fraternity
                        Coastal Conservation Association
                        Biomedical Science Association Fall 2010–present
                        Texas A&M Pre-Med Society Fall 2012

AWARDS

Sigma Nu Fraternity

- Commander's Award, Fall 2013–Spring 2014. The chapter president recognizes one exceptional individual within the chapter.

- Man of the Year, Fall 2013–Spring 2014. The fraternity recognizes the individual who has demonstrated excellence in the area of leadership.

- Scholar of the Year, Fall 2012–Spring 2013. The fraternity recognizes the individual who has demonstrated excellence in the area of scholarship.

- Big Brother of the Year, Fall 2011–Spring 2012. The fraternity recognizes an exceptional Big Brother in the chapter.

Study Abroad Fellowship and International Education Scholarship

National Scholar's Honor Society–inducted Fall 2013

References Furnished Upon Request

FIGURE 12–3 continued

team building, communication, research, sales, production, administration, and consulting.

To prepare a functional résumé, start by brainstorming. List some categories that you think might nicely describe your experiences. Brainstorm further by listing your experiences in each of those categories. Next, select the categories and experiences that make you a particularly compelling job candidate. Remember, you don't have to display everything you've ever done, just those things that might strike a potential employer as valuable. Finish off the functional résumé with a brief reverse chronological work history and insert a dateline, as in the chronological résumé.

For example, Lawrence's chronological résumé (Figure 12–2) is impressive, but he realized it would not help him in applying for a research assistant position in entomology that had been advertised on campus. He developed a functional résumé (Figure 12–3) and submitted it to the entomology professor who needed the research assistant. The functional résumé emphasized what Lawrence knew about entomology and research, which was exactly what the professor wanted, and Lawrence got the job. He finished that research position but needed a new job for the summer. Lawrence revised his résumé again (Figure 12–4), adding what he had learned from the first research position, and submitted it for the second position. This new job would require him to work with two professors, one in crop sciences and one in entomology. Both were impressed, and Lawrence was again hired.

Ordinarily, you will prepare your résumé as a traditional formatted document, but you might instead be asked to fill in fields at a job site online that will automatically generate your résumé according to a standardized template.

Traditional formatted résumés. As Figures 12–2, 12–3, and 12–4 illustrate, in a traditional résumé you use variations in type and spacing to emphasize and organize information. Make the résumé look easy to read—leave generous margins and white space. Use distinctive headings and subheadings. The use of a two-column spread is traditional (heading on the left and information on the right), as is the use of bold for headings. You might use a 12-point type like Arial for headings and a 10-point type like Times for the text. Be sparing, however, with the typographical variation: you want to organize the page of information visually without pulling the reader's attention away from your abilities. More and more often, however, résumés are prepared in a single-column format with headings above the pertinent information for each category: the single-column format accommodates screen readers for people with limited vision as well as scanning to a database and processing for keywords.

If you are printing your résumé, use a high-quality bond paper in white or off-white. Your letter and résumé should be on matching paper. If you are attaching the résumé as an electronic file to an e-mail message or uploading it to a job site, again make sure you submit it as a read-only file in a widely used format (e.g., doc, docx, pdf, or rtf) for easy access by the recipient.

Who is the recipient of your letters and résumés? When answering an advertisement, you should follow whatever instructions are given there. You might be

Lawrence T. Morgan
ltmorgan@tamu.edu

Local Address
1221 Locust Street
College Station, TX 77840
(713) 515-8601

Permanent Address
2305 18th Street
Lubbock, TX 79401
(806) 543-6039

EDUCATION **Texas A&M University, College Station TX**
Bachelor of Science Degree
Double Major: Biomedical Science and Entomology
Graduated: December 2014 GPA: 3.3/4.0

LABORATORY EXPERIENCE & QUALIFICATIONS

Texas A&M University, Department of Entomology–Dr. Michel Slotman

- Polymerase Chain Reaction
- Gel Electrophoresis
- Ligation & Cloning
- DNA Extraction
- Basic Lab skills and safety

Collegiate Science Coursework

- General Chemistry w/lab I & II
- General Biology w/lab I & II
- Physics w/lab I & II
- Organic Chemistry w/lab I & II

- Biochemistry
- Microbiology w/lab
- Basic knowledge of diseases

Collegiate Entomology

- Identifying Species of Insects
- Collection of Insects
- Forensic Entomology
- Medical Entomology w/lab

- Veterinary Entomology w/lab
- General Entomology w/lab
- Insect Morphology w/lab

Other Skills

- Microsoft Office–Word, Excel, PowerPoint, and Access

EXPERIENCE

The History of Medicine Study Abroad, Germany, Summer 2013

- Shadowed and observed a triple bypass surgery, North Rhine Westfalia Heart and Diabetes Center

continued

FIGURE 12–4 Functional Résumé

Lawrence T. Morgan   2

- Visited German veterinary cloning facilities in Bayer Healthcare Target Research Facilities, Chemical Research Facilities, and fermentation facilities

- Participated in research on "Nazi medicine"

St. Joseph's Hospital, Bryan TX

Day Surgery Tower Volunteer, August 2011–present

- Assisted patients and their families while at the hospital

- Aided doctors, nurses, and staff in various duties

International Student Volunteers Conservation Project

- Traveled to Australia to work with a team on trail building in national forests, planting over 1,000 trees, sages, and grasses and removing invasive species from devastated swamp lands

ACTIVITIES   Sigma Nu National Fraternity
                        Coastal Conservation Association
                        Biomedical Science Association Fall 2010–present
                        Texas A&M Pre-Med Society Fall 2012

AWARDS

Sigma Nu Fraternity

- Commander's Award, Fall 2013–Spring 2014. The chapter president recognizes one exceptional individual within the chapter.

- Man of the Year, Fall 2013–Spring 2014. The fraternity recognizes the individual who has demonstrated excellence in the area of leadership.

- Scholar of the Year, Fall 2012–Spring 2013. The fraternity recognizes the individual who has demonstrated excellence in the area of academics.

- Big Brother of the Year, Fall 2011–Spring 2012. The fraternity recognizes an exceptional Big Brother in the chapter.

Study Abroad Fellowship and International Education Scholarship

National Scholar's Honor Society–inducted Fall 2013

References Furnished Upon Request

FIGURE 12–4 *continued*

directed, for example, to a company-wide job site to submit your application electronically. Otherwise, direct the materials, if at all possible, to the person in the organization who directly supervises the position for which you are applying. This person often has the power to hire for the position. Your research into the company may turn up the name you need. If not, don't be afraid of calling the company switchboard and asking directly for a name and title. Whatever you do, write to a specific individual by name.

Sometimes, of course, you may gain an interview without having submitted a letter of application—for example, when recruiters come to your campus. Bring a résumé with you, and give it to the interviewer at the start of the interview. The résumé gives the interviewer a starting point for questions and often helps to focus the interview on your qualifications for the job.

Online-generated résumés. A lot of companies allow you to submit résumés as documents generated electronically by filling in the fields on a series of pages at the site. The fields solicit the details of your address, credentials, and objectives and allow you to identify the job or jobs for which you are applying. This information is automatically compiled in a standardized format for access by hiring managers. Because you can't organize the information yourself to emphasize your knowledge and experience, using keywords to highlight your suitability for the job will be critical. Think carefully about the words that would be used to find the ideal candidate for the position. Make sure you include these words in describing your skills, your previous jobs, and your academic training.

Follow-up letters.   Write *follow-up letters* (1) if after 2 weeks you have received no answer to your letter of application, (2) after an interview, (3) if a company fails to offer you a job, and (4) to accept or refuse a job.

No answer.   If a company doesn't acknowledge receipt of your original letter of application within 2 weeks, write again with a gracious inquiry such as the following:

Dear Mr. Liao:

On April 12 I applied for a position with your company. I have not heard from you, so perhaps my original letter and résumé have been misplaced. I enclose copies of them.

If you have already reached a decision regarding my application or if there has been a delay in filling the position, please let me know at your earliest opportunity.

I look forward to hearing from you.

Sincerely yours,

After an interview.   Within a day's time, follow up your interview with a letter. Such a letter draws favorable attention to yourself as a person who understands business courtesy and good communication practice. Express your appreciation for the interview. Emphasize any of your qualifications that seemed to be especially important to the interviewer. Express your willingness to live with any special conditions of employment, such as relocation. Make clear that you want the job and feel qualified to do it. If you include a specific question in your letter, it may hasten a reply. Your letter might look like this one:

Dear Ms. Cooper:

Thank you for speaking with me last Tuesday about the computer security position you have open.

Working in computer security relates well to my experience and interests. The job you have available is one I am qualified to do. I am working this semester on a research report that examines the legal and ethical issues regarding electronic retrieval of patient records in public hospitals. May I send you a copy next week when it is completed?

I understand that the position would include working alternating weekends. This requirement presents no difficulty for me.

I look forward to hearing from you soon.

Sincerely yours,

After being refused a job. When a company refuses you a job, good tactics dictate that you acknowledge the refusal. Express thanks for the time spent with you, and state your regret that no opening exists at the present time. If you like, express the hope of being considered in the future. You never know—a new job might become available at a later time. In any case, you want to maintain a good reputation with this employer and its representatives. These are people working in the same industry as you: you may encounter them at professional conferences or meetings of the local chapter of your professional association. They may later come to work at your same company as your colleague, your supervisor, or your subordinate. The few minutes you devote now to a courteous reply will create a lasting impression on them of your extraordinary professionalism.

Accepting or refusing a job. Writing an acceptance letter presents few problems. Be brief. Thank the employer for the job offer, and accept the job. Determine the day and time you will report for work, and express pleasure at the prospect of joining the organization and working with your new colleagues. A good letter of acceptance might read as follows:

Dear Mr. Simmons:

Thank you for offering me a job as research assistant with your firm. I happily accept. I will report to work on July 1 as you have requested.

I look forward to working with Price Industries and particularly to the opportunity of doing research with Dr. Martinez.

Sincerely yours,

Writing a letter of refusal can be difficult. Be as gracious as possible. Be brief but not so brief as to suggest rudeness or indifference. Make it clear that you appreciate the offer. If you can, give a reason for your refusal. The employer who has spent time and money on interviewing you and corresponding with you deserves these courtesies. And, of course, your own self-interest is involved. In the future you may wish to work for this organization even though you must decline the opportunity at this time. A good letter of refusal might look like this one:

Dear Ms. Okujeni:

I enjoyed my visit to the research department of your company. I would very much have liked to work with the people I met there. I thank you for offering me the opportunity to do so.

After much serious thought, I have decided that the research opportunities offered to me in another job are closer to the interests I developed at Virginia State University. Therefore, I have accepted that job and regret that I cannot accept yours.

I appreciate the great courtesy and thoughtfulness that you and your associates extended to me during the application and interview process. The individual you do hire will be blessed with exceptional colleagues.

Sincerely yours,

INTERVIEWING

The immediate goal of all your preparation and letter and résumé writing is an interview with a potential employer.

The interview. If you have prepared properly, you should show up at the interview knowing almost everything about the organization, including its basic history, chief locations, key products and services, financial situation, and mission.

You should have the following with you: your résumé; a portfolio of your work, if appropriate; pen and notebook; a list of your references; and your business card.

The interviewer will probably start with a bit of social conversation to put you at ease before proceeding to questions aimed at assessing your skills and interests and how you might be of value to the organization. You might anticipate questions such as the following:

- What can you tell me about yourself?
- What are your strengths and weaknesses?
- What do you want to be doing 5 years from now?
- Why would you want to work for us?
- Why should we hire you?
- What accomplishment are you most proud of?
- What is the biggest problem you've encountered and how did you solve it?

To the question "What can you tell me about yourself?" the interviewer really doesn't expect an extended life history. This question provides you the opportunity to talk about your work, educational experiences, and skills. Try to relate your skills and experience to the needs of the organization. Don't overlook the people and communication skills essential to nearly every professional job.

In your answer to this and other questions, be specific in your examples. If you say something like "I have good managerial skills," immediately back it up with an occasion or experience that supports your statement. Focus on offering the specific evidence that will lead the interviewer to the right conclusions about you. Such specifics are more memorable than unsupported claims and will help you stick in the interviewer's mind.

In answering questions about your strengths and weaknesses, be honest but don't identify weaknesses that could eliminate you from consideration. Admit to weaknesses that would have no impact on the specific job for which you are interviewing: for example, if the job has no managerial duties, it would be okay to acknowledge that your managerial skills are undeveloped.

To the question "What do you want to be doing 5 years from now?" a good answer would identify a job with greater responsibilities at the same company.

The question "Why do you want to work for us?" allows you to display what you have learned about the organization. In answering this question, you should again show that what you have to offer fits with what the company needs.

In the final portion of the interview you will typically be given a chance to ask some questions of your own. It's a good time to get more details about the job or the working environment. Ask about the organization's goals. "What is the company most proud of?" is a good question. Or "What are you looking for in the candidate for this position?" "What will you want him or her to accomplish in the first year on the job?" Don't ask naive questions—questions you ought to know the answers to already from your research—like the size of the company or the number of employees. Don't ask questions just to ask questions. The interview is a good time for you to find out if you really want to work for this organization. Not every organization is going to be a good fit for what you have to offer and what you want to do. Avoid questions about salary and benefits unless the interviewer has raised this subject.

If you really want to work for the organization, make that clear before the interview ends. But don't allow your willingness to appear as desperation. Companies don't hire desperate people. At some point in the interview, be sure to get the interviewer's name (spelled correctly), title, address, phone number, e-mail address, and fax number. You'll need them for later correspondence. When the interviewer thanks you for coming, thank him or her for seeing you and leave. Don't drag the interview out when it's clearly over or linger at the door.

If you are invited for a site visit as a finalist for the position, you will find yourself continuously interviewed—from the minute you arrive until the minute you leave—by a variety of potential colleagues and supervisors during lunches or dinners, while waiting for elevators, while walking from office to office, or while driving to and from the airport. For this series of informal interviews and conversations—each of which could contribute to the hiring decision—prepare 1-minute answers to each of the formal interview questions: that is, abbreviate your usual 3-minute or 5-minute answer in a cogent emphasis on your key selling point. If time allows, you could always elaborate, but if the conversation is interrupted or the subject abruptly changes, you will at least have made your point and proven yourself articulate.

Negotiation. Interviewers rarely raise the subject of salary and benefits until they either see you as a good prospect or are sure they want to hire you. If they offer you the job, the negotiation is sometimes conducted by others in a separate interview.

Sometimes, the negotiator may offer you a salary. At other times, you may be asked to name a salary. Now is the time to put to good use the information you may have received through your networking activities. Or check the online job-finding services or specialized services such as Salary.com (www.salary.com) for their estimates of earnings for different positions in different industries.

Your research in these sources will give you not a specific salary but a salary range. If asked to name a salary, do not ask for the bottom of the range. Ask for as near the top as you believe is reasonable given your education and experience. The negotiator will respect you the more for knowing what you are worth. However, balance the compensation package—vacations, pension plans, healthcare, educational opportunities, and so forth—against the salary. Some compensation packages are worth a good deal of money and may allow you to accept a lower salary.

The location of the job is also a critical factor to consider. Online services such as Salary.com offer cost-of-living calculators that assess the comparative prices for housing, food, clothing, transportation, and basic services in different locations. For example, $50,000 in New York would be the equivalent of $33,000 in Chicago, $24,000 in Houston, or $40,000 in Los Angeles.

Before and after the interview. If you have not participated in job interviews before, you would be wise to practice with several friends. Using the information that we have given you and that you have gathered for yourself, practice several interviews, with two of you as interviewer and interviewee and the remainder as observers. Ask the observers to look for strengths and weaknesses in your answers, diction, grammar, and body language and to give you a candid appraisal. Practice until you feel comfortable with the process.

After you finish each interview, write down your impressions as soon as possible. How did your clothes compare to the interviewer's? Were there unexpected questions? How good were your answers? What did you learn about the organization? What did you learn about a specific job or jobs? Did anything make you uncomfortable about the organization? Do you think you would fit in there? By the next day, get a thank-you note (letter, e-mail, or fax) off to the interviewer.

JOB SEARCH CHECKLIST

The Letter and Résumé

Planning

☐ Your letter of application:
- What position are you applying for?
- How did you learn about this position?
- Why are you qualified for this position?
- What interests you about this company?
- What will you do for the organization?
- How will the employer reach you?

☐ Your résumé:
- Do you have all the necessary details about your education and experience (e.g., dates, job descriptions, schools, majors, degrees, extracurricular activities)?
- Which résumé format will suit your experience and abilities, chronological or functional?
- Do you have permission to use the names of at least three people as references?

Revising

☐ Your letter of application:
- Do you seem self-confident but not arrogant or boastful?
- Does your letter show how you could be valuable to the employer?
- Does your letter reflect interest in a specific job?
- Have you emphasized the education and experience that suit you for this job?
- Have you made it clear you would like to be interviewed? Have you made it easy for the employer to arrange the interview?
- Is your letter completely free of grammatical and spelling errors? Is it designed for easy reading and skimming?

☐ Your résumé:
- Have you picked the résumé type that suits your experiences and qualifications?
- Have you listed your education and job experience in reverse chronological order?
- Have you used active verbs and appropriate keywords to describe your education and experience?
- Is your résumé completely free of grammatical and spelling errors? Is it designed for easy reading and skimming?

JOB SEARCH CHECKLIST ✔ *continued*

The Interview

Planning

☐ Have you completed your research about the organization? Are you familiar with its history, goals, products, locations, and reputation in the industry and region?

☐ Have you practiced your answers to likely questions?

☐ Do you have good questions to ask the interviewer?

Reviewing

☐ Did you answer all the questions effectively? Which questions could you have answered better?

☐ Were you asked questions that you didn't expect?

☐ How did the interviewer answer your questions? Did he or she seem to think you asked good questions?

☐ Do you think you will be offered the job? Why or why not?

☐ Do you think you are a good fit for this organization? Why or why not?

EXERCISES

1. Investigate a potential employer of majors in your field. Locate information available from the employer's website as well as online job-finding services. Are all the profiles of this employer basically similar? Do you notice important differences? Locate additional information available in newspapers, magazines, and government documents about this employer. Do a search on Twitter for comments about this employer and its products and services. What is the public's perception? What is its reputation in the industry? Interview employees or managers for their perspectives on the advantages and disadvantages of working for this employer. Compile the findings of your research in a slide presentation for majors in your field.

2. Given the findings of your investigation of a major employer in your field, list the characteristics of this employer's ideal job candidate. What would be the ideal job candidate's education and experience? What would be his or her special skills, dispositions, and achievements? How are you similar to this ideal job candidate? How do you differ?

3. Given the findings of your investigation of a major employer in your field and your inventory of the ideal job candidate, tailor your typical letter of job application and résumé to the needs and interests of this employer. What changes will you make in your usual version of each document? How will these changes increase the likelihood of your being interviewed for a job?

4. If a friend of yours were applying for a position as a technical writer at Google, which of the following two résumés (see Figures 12–5 and 12–6) would you advise she include with the letter of application? If you were given 15 more minutes to make this résumé more effective, what changes would you make? How would your advice differ if your friend were applying for a job at IBM?

Jessica Perry

19309 Magnolia Avenue
Long Beach CA 90802
562-555-4433
jessica.perry@uc.edu

Employment Objective	Position as a Technical Communicator, with opportunities to apply my experience and skills in technical editing and document design
Education *University of California* Long Beach, CA	BA, Technical Communication, May 2014 GPA: 3.5 GPA in Major: 3.75

Employment Experience

Crisp Communications Long Beach, CA 2013–2014	• Compose, edit, and design a variety of documents for clients of this consulting firm, including brochures, newsletters, reports, and manuals • Supervise small projects • Coordinate project staff and schedule • Maintain billing records
Mercury Computers Los Angeles, CA 2012–2013	• Provided customer support for hardware and software products • Researched customer problems, identified appropriate solutions, and replied to customers • Served average of 50 customers daily
Special Skills	• Proficient in all Adobe applications • Read/write/speak Spanish
Scholastic Honors	• Sigma Pi Delta Honor Society, 2014 • R.T. Brookings Scholarship Recipient, 2013 • Dean's Academic Honor Roll, 2012–2014
Professional Memberships	IEEE Professional Communication Society Society for Technical Communication (STC)
References	Available upon request from the Career Planning and Placement Center, University of California, Meredith Hall 17, Long Beach, CA 90815

February 2014

FIGURE 12–5 Document for Exercise 4

Jessica Perry

19309 Magnolia Avenue
Long Beach, CA 90802
562-555-4433
jessica.perry@uc.edu

Employment Objective	Position as a Technical Communicator specializing in technical editing and document design
Education	University of California, Long Beach, CA BA, Technical Communication, May 2014
Communication Skills	• Compose, edit, and design a variety of documents, including brochures, newsletters, reports, and manuals • Create projects using all Adobe applications • Provide customer support for hardware and software products • Research customer problems, identify appropriate solutions, and reply to customers by telephone or e-mail message • Read/write/speak Spanish • Graduating with honors: GPA: 3.5; GPA in major: 3.75
Project Management	• Supervise small document design projects from initial meeting with client through publication and delivery • Coordinate project staff and schedule • Maintain billing records
Professional Involvement	• Sigma Pi Delta Honor Society • Association for Computing Machinery • IEEE Professional Communication Society • Society for Technical Communication
Work Experience	• Crisp Communications, Long Beach, CA, 2013–2014, document design • Mercury Computers, Los Angeles, CA, 2012–2013, customer support
References	Available upon request from the Career Planning and Placement Center, University of California, Meredith Hall 17, Long Beach, CA 90815

February 2014

FIGURE 12-6 Document for Exercise 4

Appendix A

▬▬ ▬▬ ▬▬▬▬ ▬▬ ▬▬ ▬▬ ▬▬ ▬▬

Brief Guide to Grammar, Punctuation, and Usage

ab ♦ ABBREVIATIONS

Every field and organization uses standard and specialized terms that are abbreviated for convenience and conciseness.

First, determine whether your audience is familiar enough with the unabbreviated term to allow you to use it without defining it. Second, determine whether your audience is familiar enough with the abbreviation for you to use it without spelling it out.

If you decide that an abbreviation is appropriate and must be explained, place the abbreviation in parentheses following your first use of the unabbreviated term. Thereafter, use the abbreviation by itself. If necessary, after the parentheses, provide a definition of the term. Make the definition as brief as possible but as detailed as necessary for your purpose and audience:

> Fluid catalytic cracking (FCC) changes crude oil to gasoline by breaking the long-chain molecules that are characteristic of hydrocarbon liquids. The process involves exposing the oil to a special chemical agent under high temperature and pressure. FCC is the key conversion process at oil refineries.

Use Latin abbreviations like *i.e.* (that is) and *e.g.* (for example) in parenthetical explanations or in tables and figures if space is limited.

Check the appropriate style guide or publication manual in your field for specific guidelines on abbreviations.

apos ♦ APOSTROPHES

Use apostrophes to indicate the possessive case of nouns (e.g., singular: supervisor's; plural: supervisors') and the missing letters in contractions (would've: would have).

With coordinated nouns, use apostrophes to make all possessive in order to indicate individual ownership. To indicate joint ownership, make only the final noun possessive.

• Maria's and Craig's proposals are late. (i.e., The proposal written by Maria and the proposal written by Craig are late.)

- Maria and Craig's proposals are late. (i.e., The proposals written together by Maria and Craig are late.)

Also use apostrophes to indicate the possessive case of indefinite pronouns ending in -one or -body as well as *either, neither, another,* or *other*:

- Everyone's utility bill is incorrect because of a computer error.
- The costly error probably raised somebody's chances of getting fired.
- I asked both supervisors and either's approval is enough to proceed.
- I asked both supervisors but neither's approval was forthcoming.
- We admired each other's abilities.
- He preferred another's answer.

acro ◆ ACRONYMS

Acronyms are names for objects or entities formed from a combination of the initial letters of the words in a title or phrase. Unlike abbreviations (e.g., UN for United Nations), acronyms are pronounced as words (e.g., UNICEF for United Nations International Children's Emergency Fund).

First, determine whether your audience is familiar enough with the acronym to allow you to use it without showing the phrase from which it is derived. Second, determine whether the phrase from which it is derived is itself a sufficient definition or explanation of the acronym.

If you decide that an acronym is appropriate, place the phrase from which it is derived in parentheses following your first use of the acronym. If necessary, supply additional explanation immediately following. Thereafter, use the acronym by itself:

The mysterious illness was identified as SARS (severe acute respiratory syndrome), a disease similar to pneumonia but viral in nature and thus impervious to antibiotics. SARS killed 299 people in Hong Kong in a 2003 epidemic.

brackets ◆ BRACKETS

Use brackets inside quotations or parentheses to insert a clarifying note or comment:

- "This is the highest level of recognition in the organization [2.5% of 475,000] and certifies exceptional professional achievement."
- If you would like more information about this opportunity, please call my office (M–F 8–5 [ET], 866-555-7423).

cap ◆ CAPITALIZATION

Capitalize months but not seasons (e.g., February was unusually dry this winter).

Capitalize geographic areas but not directions (e.g., The blight started in the Southwest but quickly traveled north and east).

Capitalize titles preceding names but not following or in isolation:

- The invitation is from President Juanita Rodriguez.
- Juanita Rodriguez, president of the company, arrived earlier this morning.
- The president of the company is Juanita Rodriguez.

colon ◆ COLON

Use a colon to introduce quotations, lists, or supporting statements:

- Dr. Raju's assertion was unequivocal: "This test has no predictive value, and it is unethical to advertise it as though it does."
- The company has five international locations: Buenos Aires, Cairo, Hong Kong, Mumbai, and Rio de Janeiro.
- The instructor explained the maieutic method: that is, the practice of teaching by asking questions that stimulate thinking and elicit new ideas.

Never use a colon immediately following a verb or a preposition:

WRONG: The five stores are located in: Buenos Aires, Cairo, Hong Kong, Mumbai, and Rio de Janeiro.
WRONG: The locations of the five stores are: Buenos Aires, Cairo, Hong Kong, Mumbai, and Rio de Janeiro.
RIGHT: The stores are in five locations: Buenos Aires, Cairo, Hong Kong, Mumbai, and Rio de Janeiro.
RIGHT: The stores are located in the following five cities: Buenos Aires, Cairo, Hong Kong, Mumbai, and Rio de Janeiro.

Use a colon to separate two independent clauses if the second clause explains or amplifies the first:

- The ability to find information is important but insufficient: we must also develop skills in synthesizing and analyzing information.
- The new map uses a different color for each section of the zoo: for example, green for the forest animals and brown for the desert animals.

Use a colon following the salutation of a business letter.

Dear Ms. Banerjee:

c ◆ COMMA

Use a comma before a coordinating conjunction (*and, but, or, nor, for, yet*) that joins two independent clauses:

- The investigators made their recommendation, and the client supported their judgment.
- The investigators made their recommendation, but the client questioned their judgment.

Use a comma after an introductory word, phrase, or clause:

- Unfortunately, the experiment did not yield statistically significant findings.
- In appreciation of their 25 years of service to the hospital, the president awarded each of the nurses a $1,000 prize.
- After you inspect the vehicle and verify that its condition is satisfactory, you must complete and sign a lease agreement.

Use commas to separate nonrestrictive modifiers from the remainder of the sentence.

RESTRICTIVE: Cotton farming that is conducted without artificial irrigation lowers the cost of production but raises the risk of a poor crop.
NONRESTRICTIVE: Cotton farming, which typically involves the use of fertilizers and defoliants as well as herbicides and pesticides, has high costs for the environment.

A *restrictive modifier* is necessary to the meaning of the sentence. The writer has thus restricted the subject of cotton farming to only that which is conducted without artificial irrigation. The writer here is making a single claim: cotton farming without artificial irrigation has lower costs but higher risks.

A *nonrestrictive modifier* supplies additional information about the subject without restricting it. The writer here is making two claims: cotton farming usually has high costs for the environment, and cotton farming usually involves fertilizers, defoliants, pesticides, and herbicides.

Use commas to separate items in a series of words, phrases, or clauses.

- This unit of the facility produces ethylene oxide, ethylene dichloride, and polyethylene.
- Ethylene contributes to the ripening of fruit, the opening of flowers, and the shedding of leaves.
- Florists complain about the impact of ethylene exposure on the life of their flowers, grocers worry about high-ethylene products like bananas, and shippers focus their efforts on good temperature controls to minimize the chemical's release.

Use commas to separate dates, geographical locations, and titles:

- On Tuesday, June 21, 2014, the case was closed.
- The artifact was shipped from Prague to New York, transported by train to Chicago, and delivered by bicycle messenger to the museum.
- The new members of the grievance committee are Kris Kelly, Ph.D., and Angela Pierce, J.D.

dm ♦ DANGLING MODIFIER

A dangling modifier occurs whenever a modifying word or phrase is used without a suitable noun for it to modify:

WRONG: Fearing the infectious disease might jump species, the chickens were quarantined.
RIGHT: Fearing the infectious disease might jump species, the veterinarian quarantined the chickens.

dash ◆ DASH

The dash serves the same function as parentheses (i.e., separating a tangential or explanatory comment from the remainder of the sentence), but the dash does so with greater emphasis—like a shout instead of a whisper.

ell ◆ ELLIPSIS

Use three spaced periods to indicate that words have been omitted from a quotation. If the quotation is at the end of a sentence, use four periods (i.e., three spaced periods for the ellipsis and a period for the sentence).

- "On the job, most readers . . . will feel no obligation to read what you write unless your messages are useful to them in making decisions or taking action."
- "On the job, most readers, coming from a variety of professional and cultural backgrounds, will feel no obligation to read what you write unless your messages are useful to them . . . ."

If the quotation is obviously incomplete, skip the ellipsis.

- According to the investigator, their actions were "deliberate and malicious."

exc ◆ EXCLAMATION POINT

Avoid using exclamation points in the writing of reports, letters, memos, and e-mail messages so that you never come across as easily excited or agitated. If writing instructions, use exclamation points as necessary with cautions and warnings.

frag ◆ FRAGMENT

A *sentence fragment* is a phrase or a clause punctuated as though it were a complete sentence. This error is usually fixed by connecting the fragment to the preceding sentence.

WRONG: The results were exactly as expected. Although we were hoping for better news.
RIGHT: The results were exactly as expected, although we were hoping for better news.
WRONG: The recommendations were made by the special investigating committee. Which included two forensic scientists and the city's medical examiner.

RIGHT: The recommendations were made by the special investigating committee, which included two forensic scientists and the city's medical examiner.

WRONG: The project supervisor decided to change the vacation policy. Surprising all of us at the meeting.

RIGHT: The project supervisor decided to change the vacation policy, surprising all of us at the meeting.

hyphen ◆ HYPHEN

Use a hyphen with the following suffixes and prefixes:

- -elect: president-elect
- -in-law: sister-in-law
- all-: all-around athlete
- ex-: ex-employee
- quasi-: quasi-empirical study
- self-: self-confident

Also use hyphens with prefixes preceding capitalized words (e.g., pre-Olympics competition), prefixes preceding numbers (e.g., post-9/11 security), and prefixes preceding acronyms and abbreviations (e.g., anti-HIV medication).

In addition, use a hyphen between words joined together to modify a word:

- The job was a time-intensive effort.
- We introduced the jazz-inspired designs at the music festival.
- He makes six-, eight-, and twelve-string guitars.
- She is a well-connected representative for the organization.

ital ◆ ITALICIZATION

Italicize titles of books, journals, magazines, plays, films, radio and television programs, sculptures and paintings, ships, aircraft, and spacecraft.

Italicize words that are considered foreign to the English language.

Italicize Latin words for genus and species (e.g., The zoological designation for the American toad is *Bufo americanus*).

Italicize words, letters, and numbers identified as words, letters, and numbers.

The words *compose* and *comprise* have opposite meanings.
His middle initial is *Q*.
She said it scored *75* on the grading scale.

Italicize letters used as statistical symbols:

- *M* (i.e., mean)
- *N* (i.e., number in the population studied)

- *p* (i.e., probability)
- *SD* (i.e., standard deviation)

mm ◆ MISPLACED MODIFIER

A *misplaced modifier* is a word, phrase, or clause that is in the wrong position in a sentence; as a consequence, it modifies a word that the writer never intended it to and creates ambiguity.

> WRONG: The machine was repaired almost at no cost to us.
> RIGHT: The machine was repaired at almost no cost to us (*or* The machine was almost repaired but at no cost to us).
> WRONG: He was trying to find a job for two years.
> RIGHT: He was trying for two years to find a job (*or* He was trying to find a two-year job).
> WRONG: A great museum in the city that he visits all the time is the Art Exchange on 3rd and Elm.
> RIGHT: A great museum that he visits all the time in the city is the Art Exchange on 3rd and Elm (*or* In the city that he visits all the time, a great museum is the Art Exchange on 3rd and Elm).

np/ag ◆ NOUN–PRONOUN AGREEMENT

In all formal communication, collective nouns such as *each, everyone, either, neither, anybody, somebody, everybody,* and *no one* use singular pronouns and singular verbs:

> FORMAL: Each has his or her book.
> FORMAL: Everybody was given his or her ticket.

For informal communications, the plural pronoun is widely used:

> INFORMAL: Each has their book.
> INFORMAL: Everybody was given their ticket.

A better solution is to revise the wording:

- All have their books.
- Each has a book.
- All were given their tickets.
- Everybody was given a ticket.

The same issue arises with nouns used in a generic sense, such as the word *employee* in the following:

> FORMAL: The employee must complete his or her training within one week of being hired.
> INFORMAL: The employee must complete their training within one week of being hired.

Again, a better solution is to revise the wording:

- All employees must complete their training within one week of being hired.
- The employee must complete the training within one week of being hired.

With collective nouns such as *team*, *group*, *class*, or *committee*, use the singular pronouns *it/its* if you are referring to the unit and the plural pronouns *they/their/them* if you are referring to the constituents.

- The hiring committee is using its rigorous criteria to find job candidates it believes are qualified.
- The security squad must make sure that they keep their uniforms clean and in good repair.

num   ◆   NUMBERS

Ordinarily, use words for numbers from zero to nine and numerals for numbers 10 and higher.

In a series of numbers, if any one of the numbers is 10 or higher, use all numerals:

- The collision damaged seven vehicles and injured four passengers.
- The collision damaged 7 vehicles and injured 16 passengers.

Never start a sentence with a numeral: either use words for the number or revise the sentence to shift the number from the opening position.

- Twenty-five engineers were involved in the project.
- The project involved 25 engineers.

With two adjacent numbers, use words for one and numerals for the other in order to avoid confusion.

- He ordered twelve 50-liter canisters.
- The list includes 15 eight-digit passwords.

Always use numerals for dates, addresses, exact time, exact sums of money, measurements, and cross-references.

- 1 October 2015 or October 1, 2015
- 2305 18th Street
- 2:00 P.M. (but two o'clock)
- $7,889.97 (but about eight thousand dollars)
- 16 km
- 36 g
- see page 85
- see Table 3

paral ◆ PARALLELISM

Use the same grammatical structure for items in a series:

> WRONG: A candidate for this job must be skilled in writing proposals, publication management, usability tests, and designing websites.
> RIGHT: A candidate for this job must be skilled in writing proposals, managing publications, running usability tests, and designing websites.
> RIGHT: A candidate for this job must be skilled in proposal writing, publication management, usability testing, and website design.

paren ◆ PARENTHESES

Use parentheses to separate a tangential or explanatory comment.

Never put any mark of punctuation before an opening parenthesis, but put any required marks of punctuation inside the parentheses and after the closing parenthesis:

- Use parentheses to separate a tangential or explanatory comment (including definitions, clarifications, examples, quotations, or statistics) from the remainder of a sentence.
- Latin words are used to designate genus and species in zoological classification (e.g., *Bufo americanus* for the American toad). (Note that these scientific names are always displayed in italics.)

pron ◆ PRONOUN

Use the subjective case of a pronoun (*I, he, she, we, they*) if it serves as the subject of a clause:

- She and I will work on this project.
- He visited the manufacturing facility yesterday, even though she and I made the same trip a week ago.

Use the objective case of a pronoun (*me, him, her, us, them*) if it is the object of a verb or the object of a preposition:

- The supervisor praised them and us for working together to find a solution.
- The grievance committee interviewed him and her at separate times.
- The medical examiner reported the results of the test to them and me.
- This project is a great opportunity for her and me as new scientists.

quot ◆ QUOTATION MARKS

Use quotation marks for brief quotations:

- As Simon Miller has noted, "The only solution is to limit the population of the species."

• According to Gloria Martinez, "Trillions of nurdles are made annually in factories on every continent. This growing epidemic of plastic pellets is a looming disaster for the world's environment."

For quotations of four or more lines, skip the quotation marks and indent and single space the quotation.

Use quotation marks around titles of articles from journals and magazines:

• The article "Polymers are Eternal" was originally published by Eduardo Quesada in 1989.
• Emily Kolosov's "The Geologic Heritage in Dispute" claims that crop diversity is diminishing worldwide.

Always put commas and periods inside the quotation marks:

• According to Emily Kolosov's "The Geologic Heritage in Dispute," crop diversity is dwindling worldwide.
• Eduardo Quesada is the author of "Polymers are Eternal."

Put all other marks of punctuation outside the quotation marks unless they are part of the quotation.

• In 1986, Eduardo Quesada published his article "Polymers are Eternal": it was greeted with derision by the plastics industry.
• Emily Kolosov's "The Geologic Heritage in Dispute" asked the question, "What are the avoidable contributors to this growing environmental crisis?"

run-on ◆ RUN-ON SENTENCE

A *run-on sentence* is two independent clauses joined with only a comma or with no punctuation at all.

Two independent clauses may be punctuated as two separate sentences (each with a period) or may be joined with a semicolon, a colon, or a comma and a conjunction (*and, but, for, nor,* or *yet*):

WRONG: The instructions manual has legal and ethical implications, it must assure safe and efficient operation of the product and minimize liability for the manufacturer.
RIGHT: The instructions manual has ethical and legal implications. It must assure safe and efficient operation of the product and minimize liability in case of damage or injury.
RIGHT: The instructions manual has ethical and legal implications; specifically, it must assure safe and efficient operation of the product and minimize liability in case of damage or injury.
RIGHT: The instructions manual has ethical and legal implications, and thus it must assure safe and efficient operation of the product and minimize liability in case of damage or injury.

A run-on sentence may also be repaired by changing one or the other of the independent clauses to a dependent clause or phrase:

RIGHT: The instructions manual has ethical and legal implications because it must assure safe and efficient operation of the product and minimize liability in case of damage or injury.

RIGHT: The instructions manual has ethical and legal implications, making it necessary to assure safe and efficient operation of the product while minimizing liability in case of damage or injury.

semi ◆ SEMICOLON

Use a semicolon to join related independent clauses (e.g., Victims of the mysterious disease exhibited symptoms of pneumonia; however, unlike typical pneumonia, their disease was viral and unresponsive to antibiotics).

Use a semicolon to separate items in a series if the items are internally punctuated with commas (e.g., The budget for new equipment included security cameras for all exits; projectors, computers, and screens for 12 of the 18 classrooms; and one desk, one desk chair, two guest chairs, three bookshelves, and one filing cabinet for each of 35 faculty and administrative offices).

av/ag ◆ SUBJECT–VERB AGREEMENT

Note that the words *each*, *either*, and *neither* take a singular verb regardless of intervening words:

- Each is qualified.
- Each of the job candidates who applied for the supervisor positions is qualified.
- Either is qualified.
- Either of the job candidates who were interviewed yesterday by the hiring committee is qualified.
- Neither is qualified.
- Neither of the job candidates who applied for the engineering positions is qualified.

In a coordinated subject joined by *or* or *nor*, make the verb agree with the closer noun or pronoun:

- The original painting or digital reproductions are usually on display.
- Digital reproductions or the original painting is usually on display.
- Neither the original painting nor digital reproductions are on display at this time.
- Neither digital reproductions nor the original painting is on display at this time.

In a coordinated subject joined by *and*, the verb is always plural:

- The original painting and digital reproductions are available for examination.
- Digital reproductions and the original painting are available for examination.

With collective nouns such as *team*, *group*, *class*, and *committee*, use a singular verb if you are referring to the unit and a plural verb if you are referring to the constituents.

- The marketing team is flying to the annual meeting.
- The marketing team are flying together to the annual meeting.

WW ◆ WRONG WORD

The following words are often confused and misused. Here are examples of correct usage:

I **accept** your resignation.
She liked every job candidate we interviewed **except** the third.

His engineering experience will **affect** [influence] his decisions about this project.
Every delay on the bridge project will **effect** [cause] a loss in sales and profits for local businesses.

The proposal was **already** submitted.
The proposals were **all ready** to be submitted.

The clients were **all together** [united] in their support of the policy.
The clients were **altogether** [completely] supportive of the policy.

I don't make paper copies of the annual reports **anymore**.
I don't need **any more** paper copies of the annual reports.

She visited **awhile** with each of the job candidates.
She visited for **a while** with each of the job candidates.

The engineers believe that **canvas** is the right material for this product.
I will **canvass** the engineers for their opinions about this product.

She was pleased to **cite** each source of information.
He inspected the **site** of the accident.

The advertising through social media **complemented** the television commercials.
The project manager **complimented** the scientists and praised their efficiency.

A chemical engineer, a historian, and a psychologist **compose** the hiring committee.
The hiring committee **comprises** a chemical engineer, a historian, and a psychologist.

She is **continually** [repeatedly] making changes to the report.
She worked **continuously** [without interruption] all afternoon to finish the report.

The **council** [committee] of advisors made five recommendations.
He appreciated his mentor's **counsel** [advice].

She was always **discreet** [prudent] in discussing clients and colleagues.
The two projects are equally important but **discrete** [separate].

The **eminent** scientists arrived early to the meeting.
The arrival of the scientists is **imminent**.

Their **everyday** routine includes a morning review of international news.
He is in the office **every day** by 8 o'clock.

The train station is five kilometers **farther** [distance].
He will investigate this case **further** [degree].

He recruited **fewer** [number] new clients this week.
He is devoting **less** time [amount] to recruiting new clients.

The company is moving **forward** with its new president.
The report includes a **foreword** by the president of the company.

My supervisor **implied** [suggested] that I would be promoted.
My supervisor **inferred** [concluded] from my record of success that I
deserved a promotion.

It's [it is] a historic building.
Its history is impressive.

The walls were covered in **lead** paint.
The assistant manager **led** us on a walking tour of the facility.

The doctors identified the **principal** cause of the disease.
The **principal** of my loan is slowly shrinking.
She operated on the **principle** of people over profits.

He located the source of the **quotation**.
He will **quote** the president in this report.

The police officer talked **respectfully** to the victim about the details of the crime.
The awards were announced for arts and sciences on April 2 and 3, **respectively**.

The instructions for my **stationary** bicycle are clear and simple.
She printed the letters of inquiry on the company's **stationery**.

The report that **they're** writing has a Friday deadline.
I was in **their** office earlier this week.
She is taking the train **there** this evening.

The **weather** is a safe topic of conversation.
I asked him **whether** he could join the meeting.

He never said **whose** jobs were at risk.
My manager knows **who's** being hired.

I noticed that **you're** writing this proposal.
I appreciate **your** willingness to write this proposal.

Appendix B

Using Sources of Information

Clear, credible, pertinent, and sufficient information is the basis of all effective technical communication. To deliver the right information at the right time, you must develop the six skills of information literacy:

1. *Locate sufficient sources of information*: Given your audience and purpose, you must determine the nature and scope of potential information sources: interviews, surveys, books, articles, blogs, websites, videos, and government documents.

2. *Examine sources for pertinent information*: After you have compiled the potential sources of information, you must review each for information related to your subject. This effort includes skimming tables of contents and lists of illustrations, searching for keywords, and navigating a subject index.

3. *Evaluate sources for credibility*: You must judge the accuracy of the information in your sources. Ordinarily, the more reliable the source or the more sources offering the same information, the more likely it is you have information that deserves to be trusted. No source is infallible, but credible sources typically identify their sources of information, have a track record of offering valid and reliable information, or publish their information only after rigorous review by editors and subject specialists.

4. *Summarize information in sources clearly and correctly*: You will have to make sure that your abridged version of the information in a source is neither misleading nor mystifying. If you are unsure of the right wording, consider integrating quotations judiciously.

5. *Synthesize information across sources*: You must piece together the information from each of your sources so that it creates a coherent picture of your subject. This job includes explaining the relationships among all the pieces of pertinent information that you have located (e.g., cause/effect, chronological, comparison/contrast) and making sure that your assertions about the relationships are supported by evidence and clarified by examples.

6. *Cite sources of information appropriately*: According to your audience and purpose as well as the practices of your organization, choose a formal or informal method of identifying your information sources. In the majority of documents that you write (e-mail messages, blog postings, letters, and memos), informal citation will be satisfactory (e.g., "According to the book *Risk Analysis for Civil Engineers* . . ." or "The video at www.worldnews.com/112014/ solarpower.html makes clear that . . ."). In proposals and reports using multiple sources of information, especially if circulated outside your organization,

a formal system of documentation will be necessary to assist readers in weighing the merits of your research and keeping track of your sources; that is, in addition to specifying which sources supplied which information, a formal system makes evident the number of sources you used, the variety of your sources, and the frequency with which you relied on each source.

Applications such as EndNote (www.endnote.com), RefWorks (www.refworks .com), and Zotero (www.zotero.org) will assist you with managing, citing, and annotating your sources, but your conscientious efforts are critical to developing the skills of information literacy. No application is a substitute for your cautious judgment about information that is clear, credible, pertinent, and sufficient.

THE MLA SYSTEM

In the MLA (Modern Language Association) system of documentation, cite sources parenthetically by author's last name (or by authoring organization or by title if no author is identified). If you cite a specific passage or quote from a source, include specific pages:

> With its officious language and its emphasis on damage to buildings and vehicles, this accident report "silences the horror of human suffering and the loss of human life" (Sauer 72).

In a list titled "Works Cited," detail your sources in alphabetical order by the author's last name (or by authoring organization or by title if no author is identified). If several of your sources are by the same author, list the items in alphabetical order by title and substitute three hyphens for the author's name on the second and subsequent entries.

If you used a paper copy of the source, close the entry with the word *Print*. If you accessed the source through the Internet, close the entry with the word *Web* and specify the date of access (include the URL, in angled brackets, only if it is important to finding the source). If neither print nor web, specify other media as appropriate (e.g., *E-mail, Television, Radio, Performance, DVD*).

Book:
Appadurai, Morarji. *The Rhetoric of Multinational Corporations*. 3rd ed. New York: Dolphin, 2014. Print.

Collection:
Williams, Miriam F., and Octavio Pimentel, eds. *Communicating Race, Ethnicity, and Identity in Technical Communication*. Amityville: Baywood, 2013. Print.

Essay or article in a book:
Still, Joyce L., Brian T. Carter, and Maria V. García. "The Risk and Rhetoric of Statistical Illustrations." *Visual Communication: Investigations in Theory and Practice*. Ed. Mary M. Staggers and William B. Cook. Chicago: Radius, 2008. 170–205. Print.

Article in a professional journal:

Sauer, Beverly. "Sense and Sensibility in Technical Documentation: How Feminist Interpretation Strategies Can Save Lives in the Nation's Mines." *Journal of Business and Technical Communication* 7 (1993): 63–83. Print.

Zoetewey, Meredith W. "The Rhetoric of Free: Open Source Software and Technical Communication During Economic Downturns." *Technical Communication Quarterly* 23.4 (2013): 323–42. Web. 27 Sept. 2013.

Article in a monthly publication:

Cook, Andrew. "Risk Communication and Vaccines." *Medicine* Oct. 2012: 49–53. Print.

Article in a daily newspaper, anonymous:

"Bike-Friendly Culture Inspires Business Community." *The Long Beach Daily* 17 Jan. 2014. Web. 3 Sept. 2014.

Government publication:

United States. Department of Transportation. National Highway Traffic Safety Administration. *Blueprint for Ending Distracted Driving.* DOT HS 811 629. Washington: GPO, 2012. Web. 18 Sept. 2013.

Article in an online news source:

Luo, Leilei. "Broccoli and the Brain." INN.com. International News Network. 17 Dec. 2013. Web. 19 Dec. 2013.

Page of a website:

"Medical Issues." Doctors Without Borders. 2014. Web. 11 October 2014.

Online multimedia source:

Luo, Leilei. "Broccoli and the Brain." Video. INN.com. International News Network. 17 Dec. 2013. Web. 19 Dec. 2013.

Posting to an e-mail distribution list:

Vasquez, Eduardo. "Research Ethics Question." Message to ATTW Distribution List. 24 Sept. 2013. E-mail.

Posting to a blog or bulletin board:

Taylor, Michael R. "Defining Boundaries for Caffeine in Today's Marketplace." *FDA Voice.* 26 Aug. 2013. Web. 9 Sept. 2013.

Tweet:

Rickard, Lydia (lydiarickard). "FDA regulations promote innovation in devices intended to treat obesity." 22 Jan. 2014, 11:06 a.m. Tweet.

THE CHICAGO SYSTEM

The Chicago (Chicago Manual of Style) notes system of documentation includes a numerical list of citations (i.e., notes) in the order cited as well as a list of sources (i.e., bibliography) in alphabetical order by author's last name (or by authoring organization or by title if no author is identified). For online sources, include the Digital Object Identifier (DOI), a unique alphanumeric designator of the source's fixed location on the Internet. If no DOI is available, include the URL.

For each source cited, insert a superscript number in the text corresponding to the appropriate source in the numerical list:

> With its officious language and its emphasis on damage to buildings and vehicles, this accident report "silences the horror of human suffering and the loss of human life."[2]

Notes:

1. Morarji Appadurai, *The Rhetoric of Multinational Corporations*, 3rd ed. (New York: Dolphin, 2014), 158.
2. Beverly Sauer, "Sense and Sensibility in Technical Documentation: How Feminist Interpretation Strategies Can Save Lives in the Nation's Mines," *Journal of Business and Technical Communication* 7, no. 1 (1993): 72.
3. Appadurai, *Rhetoric of Multinational Corporations*, 184.
4. Sauer, "Sense and Sensibility," 75.

Following are a variety of sources as each would be displayed in numerical notes and in the alphabetical bibliography:

Book:

1. Morarji Appadurai, *The Rhetoric of Multinational Corporations*, 3rd ed. (New York: Dolphin, 2014), 158.

Appadurai, Morarji. *The Rhetoric of Multinational Corporations*. 3rd ed. New York: Dolphin, 2014.

Collection:

2. Miriam F. Williams and Octavio Pimentel, eds., *Communicating Race, Ethnicity, and Identity in Technical Communication* (Amityville, NY: Baywood, 2013), ii–iii.

Williams, Miriam F., and Octavio Pimentel, eds. *Communicating Race, Ethnicity, and Identity in Technical Communication*. Amityville, NY: Baywood, 2013.

Essay or article in a book:

3. Joyce L. Still, Brian T. Carter, and Maria V. Garcia, "The Risk and Rhetoric of Statistical Illustrations," in *Visual Communication: Investigations in Theory and Practice*, ed. Mary M. Staggers and William B. Cook (Chicago: Radius, 2008), 124.

Still, Joyce L., Brian T. Carter, and Maria V. García. "The Risk and Rhetoric of Statistical Illustrations." In *Visual Communication: Investigations in Theory and Practice*, edited by Mary M. Staggers and William B. Cook, 117–131. Chicago: Radius, 2008.

Article in a professional journal:

4. Beverly Sauer, "Sense and Sensibility in Technical Documentation: How Feminist Interpretation Strategies Can Save Lives in the Nation's Mines," *Journal of Business and Technical Communication* 7, no. 1 (1993): 72.

Sauer, Beverly. "Sense and Sensibility in Technical Documentation: How Feminist Interpretation Strategies Can Save Lives in the Nation's Mines." *Journal of Business and Technical Communication* 7, no. 1 (1993): 63–83.

5. Meredith W. Zoetewey, "The Rhetoric of Free: Open Source Software and Technical Communication During Economic Downturns," *Technical Communication Quarterly* 23, no. 4 (2013): 323–342, doi:10.1080/10572252.2013.794090.

Zoetewey, Meredith W. "The Rhetoric of Free: Open Source Software and Technical Communication During Economic Downturns." *Technical Communication Quarterly* 23, no. 4 (2013): 323–342. doi:10.1080/10572252.2013.794090.

Article in a monthly publication:

6. Andrew Cook, "Risk Communication and Vaccines," *Medicine*, October 2012, 51.

Cook, Andrew. "Risk Communication and Vaccines." *Medicine*, October 2012: 49–53.

Article in a daily newspaper, anonymous:

7. "Bike-Friendly Culture Inspires Business Community," *The Long Beach Daily,* January 17, 2014, http://www.lbdaily/com/011411/bike.html.

"Bike-Friendly Culture Inspires Business Community." *The Long Beach Daily,* January 17, 2014. http://www.lbdaily/com/011714/bike.html.

Government publication:

8. US Department of Transportation, National Highway Traffic Safety Administration, *Blueprint for Ending Distracted Driving*, DOT HS 811 629 (Washington, DC: US Government Printing Office, 2012), 19, http://www.distraction.gov/blueprint-2012.pdf.

US Department of Transportation, National Highway Traffic Safety Administration. *Blueprint for Ending Distracted Driving*. DOT HS 811 629. Washington, DC: US Government Printing Office, 2012. http://www.distraction.gov/blueprint-2012.pdf.

Article in an online news source:

9. Leilei Luo, "Broccoli and the Brain," *INN.com*, December 17, 2013, http://www.inn.com/2013/SCIENCE/12/17/broccoli/index.html.

Luo, Leilei. "Broccoli and the Brain," *INN.com*, December 17 2013. http://www.inn.com/2013/SCIENCE/12/17/broccoli/index.html.

Page of a website:

10. "Medical Issues," *Doctors Without Borders*, accessed October 11, 2014, http://www
.doctorswithoutborders.org/news/issues.cfm.

"Medical Issues." *Doctors Without Borders*. Accessed October 11, 2014. http://www
.doctorswithoutborders.org/news/issues.cfm.

Online multimedia source:

11. Leilei Luo, "Broccoli and the Brain," Video. *INN.com*, December 17 2013, http://www
.inn.com/2013/SCIENCE/12/17/broccoli/index.html#innSTCVideo.

Luo, Leilei. "Broccoli and the Brain," Video. *INN.com*, December 17 2013. http://www
.inn.com/2013/SCIENCE/12/17/broccoli/index.html#innSTCVideo.

Posting to e-mail distribution list:

12. Eduardo Vasquez, "Research Ethics Question," *ATTW-L* (e-mail distribution list), September 24, 2013, http://attw.interversity.org/read/messages?id100130924.

Vasquez, Eduardo. "Research Ethics Question." *ATTW-L* (e-mail distribution list). September 24, 2013. http://attw.interversity.org/read/messages?id100130924.

Posting to a blog or bulletin board:

13. Michael R. Taylor, "Defining Boundaries for Caffeine in Today's Marketplace," *FDA
Voice* (blog), August 26, 2013, http://blogs.fda.gov/fdavoice/index.php/2013/08/defining
-boundaries-for-caffeine-in-todays-marketplace/.

Taylor, Michael R. "Defining Boundaries for Caffeine in Today's Marketplace." *FDA
Voice* (blog). August 26, 2013. http://blogs.fda.gov/fdavoice/index.php/2013/08/defining
-boundaries-for-caffeine-in-todays-marketplace/.

Tweet:

14. Lydia Rickard, "FDA regulations promote innovation in devices intended to treat obesity," *Twitter*, January 22, 2014, http://twitter.com/lydiarickard/status/192837465001.

Rickard, Lydia. "FDA regulations promote innovation in devices intended to treat obesity"
Twitter. January 22, 2014. http://twitter.com/lydiarickard/status/192837465001.

THE APA SYSTEM

The APA (American Psychological Association) system of documentation cites
sources parenthetically using the author's last name (or authoring organization or
title if no author is identified) and the year of publication, separated by a comma.
If emphasizing or quoting a specific passage, also cite specific pages using the abbreviation *p.* or *pp.* as necessary:

> With its officious language and its emphasis on damage to buildings and vehicles, this accident report "silences the horror of human suffering and the loss of human life" (Sauer, 1993, p. 72).

In a list titled "References" and organized alphabetically by author's last name (or by authoring organization or by title if no author is identified), detail your sources. If available, include the Digital Object Identifier (DOI), a unique alphanumeric designator of the source's fixed location on the Internet. If the source was accessed online but has no DOI, include the URL. If the online source is subject to change (e.g., wiki, website), include the date of retrieval.

Book:
Appadurai, M. (2014). *The rhetoric of multinational corporations* (3rd ed.). New York, NY: Dolphin.

Collection:
Williams, M. F., & Pimentel, O. (Eds.). (2013). *Communicating race, ethnicity, and identity in technical communication.* Amityville, NY: Baywood.

Essay or article in a book:
Still, J. L., Carter, B. T., & García, M. V. (2008). The risk and rhetoric of statistical illustrations. In M. M. Staggers & W. B. Cook (Eds.), *Visual communication: Investigations in theory and practice* (pp. 117–131). Chicago, IL: Radius.

Article in a professional journal:
Sauer, B. (1993). Sense and sensibility in technical documentation: How feminist interpretation strategies can save lives in the nation's mines. *Journal of Business and Technical Communication, 7,* 63–83.

Zoetewey, M. W. (2013). The rhetoric of free: Open source software and technical communication during economic downturns. *Technical Communication Quarterly, 23,* 323–342. doi:10.1080/10572252.2013.794090

Article in a monthly publication:
Cook, A. (2012, October). Risk communication and vaccines. *Medicine, 23,* 49–53.

Article in a daily newspaper, anonymous:
Bike-friendly culture inspires business community (2014, January 17). *The Long Beach Daily.* Retrieved from http://www.lbdaily/com/011714/bike.html

Government publication:
United States. Department of Transportation. National Highway Traffic Safety Administration. (2012). *Blueprint for Ending Distracted Driving* (DOT HS 811 629). Washington, DC: US Government Printing Office. Retrieved from http://www.distraction.gov/blueprint-2012.pdf

Article in an online news source:

Luo, L. (2013, December 17). Broccoli and the brain. *INN.com*. Retrieved from http://www.inn.com/2013/SCIENCE/12/17/broccoli/index.html

Page of a website:

Doctors Without Borders. (2014). *Medical Issues*. Retrieved October 11, 2014, from http://www.doctorswithoutborders.org/news/issues.cfm

Online multimedia source:

Luo, L. (2013, December 17). Broccoli and the brain. [Video]. *INN.com*. Retrieved from http://www.inn.com/2013/SCIENCE/12/17/broccoli/index.html#innSTC Video

Posting to e-mail distribution list:

Vasquez, E. (2013, September 24). Research ethics question. [Electronic mailing list message]. Retrieved from http://attw.interversity.org/read/messages ?id100130924

Posting to a blog or bulletin board:

Taylor, M. R. (2013, August 26). Defining boundaries for caffeine in today's marketplace [Blog post]. Retrieved from http://blogs.fda.gov/fdavoice/index.php/2013/08/defining-boundaries-for-caffeine-in-todays-marketplace/

Tweet:

LydiaRickard. (2014, January 22). FDA regulations promote innovation in devices intended to treat obesity [Twitter post]. Retrieved from http://twitter.com/lydiarickard/status/192837465001

Appendix C

Report for Study and Analysis

At ABC University, premedical majors may pursue an internship at a hospital during two summer terms. At the end of their internship, they write a formal report on a disease or condition that they believe to have major importance to medicine. Internship allows them to spend some of their time addressing a major research/medical topic. They have to first propose a topic at the end of their first summer internship, then submit a progress report during the next year, and then submit a final report at the end of their second summer internship.

Hannah Starke chose Alzheimer's as she saw the ravages of this disease during her internship.

MEMORANDUM

To: Elizabeth Tebeaux
From: Hannah Starke
Date: 17 April 2012
Subject: Transmittal of the Final Report on the Mechanics and Treatment of
             Alzheimer's Disease

Attached to this memo is my final report on Alzheimer's disease. This report provides an overview of the current state of research on Alzheimer's disease on the following topics:

- Alzheimer's effect on the affected individual
- Current physiological hypotheses on disease development
- Risk factors for developing the disease
- Preventative measures
- Current treatment options and their effectiveness

Alzheimer's slowly destroys the brain until the individual is left with no memories and an inability to function independently. Alzheimer's is a terrible fate. I have watched two people I love dearly succumb to this disease, and I can imagine no worse a fate. Their helplessness, confusion, and loss of self rip your heart in two; and I hope that at some point in my career, I will help those who suffer from this disease or prevent them from developing it at all.

The purpose of this report is to inform the reader of all the basic mechanisms that affect Alzheimer's development in individuals. In addition to that, I want to make sure people know that despite the fact that Alzheimer's cannot be cured or reversed in any way, individuals can take several actions to reduce their risk of disease development.

Thank you, Dr. Tebeaux for your instruction, guidance, and complete dedication to your students. Being in your class reminds me of how it feels to be in the presence of someone who loves what she does and, more importantly, cares more about the success of those she teaches. Also, thank you to the library staff for demonstrating how to use the extensive database network available and fielding all questions I asked them.

Sincerely,

Hannah Starke

Final Report on the Mechanics and Treatment of Alzheimer's Disease

17 April 2012

Hannah Starke, Author	Date

Abstract

This report serves to provide the reader with an overview of the current research on risk factors, disease development, and treatment options for Alzheimer's disease. In addition to overall summaries of the listed topics, this report highlights the possible future development associated with the research developments. The report itself contains a summary, works cited list, and original proposal documents.

Keywords:

Alzheimer's, Amyloid cascade hypothesis, Mitochondrial cascade hypothesis, Glucocorticoid hypothesis, Neurofibrillary tangles, Amyloid β protein

Final Report on the Mechanics and Treatment of Alzheimer's Disease

Table of Contents

List of Figures

Executive Summary

Report Purpose

I hope to educate the reader on all aspects of Alzheimer's disease and steps that he or she can take to try to reduce the risk of developing the disease. The incidence of Alzheimer's has increased in recent years. Actions that people take early in life can greatly influence their risk of developing a disease that changes everything. Understanding the actions you can take to decrease your chances of developing this disease may help you avoid it.

Research Methods

To collect information for the report, I first used medical help sites such as WebMD and the official websites of the Alzheimer's Association and National Institute of Aging to gain basic understanding of Alzheimer's disease and its progression. Next I used the TAMU library databases and free online journals such as *PLoS One* to gain a greater understanding of the mechanisms and technical problems associated with Alzheimer's. Despite continued efforts, I was unable to communicate with any neurological specialists at Baylor College of Medicine or any other institution.

Findings

Alzheimer's begins by attacking and killing brain cells. The brain damage causes horrible memory loss, personality change, and motor skill degeneration. An individual can begin having internal physiological signs of the disease two decades before any noticeable symptoms. Brain scans can detect these early developments, but most of the time detection and preventative measures come too late. Once the disease starts, nothing can stop its progression.

We do not know the exact mechanism of disease formation. As of now, no effective treatments exist to stop disease progression, but research continues to investigate the catalytic events responsible for the disease. With a more comprehensive understanding of what initiates disease formation, scientists can develop more effective treatments. These treatments work by targeting the central problem and stopping the disease before it begins.

Anyone can develop this disease. Genetic predispositions do exist and contribute to the likelihood of development, but genetics is not the deciding factor. Researchers have found several preventative measures that at-risk individuals can take to reduce the risk or slow down the progression of Alzheimer's. Several of these methods include dieting, exercise, and education.

Conclusions

My research and personal experiences have led me to firmly believe that Alzheimer's is one of the worst diseases a person can have. Everyone needs to have a general understanding of the risk factors associated with this disease and the small measures that one can take to reduce the risk of developing this disease. Avoiding any development and catching it early are the two most important steps. As someone who has several of the risk factors associated with developing Alzheimer's, I can assure you that I will do everything reported here. In future years, research advances will lead to effective treatments aimed at eliminating the progression of the disease.

Final Report on the Mechanics and Treatment of Alzheimer's Disease

Introduction

Alzheimer's is the most common form of dementia, a blanket term for age-associated cognitive decline. Progression of Alzheimer's disease leads to decreases in memory storage, cognitive function, and basic task performance. To a family, Alzheimer's takes a loved one; but to the affected, the disease takes everything. Anyone who has ever watched a loved one struggle with this disease needs no reminder or explanation of the toll it takes on everyone involved.

With the aging of the baby boomer generation comes an increase in the number of individuals entering the high-risk age group. In 2010, about 5.1 million individuals suffered from mild to severe Alzheimer's; and by 2050, that number will rise to an estimated 19 million [1]. The disease takes another toll as well. In this year alone, the United States will spend about $200 billion on the treatment and care of Alzheimer's [2].

Due to these enormous medical and economic impacts on the nation, institutions have increased research to accelerate the creation of effective preventative treatments. Understanding how the disease begins and progresses helps guide this new research. Knowing how the disease advances allows for treatments aimed at stopping the disease before it begins. Through research scientists have developed three main hypotheses explaining how the disease develops. Each one of these hypotheses describes a different physiological mechanism responsible for the initial disease development.

No matter the catalyst, a few neurological changes characterize Alzheimer's:

- Protein aggregations in the brain called "plaques"
- Fibrous networks within brain cells
- Breaking down of synaptic (communicative) connections in the brain
- Neuron (brain cell) death

The progression of these neurological alterations causes the cognitive, psychological, and behavioral symptoms associated with Alzheimer's.

To date, no treatment effectively stops the progression of the disease, but individuals can still reduce their risk of disease development through a few simple actions. A healthy lifestyle, aerobic exercise, and continued educational stimulation all correlate with a reduced risk of Alzheimer's.

This report outlines the following aspects of Alzheimer's:

- General progression
- Developmental hypotheses
- Associated risk factors
- Treatment options available

The most important point to take away from this report is that anyone can develop Alzheimer's disease. Genetic predispositions play a role in who develops this disease, but many other factors contribute. Despite the fact that everyone is at risk, individuals can improve their chances of not developing the disease in several ways.

Alzheimer's Disease

What Is Alzheimer's Disease?

Alzheimer's is a neurodegenerative disease facilitated by neuronal destruction. As the destruction progresses, the affected individual experiences increased memory loss, decreased cognitive function, personality change, and, eventually, inability to care for oneself. The characeristcs of Alzheimer's, the buildup of protein plaques in the brain and fibrous networks (neurofibrillary tangles or NTs) in the neurons, cause cell destruction (Figure 1). Plaques of amyloid β (Aβ) protein build up in the spaces between neurons and cause cell death mainly by destroying cellular communication sites called "synapses." Synapse destruction inhibits message and nutrient transfer between neurons. Scientists are unsure if these plaques are the cause or a by-product of the disease, as discussed later in this report. Within neurons NTs occur when a protein named "tau" begins interacting uncharaceristically with itself [1]. Most likely, Aβ facilitates this unhealthy interaction.

Alzheimer's Brain Cells

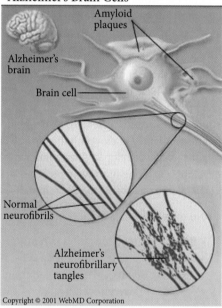

Copyright © 2001 WebMD Corporation

Figure 1: **Neurofibrillary Tangles and Amyloid Plaques.** Courtesy of: The Doctor Weighs In http://www.thedoctorweighsin.com/alzheimers-disease-is-there-daylight-at-the-end-of-the-tunnel-or-is-it-an-oncoming-train/

The first site of deterioration in the brain is the hippocampus, the area responsible for memory creation and storage [1]. Accordingly, memory loss is the first symptom of Alzheimer's. Brain damage continues beyond the hippocampus into sites responsible for judgment, personality, speech, and basic motor function. The widespread destruction in the brain leads to progressively less functional individuals (Figure 2). Doctors characterize a person's disease severity through a ranking system. The seven stages of Alzheimer's describe the relative impairment as the disease progresses and affects different areas of the brain [3].

- Stage 1: no signs of disease development in the brain but no cognitive impairment
- Stage 2: very mild cognitive impairment—memory loss of small facts
- Stage 3: mild cognitive impairment—more noticeable memory loss detectable by others, especially short-term memory loss
- Stage 4: moderate cognitive decline—higher cognitive function deterioration begins
- Stage 5: moderately severe cognitive decline—long-term memory deterioration and loss of judgment, moderate assistance necessary
- Stage 6: severe cognitive decline—personality change and constant assistance needed
- Stage 7: very severe cognitive decline—inability to interpret surroundings, loss of motor movement and muscle coordination, eventual death

Due to the extensive brain damage, Alzheimer's eventually kills. The degeneration of the brain leads to impaired ability to perform basic motor skills including bathing, talking, eating, swallowing, and controlling bladder and bowel elimination. Without these basic skills, the affected individual becomes

Figure 2: **Normal Brain vs. Alzheimer's Brain.** Courtesy of: Highlight Health http://www.highlighthealth.com/neurological-disorders/amyloid-deposits-in-cognitively-normal-people-may-predict-risk-for-alzheimers-disease/

susceptible to contracting a deadly infection, most commonly pneumonia. If infection does not kill the individual first, the disease will eventually progress to the point where the affected person can no longer breathe without mechanical assistance [4].

Why Is Alzheimer's Disease Important?

Today, Alzheimer's is the 6th leading cause of death in the United States. It is the only one in the top 10 which cannot be prevented, cured, or significantly slowed [2]. The detrimental effects of Alzheimer's reach beyond the loss of life however. For families, caring for an affected individual takes great emotional and psychological tolls. Watching someone you love forget who you are and lose every aspect of his or her personality are two of the most painful human experiences.

For the country, a great economic burden associated with caring and curing the disease exists. The economic toll comes from two sources. First, many family members give up careers to provide constant care for affected loved ones. Caring for an individual with Alzheimer's requires money for medication, doctors' visits, and general care. The loss of wages leads to economic instability for the family [5]. Second, the United States spends an enormous amount of money on caring for those with Alzheimer's. The Alzheimer's Association expects the United States to spend $200 billion in 2012 and $1.1 trillion in 2050 on Alzheimer's care [1]. As seen in Figure 3, the majority of that money is spent on Medicare and Medicaid programs, which facilitate nursing home enrollment and long-term care. This incredible amount of money adds to the ever-worsening economic state of the country.

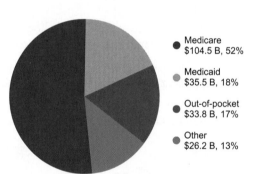

Total cost: $200 Billion (B)

- Medicare $104.5 B, 52%
- Medicaid $35.5 B, 18%
- Out-of-pocket $33.8 B, 17%
- Other $26.2 B, 13%

Figure 3: **Total US Spending on Alzheimer's, 2012.** Courtesy of the Alzheimer's Association [2].

Lastly, for the generation following the affected individual, the genetic link to Alzheimer's takes a cognitive toll. You do not need to have memory loss to be affected by Alzheimer's. You can see physiological findings consistent with Alzheimer's on a magnetic resonance image two decades before any symptoms arise. Having genetic risk factors of Alzheimer's alters cognition even in children. Children who expressed the gene *ApoE4* (discussed later) and had a family history of Alzheimer's scored significantly lower on cognitive tests than those lacking either risk factor [6]. Also, some people who develop Alzheimer's report having reduced memorization or cognitive skills throughout their lives. Finding effective prevention treatments could reduce all three of these detrimental effects.

Physiological Hypotheses

Several hypotheses exist to explain Alzheimer's development, each based on different physiological mechanisms leading to the morphological changes characteristic of Alzheimer's. These hypotheses differ in their explanation of the first step of disease development but not the physical changes of neural anatomy. The following sections highlight the basics of each of these hypotheses and the possible treatment importance of each.

Amyloid Cascade Hypothesis

Since 1992, the amyloid cascade hypothesis (ACH) has become the most widely accepted hypothesis of Alzheimer's development. The ACH proposes that the deposition of Aβ protein, the major component in protein plaques, precedes the development of NTs and neuron death [7]. The theory starts with the synthesis of excess Aβ. Several factors, including genetic ones, contribute to the synthesis of a protein called "APP," which later forms Aβ. This newly synthesized protein aggregates outside the cell or breaks into smaller fragments, enters the cell, and alters cellular processes (Figure 4) [8]. In addition to its role in synaptic degradation discussed earlier, Aβ initiates events in the cell, including tau alteration and mitochondrial dysfunction, by interacting with receptors on the cells. The Aβ protein's interaction with extracellular receptors activates enzymatic proteins within the cell responsible for morphological changes. Scientists have not yet identified the exact mechanism for

Figure 4: **Amyloid Cascade Hypothesis.** Courtesy of C. Reitz [8].

mitochondrial dysfunction, but isolated Aβ alters the protein tau in a way that contributes to the development of NTs seen in neurons [9].

Most novel treatments revolve around this hypothetical progression. The treatments target regulatory mechanisms of Aβ through receptor interactions or destruction of free Aβ. Scientists recently discovered a breakthrough treatment utilizing protein antibodies specific for a related protein. I discuss the details later in this report.

Mitochondrial Cascade Hypothesis

The mitochondrial cascade hypothesis (MCH) challenged the ACH by proposing that mitochondrial dysfunction precedes Aβ protein aggregation [10]. Mitochondria generate usable energy (in the form of adenosine triphosphate) necessary for normal cellular function. Neurons require a lot of energy for communication, transport, and growth. The brain only constitutes about 2% of a person's body weight but requires at least 20% of all energy generated. The DNA which resides in the mitochondria regulates function through protein synthesis. If mutations exist in DNA regions encoding vital proteins, the mitochondria will produce malfunctioning proteins which decrease energy production. Proteins outside the mitochondria can alter activity as much as genetics. The Aβ precursor APP blocks nutrient transport channels leading into the mitochondrial matrix, hindering energy production.

There exists an activity threshold required to maintain normal cellular function. When activity drops below the threshold, energy generation decreases and neuronal dynamics deteriorate (Figure 5). Mitochondrial dysfunction causes neuron death by disrupting synaptic connections long before plaques begin developing; neuron death marks the beginning of Alzheimer's. In addition to cell death, a decrease in mitochondrial activity allows an increase in Aβ production and tau modification in neurons. The glucocorticoid cascade hypothesis argues that the increase in Aβ does

Figure 5: **Mitochondrial Cascade Hypothesis.**
Courtesy of Swerdlow, R.H. [10]

not start the cascade, as argued by the ACH; it merely contributes to continued degeneration.

Treatment options pinpointing mitochondrial dysfunction are difficult to test and implement. Altering mitochondrial activity affects every part of the body. Recent research shows that some of the dysfunctions found in neurons affect other cells in the body, most notably platelets. Though the following theory has not been tested extensively, it may propel future studies: Alzheimer's may not be just a neurological disease. Dysfunction in other cells and tissues may contribute to disease development in unknown ways. Once identified, more extensive intracellular relationships may provide researchers targets for experimental treatments.

Glucocorticoid Cascade Hypothesis

The glucocorticoid cascade hypothesis (GCH), the last of the major hypotheses, proposes that long-term depression and exposure to high levels of stress hormones (glucocorticoids) initiate hippocampal damage. In a normal body, stress hormones regulate their own release by binding to receptors in the brain and signaling for the cessation of hormone release. However, long-term elevated stress hormone levels cause destruction of the regulatory receptors, thereby causing hypersecretion. General hormone insensitivity also accompanies prolonged exposure to high cortisol and contributes to increased glucocorticoid circulation. Great increases in hormone concentrations in the brain cause damage to neurons through a variety of mechanisms. The primary mechanism involves the inhibition of glucose uptake into brain cells, which causes brain damage [11].

In addition to direct cell death, high levels of cortisol, the primary stress hormone in humans, accelerate disease progression by contributing to the toxicity and deposition of Aβ and tau proteins in the brain [12]. Hormone mechanisms are incredibly complex, so the exact manner in which they contribute to Aβ and tau deposition is not fully understood. In addition to effectively reducing cortisol levels in the blood, treatment options based on this hypothesis would target maintaining those receptors responsible for muting the stress response.

Risk Factors

Who Is at Risk?

The Alzheimer's Association reports that one in eight individuals above the age of 65 develops the disease and that about half of individuals 85 and older develop it [1]. Several genetic and health factors contribute to Alzheimer's (discussed in the following sections), but everyone is at risk for developing Alzheimer's. As a person ages, neurological decline occurs. The process occurs naturally with aging. Think of it this way: you can physically see the effects of muscle degradation in the elderly, and this degradation occurs due to protein breakdown in tissues over a lifetime. Just like muscles, internal processes depend on proteins, which also degrade. All the internal processes which keep you alive get worn out just like muscles; you just cannot see the effects on the outside. If degradation proceeds far enough, the same kind of damage which occurs in Alzheimer's happens.

Genetic Factors

Mitochondrial DNA carries genes which determine baseline function and durability [10]. When the mitochondria stop working, cell death ensues. Every person inherits mitochondria and mitochondrial DNA from his or her mother. Out of individuals with affected parents, those who had affected mothers had a greater risk of developing Alzheimer's than those who had affected fathers. Whether or not females are more susceptible to Alzheimer's than males has not been investigated thoroughly, but some initial findings do suggest gender linkage [13]. The bottom line is that individuals with mothers affected with Alzheimer's should familiarize themselves with preventative measures. The genetic link cannot be avoided.

Genes found in the human genome (as opposed to the mitochondrial genome) also contribute to disease development. Individuals carrying "the Fatso gene," a gene linked to obesity, appear to have a higher risk of developing Alzheimer's. This increase in susceptibility is because the presence of this gene appears to contribute to brain atrophy, or withering, a contributing factor in disease development [14]. This genetic linkage poses a problem in the United States since an estimated 43% of people of western European descent carry this gene [15]. Other forms of obesity also contribute to Alzheimer's development. Most likely this occurs due to the poor overall health associated with obesity.

Another genomic linkage comes from the encoding of apoproteins, proteins which circulate around the body and deliver lipids to tissues. One specific form of apoprotein E, ApoE4, binds to receptors on brain cells and initiates the generation of Aβ protein. The ApoE4 form promotes this generation more than the other forms, E2 and E3 [16]. High concentrations of E4 cause a great increase in Aβ production and buildup. Individuals who inherited the gene for ApoE4 from one parent are three times more likely to develop Alzheimer's than normal individuals, and individuals who inherited it from both parents are ten times more likely to develop Alzheimer's [17]. The presence of ApoE4 appears to correlate most closely with Alzheimer's development.

Early Life Risk Factors

Starting from birth, several factors influence a person's likelihood to develop Alzheimer's. Obesity, chronic stress, vascular disease, head trauma, and other factors increase someone's risk for developing the disease. This section will break down a few of the more prevalent contributing factors.

First, individuals who have had any kind of brain trauma have a much higher risk of developing neurodegenerative diseases including Alzheimer's. Many of these brain injuries cause irreversible brain damage by directly destroying neurons. These small clusters of dead neurons are easy targets for protein buildup and continued degeneration. Athletes, war veterans, and car accident victims with associated head trauma all have an increased risk of developing a neurodegenerative disease [18]. Many adolescent athletes have suffered at least one concussion during their athletic careers. Without realizing it, they increased their risk of developing neurodegenerative diseases with every hard hit taken. Though much of the time people cannot avoid accidents, avoiding brain injuries all together greatly reduced their risk for Alzheimer's.

As described before with the GCH, chronically depressed individuals are more at risk than others to develop Alzheimer's. Prolonged exposure to cortisol and other stress hormones causes neuron death and Aβ production in the hippocampus. Chronic depression arising during adolescence and continuing through much of adult life contributes to disease formation [12]. The longer an individual has suffered from depression, the more susceptible he or she becomes.

Individuals with vascular disease, such as high blood pressure, high cholesterol, and diabetes, have a higher risk for developing Alzheimer's than others. Some studies show that cardiovascular disease is the second most important indicator for Alzheimer's [19]. The vascular destruction that accompanies high blood pressure affects the vasculature in the brain as much as the rest of the body. Decreased blood flow to the brain causes a decrease in nutrient delivery. A decrease in nutrient delivery will eventually lead to mitochondrial dysfunction and cell death.

Lastly, lower education levels and less cognitively driven careers both seem to correlate with higher incidence of Alzheimer's. Learning, memorizing, and processing require the formation of numerous new connections between brain cells. The increased cognitive reserve resists damage [20]. It would take longer to degrade the theoretical increase in intracellular connections, delaying the progression of the disease. Though statistical studies have shown this correlation, it is difficult to prove this anatomical theory due to the inability to cut into a brain and investigate neuronal synapses.

Treatment Options

Alzheimer's has no cure. Once the disease develops, you cannot stop or reverse physiological changes. Once the catalytic event has occurred, it cannot be undone. However, several treatment options exist to help slow down disease progression or reduce the risk of initial development. The following sections dissect several of these options.

Disease Prevention

Since heart and vascular diseases contribute heavily to the development of Alzheimer's, individuals must maintain a healthy heart and vascular system. Staying healthy, avoiding diabetes, not smoking, and lowering blood pressure all promote healthy heart and vascular health and exponentially reduce the risk of developing Alzheimer's [21]. The characteristic vascular degeneration and buildup of cholesterol and other fatty components which accompany vascular disease cause reduced blood flow to the brain. This reduction correlates strongly to Alzheimer's development in elderly individuals. Maintaining a healthy lifestyle greatly reduces an individual's risk of Alzheimer's.

Though exercise is an important component of healthy living, it needs its own section. I cannot overstate the importance of exercising to maintain cognitive health. Numerous studies have found that exercise, most importantly exercise during the teenage years, decreases cognitive decline in the elderly. The probable mechanism is threefold: (1) exercise increases overall health, (2) exercise increases neuronal connections, and (3) exercise increases the concentration of a certain protein in the brain responsible for hippocampal growth and healthy brain function in adults [1]. A study investigating the effect of exercise on plaque formation in mice showed a marked decrease in plaque buildup in the brain [22]. Exercise during all stages of life helps improve every aspect of health in an individual. If you learn nothing else from this report, learn this: never stop exercising.

In addition to maintaining a healthy lifestyle, performing cognitive exercises can improve mental ability and decrease deterioration. Scientists debate the effectiveness of cognitive exercises. As discussed before, cognitive processes require new intracellular connections. Continued educational pursuits in any capacity could help prevent or delay the onset of Alzheimer's.

Over-the-Counter Remedies

Since Alzheimer's gained prominence in the medical world, people have tried to find ways to reduce occurrence with natural, inexpensive remedies. Dietary supplements and so-called medical foods have swept the Alzheimer's world. These alternatives to medical intervention appeal to those individuals wary of medicine or incapable of purchasing prescriptions.

Several companies claim that some dietary supplements will delay the onset of Alzheimer's. Some of these supplements pose no inherent risk to an individual, but some do. The following have not been found to have any significant effect on the development of Alzheimer's [23]:

- Caprylic acid and coconut oil
- Coenzyme Q
- Coral calcium
- *Ginkgo biloba*
- Huperzine A
- Omega-3 fatty acids
- Phosphatidylserine
- Tramiprosate
- Vitamin E [24]

Since dietary supplements are not subjected to the testing required by the US Food and Drug Administration, they can often do more harm than good. Several of these medicinal alternatives contain biological contaminants, interfere with prescription medications, or cause an overdose if taken too liberally. Compounds like omega-3 fatty acids have general health benefits, but they do not necessarily reduce an individual's risk of developing Alzheimer's [23]. The effectiveness of vitamin E drives one of the largest debates in Alzheimer's treatment. Its overall antioxidant and neurological properties lead physicians to insist that it slows cognitive decline. For that reason, vitamin E often supplements Alzheimer's treatment plans [1, 23].

Though most of the advertised supplements have no effect on disease prevention, some may. Scientists have found that high doses of vitamin B reduced the amount of brain atrophy in elderly individuals with mild cognitive impairment [25]. Vitamin D also appears to slow the rate of brain atrophy. Individuals deficient in vitamin D experience faster rates of decline than those with no deficiency [1]. Since Alzheimer's often correlates with heart disease, supplements which reduce cardiovascular disease could reduce Alzheimer's occurrence [25].

Medical Intervention

No cure exists, and there is no way to reverse changes once they begin; however, some medications exist for treating Alzheimer's symptoms. First, physicians prescribe medications which treat the behavioral symptoms of Alzheimer's. These include regular stimulants and tranquilizers prescribed for healthy individuals. Second, some patients qualify for medications to treat the neurological symptoms and slow down disease progression. These medications fall into one of two categories: cholinesterase inhibitors and *N*-methyl-D-aspartate (NMDA) receptor antagonists. (See Appendix C for specific name brand drug details.)

Cholinesterase inhibitors act to decrease the enzyme responsible for breaking down acetylcholine, a chemical messenger important in memory, judgment, and other thought processes. The effects of these drugs are small, but they can delay or slow symptom worsening for a period of time (typically no longer than one year). Physicians frequently prescribe cholinesterase inhibitors for the treatment of mild to moderate Alzheimer's [26]. Unfortunately, these drugs only work on about half of the medicated individuals, another indicator of disease complexity.

The NMDA receptor antagonists work by blocking NMDA receptors in the brain. They work by regulating the activity of glutamate, a chemical involved in information processing, storage, and retrieval. Glutamate plays an essential role in learning and memory by triggering NMDA receptors to allow a controlled amount of calcium to flow into a nerve cell. Excess glutamate overstimulates these receptors, allowing too much calcium to enter the cell. Too much calcium causes disruption of cellular processes and cell death. The NMDA receptor antagonists are used more frequently for the treatment of severe Alzheimer's where overstimulation of these receptors can be seen [26]. Again, these drugs do not help every prescribed individual and do not prevent disease progression. They provide short-term relief from some of the worst symptoms.

In March 2012, researchers published a study investigating a new treatment targeting Aβ protein. This new treatment utilized antibodies for a protein called "DKK1," a closely related protein to Aβ. The results showed a clear decrease in synaptic toxicity of Aβ and cessation of disease progression [27]. Scientists used rat models to test this treatment option; but with the promising results, continued research could potentiate an effective treatment option.

Summary and Conclusions

Alzheimer's is a terrible disease, and anyone can develop it. Unlike other neurological diseases, like Huntington's disease, genetic predispositions are not the necessary condition. Old age, early life injuries, heart disease, and many other, common factors exponentially increase your likelihood of developing the disease.

As the ever-growing population ages, the importance of finding treatment options increases. In the coming years, the prevalence of Alzheimer's will drastically increase, leading to larger economic and medical burdens in the United States (Figure 6). The general population does not know how to identify risk factors or successfully live a life that could ward off Alzheimer's. Individuals can get genetic testing to determine if they have genetic predispositions like ApoE4 or not. Lack of knowledge could theoretically further increase disease incidence. As this disease gains recognition, people need to be informed about the possible implications and avoidance techniques.

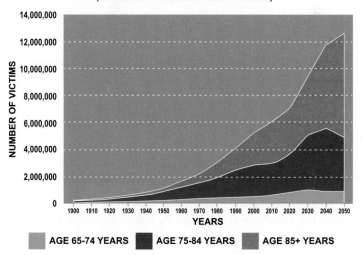

PREVALENCE OF ALZHEIMER'S DISEASE
(BY DECADES IN U.S.A. FROM 1900–2050)

AGE 65-74 YEARS   AGE 75-84 YEARS   AGE 85+ YEARS

This graph portrays how many Americans over the age of 65 are currently affected by Alzheimer's, and a projection of how many more will become affected with it as time passes.
w3.ouhsc.edu

Figure 6: **Increase in Alzheimer's through 2050.** Courtesy of Stanford University http://folding.stanford.edu/English/FAQ-Diseases

As research continues, scientists find key mechanisms to attack for more effective treatments. Like with most diseases, early detection and treatment are pivotal. Since right now no effect treatments exist to combat early stages of the disease, the most important thing you can do is prevent any development. Simple, healthier lifestyle changes can reduce the likelihood of developing Alzheimer's.

Works Cited

1. US Department of Health and Health Services. *2010 Alzheimer's Disease Progress Report: A Deeper Understanding.* National Institutes of Health, 2010, http://www.nia.nih.gov/alzheimers/publication/2010-alzheimers-disease-progress-report-deeper-understanding.

2. US Department of Health and Health Services. *2012 Alzheimer's Disease Facts and Figures.* National Institutes of Health, 2012, http://alz.org/downloads/Facts_Figures_2012.pdf.

3. "Seven Stages of Alzheimer's." Alzheimer's Association, 2012, http://alz.org/alzheimers_disease_stages_of_alzheimers.asp.

4. "How Does Alzheimer's Kill?" *Slate Magazine* (April 30, 2001), http://www.slate.com/articles/news_and_politics/explainer/2001/04/how_does_alzheimers_kill.

5. Ernst RL, Hay JW. "The US Economic and Social Costs of Alzheimer's Disease Revisited." *American Journal of Public Health* 84, no. 8 (1994): 1261–1264.

6. Bloss CS, Delis DC, Salmon DP, Bondi MW. "Decreased Cognition in Children with Risk Factors for Alzheimer's Disease." *Biological Psychiatry* 64, no. 10 (2008): 904–906.

7. Hardy JA, Higgins GA. "Alzheimer's Disease: The Amyloid Cascade Hypothesis." *Science* 256, no. 5054 (1992): 184–185.

8. Reitz C. "Alzheimer's Disease and the Amyloid Cascade Hypothesis: A Critical Review." *International Journal of Alzheimer's* (2012).

9. Jin M, Shepardson N, Yang T, et al. "Soluble Amyloid Beta-Protein Dimers Isolated from Alzheimer Cortex Directly Induce Tau Hyperphosphorylation and Neuritic Degeneration." *Proceedings of the National Academy of Sciences USA* 108 (2011): 5819–5824.

10. Swerdlow RH. "Mitochondria and Cell Bioenergetics: Increasingly Recognized Components and a Possible Etiologic Cause of Alzheimer's Disease." *Antioxidants and Redox Signaling* 16, no. 12 (2012): 1434–1455.

11. Sapolsky RM, Krey LC, McEwen BS. "The Neuroendocrinology of Stress and Aging: The Glucocorticoid Cascade Hypothesis." *Science of Aging Knowledge Environment* 38 (2002).

12. Aznar S, Knudsen GM. "Depression and Alzheimer's Disease: Is Stress the Initiating Factor in a Common Neuropathological Cascade?" *Journal of Alzheimer's Disease* 23, no. 2 (2011): 177–193.

13. Edland SD, Silverman JM, Peskind ER, Tsuang D, Wijsman E, Morns JC. "Increased Risk of Dementia in Mothers of Alzheimer's Disease Cases: Evidence for Maternal Inheritance." *Neurology* 4 (1996): 254–256.

14. Ho AJ. "A Commonly Carried Allele of the Obesity-Related FTO Gene Is Associated with Reduced Brain Volume in the Healthy Elderly." *Proceedings of the National Academy of Sciences USA* 107, no. 18 (2010): 8404–8409.

15. James F. " 'Fatso Gene' Linked to Alzheimer's." National Public Radio, April 19, 2010, http://www.npr.org/blogs/thetwo-way/2010/04/fatso_gene_linked _to_alzheimer.html?ps=rs.

16. He X, Cooley K, Chung CHY, Dashti N, Tang J. "Apolipoprotein Receptor 2 and X11α/β Mediate Apolipoprotein E–Induced Endocytosis of Amyloid β Precursor Protein and β-Secretase, Leading to Amyloid β Production." *Journal of Neuroscience* 27, no. 15 (2007): 4052–4060.

17. Oklahoma Medical Research Foundation. "Why ApoE4 Increases Alzheimer's Risk." Wellcome Trust. October 4, 2007, http://genome.wellcome.ac.uk/ doc_WTX038956.html.

18. "Brain Injury May More than Double Dementia Risk in Older Veterans." Alzheimer's Association, July 18, 2011, http://www.alz.org/aaic/monday _1230amCT_news_release_brain_injury.asp.

19. Borenstein AR, Wu Y, Mortimer JA, Schellenberg GD, McCormick WC, Bowen JD, McCurry A, Larson EB. "Developmental and Vascular Risk Factors for Alzheimer's Disease." *Neurobiology of Aging* 26, no. 3 (2005): 325–334.

20. American Academy of Neurology. "Low Education Level Linked to Alzheimer's, Study Shows." ScienceDaily, October 1, 2007, http://www.sciencedaily .com/releases/2007/10/071001172855.htm.

21. Raymond R. "Evidence Lacking to Support Alzheimer's Prevention." National Public Radio, April 28, 2010, http://www.npr.org/blogs/health/2010/04/ alzheimers_prevention_exercise.html?ps=rs.

22. Lunde A. "Preventing Alzheimer's: Exercise Still Best Bet." Mayo Clinic, March 25, 2008, http://www.mayoclinic.com/health/alzheimers/MY00002.

23. "Alternative Treatments." Alzheimer's Association, 2012, http://www.alz.org/ alzheimers_disease_alternative_treatments.asp.

24. Butler R. "Can Vitamins Prevent Alzheimer's?" Everyday Health, November 20, 2006, http://www.everydayhealth.com/specialists/senior_aging/butler/qa/ can-vitamins-prevent-alzheimers/index.aspx.

25. Russell P. "Vitamin B May Help Prevent Mental Decline in Elderly." WebMD .com, September 9, 2010, http://www.webmd.com/brain/news/20100909/ vitamin-b-may-help-preventmental-decline-elderly.

26. Morris J. "Cognitive Symptoms." Alzheimer's Association, 2012, http://www .alz.org/professionals_and_researchers_cognitive_symptoms.asp.

27. Purro SA, Dickens EM, Salinas PC. "The Secreted Wnt Antagonist Dickkopf-1 Is Required for Amyloid β Mediated Synaptic Loss." *Journal of Neuroscience* 32, no. 10 (2012): 3492–3498.

28. "Alzheimer's Disease Medications Fact Sheet." National Institute on Aging, December 13, 2011, http://www.nia.nih.gov/alzheimers/publication/ alzheimers-disease-medications-fact-sheet.

APPENDIX A: Proposal

MEMORANDUM

DATE:	February 14, 2012
TO:	Dr. Elizabeth Tebeaux
FROM:	Hannah Starke
SUBJECT:	**Proposal for Report on Alzheimer's Disease**

Project Summary

The term *dementia* describes a wide range of diseases which cause destruction of brain cells and eventual performance impairment. Alzheimer's disease is the most well known of these disorders. Alzheimer's is a neurodegenerative disease which impairs one's ability to remember, think, and perform basic tasks; and the disease leads to death. This debilitating disease changes every aspect of the lives of the afflicted individual and his or her loved ones. Though not fully known, causes for the disease have been linked to some genetic and environmental factors. The time is now to understand this disease because with an ever aging population comes the possibility for a large increase in occurrence of Alzheimer's disease.

I will consult online medical journals and consumer health articles to gain a better understanding of this disease. People have a hard time comprehending the disease in its most basic sense, and that misunderstanding is due to improper communication between the general public and those researching and treating the disease. I hope to present the information in a concise, readable manner so that readers will understand the risks and what they can do to try to prevent the disease.

This project has two main objectives:

- Investigate the biological processes and genetic factors responsible for the development of Alzheimer's disease
- Discuss recent medical and homeopathic remedies found for the prevention and treatment of Alzheimer's

Introduction and Rationale

It is estimated that today 2.4 to 5.1 million people in the United States alone suffer from Alzheimer's disease. The economic and psychological tolls that this disease has on those afflicted and those who care for the afflicted are astronomical, and those issues will only increase since recent studies estimate that the number of cases will triple by 2050.

Just as the economic impact of the disease increases every year, research on treatment options has as well. The primary at-risk age group is 65 and older, and with the baby boomer population quickly entering into that age group, research has greatly increased to try to answer questions regarding the mechanics of the disease and to prevent its development in the population.

Alzheimer's disease plays a large role in my life. My maternal grandmother was diagnosed with the disease when I was four years old, and by the time I was six she had no recollection of who I was. I remember my grandmother asking my mom, "Who is that little girl standing by the door?" every time we visited. In December 1998, Alzheimer's took her life. With recent research showing that there may be a genetic link flowing through the maternal bloodline, the time is now for me to begin to understand the steps that could be taken to reduce my risk of developing this disease. In addition, this past December, a woman so close to me that I call her my mother was diagnosed with the same disease, and she seems to be fading fast. I am a junior in college on my way to medical school, and yet she asks me and those around me if I like finally being a high school student. I have gone through caring for someone with this disease, and I never want someone I love to feel the way that I have when I realize someone I care so deeply about cannot remember who I am. I seek to shine light on this disease and all the ways that people can try to prevent it.

Research Methods

In order to begin proper research, basic information about the disease needs to be understood. There are many intricacies associated with this disease, so I must find sources that expect no prior knowledge. The best sources used in order to achieve this come from the Alzheimer's Association, the National Institute on Aging, and MedlinePlus. These sources require no prior technical knowledge on behalf of the reader in order to understand.

To gain further understanding of the mechanics of the disease, I will consult more technical journals. Databases such as the National Institute of Health's PubMed and EBSCO Medline will link me to journals such as *PLoS ONE*, the *New England Journal of Medicine, Neurology, Nature,* and other advanced scientific journals. This will provide me with an intimate familiarity of how the disease acts on the body and the biomechanics of treatment options. Gaining a more intricate understanding of the disease will allow me to synthesize all necessary data.

I hope to complete interviews with several physicians at Baylor College of Medicine in Houston. They will be able to shed light on the most recent developments and possible implications from those findings.

Tentative Outline for the Report

I. Summary/Abstract
II. Introduction
    a. What is Alzheimer's and why is it important
    b. What the reader needs to learn from this report
III. Rationale for report
    a. What the recent research is showing as the physiological issues associated with the progression of the disease
    b. Why the general population needs to be well informed on Alzheimer's
        i. Who is at risk
IV. Risk factors for the disease
    a. Genetic factors
    b. Child development
V. Current treatment options
    a. Homeopathic remedies
    b. Preventative measures
    c. Medical intervention methods and their effectiveness
VI. Summary and Conclusion

Initial Bibliography

Abbott A. "Dementia: A Problem for Our Age." *Nature* 475, no. 7355 (2011): S2–S4.

"About Alzheimer's Disease: Treatment." National Institute on Aging, 2011. http://www.nia.nih.gov/alzheimers/topics/treatment.

"Alternative Treatments." Alzheimer's Association, 2012, http://www.alz.org/alzheimers_disease_alternative_treatments.asp.

Bloss CS, Delis DC, Salmon DP, Bondi MW. "Decreased Cognition in Children with Risk Factors for Alzheimer's Disease." *Biological Psychiatry* 64, no. 10 (2008): 904–906.

Honea RA, Swerdlow RH, Vidoni ED, Burns JM. "Progressive Regional Atrophy in Normal Adults with a Maternal History of Alzheimer Disease." *Neurology* 76, no. 9 (2011): 822–829.

Jin M, Shepardson N, Yang T, et al. "Soluble Amyloid Beta-Protein Dimers Isolated from Alzheimer Cortex Directly Induce Tau Hyperphosphorylation and Neuritic Degeneration." *Proceedings of the National Academy of Sciences USA* 108 (2011): 5819–5824.

Jucker M, Walker LC. "Pathogenic Protein Seeding in Alzheimer's Disease and Other Neurodegenerative Disorders." *Annals of Neurology* 70 (2011): 532–540.

Maruszak A, Żekanowski C. "Mitochondrial Dysfunction and Alzheimer's Disease." *Progress in Neuro-Psychopharmacology and Biological Psychiatry* 35, no. 2 (2011): 320–330.

US Department of Health and Health Services. *2010 Alzheimer's Disease Progress Report: A Deeper Understanding.* National Institutes of Health, 2010, http://www.nia.nih.gov/alzheimers/publication/2010-alzheimers-disease-progress-report-deeper-understanding.

APPENDIX B: Progress Report

MEMORANDUM
To: Elizabeth Tebeaux
From: Hannah Starke
Date: 22 March 2012
Subject: Progress on Semester Project

Summary

With my report, I hope to educate readers as much as possible on the mechanics, physiological hypotheses, risk factors, and treatment options of Alzheimer's disease (AD). I have found amazing, unbelievably interesting information on AD and everything associated with it. Since I have a lot of personal connections with the disease, I hope to educate myself as much as the reader as to how I can try to prevent the development of the disease.

Everyone has risk for developing AD. Some researchers say that around half of individuals 85 and older have AD, and this number will only continue to rise in the coming decades. Many of the neurological developments associated with AD are the same as natural processes associated with aging, but different characteristics distinguish AD from natural aging.

Many different hypotheses exist to explain how AD develops, each one with a key mechanical malfunction in the brain. All of these hypotheses provide a plausible explanation of the progression of AD. I will investigate three main hypotheses. In addition to the mechanical issues associated with AD, I have found several studies discussing risk factors associated with developing AD, many of which I had never heard. Early life experiences could carry serious implications for mental health and cognitive decline.

Though no cure exists, physicians use a variety of treatments to halt the progression of symptoms. Individuals can take a variety of steps to decrease the likelihood of developing the disease.

I have found amazing sources, but much work remains. I have created a great basis off of which I can work, but I still have to find the real details and fill in the gaps.

Work Completed

Introduction/Background Information

- *2010 Alzheimer's Disease Progress Report: A Deeper Understanding*
  - This report provides a great, concise overview of the current status and basic information about AD and some basic research associated with it.
  - This is a great source. It will be used for sure.
  - US Department of Health and Health Services. *2010 Alzheimer's Disease Progress Report: A Deeper Understanding.* National Institutes of Health, 2010, http://www.nia.nih.gov/alzheimers/publication/2010-alzheimers -disease-progress-report-deeper-understanding.

- "Understanding Alzheimer's Disease: What You Need to Know."
  - The National Institute on Aging created this booklet to provide caregivers information on what to expect as loved ones develop AD. It highlights basic knowledge, beginning signs, treatment options, and other information that affected individuals and caregivers must know.
  - This is a good source; it provides basic information which may or may not help. Whether or not I use this source remains to be seen.
  - "Understanding Alzheimer's Disease: What you need to know." National Institute on Aging, June 2011, http://nia.nih.gov/sites/default/files/ UnderstandingADfinal.pdf

- "Decreased Cognition in Children with Risk Factors for Alzheimer's Disease."
  - This article describes a very early finding, so the reader must take caution. However, the study found that children who possess two major risk factors for AD, a positive family history and the presence of the allele ApoE4, had decreased performance on cognitive testing compared to peers who had one or none of the risk factors. Risk factors for AD could cause decreased cognition in young people.
  - The researchers make an interesting discovery linking childhood development to possible disease development in the future. I will discuss this point.
  - Bloss CS, Delis DC, Salmon DP, Bondi MW. "Decreased Cognition in Children with Risk Factors for Alzheimer's Disease." *Biological Psychiatry* 64, no. 10 (2008): 904–906.

Physiological Hypotheses

Amyloid Cascade Hypothesis

- "Soluble Amyloid Beta-Protein Dimers Isolated from Alzheimer Cortex Directly Induce Tau Hyperphosphorylation and Neuritic Degeneration."
  - This study provides an extremely detailed look at a very specific morphological change in neuronal components found in the brain. Doctors use the formation of amyloid β-protein plaques in the brain as a diagnostic tool. This study took that information to another level and looked at the link between these proteins and neuronal degeneration.
  - I will most likely use this report. I need to find supplementing materials to help explain the technical details, but the article itself makes good points and provides possible explanations.
  - Jin M, Shepardson N, Yang T, et al. "Soluble Amyloid Beta-Protein Dimers Isolated from Alzheimer Cortex Directly Induce Tau Hyperphosphorylation and Neuritic Degeneration." *Proceedings of the National Academy of Sciences USA* 108 (2011): 5819–5824.

- "Alzheimer's Disease: The Amyloid Cascade Hypothesis."
  - This review provides a simple overview of what the amyloid cascade hypothesis states. It also connects it to events which could have occurred earlier in life. Very well done. Will certainly be used.
  - Hardy JA, Higgins GA. "Alzheimer's Disease: The Amyloid Cascade Hypothesis." *Science* 256, no. 5054 (1992): 184–185.

Mitochondrial Cascade Hypothesis

- "Mitochondrial Dysfunction and Alzheimer's Disease."
  - This article highlights a lesser-known hypothesis of AD progression called the "mitochondrial cascade hypothesis." This hypothesis postulates that mitochondrial degeneration plays a large role in the progression of AD through a lack of respiration and energy production. In addition to that, mitochondrial baseline function and durability may be genetically determined. This article outlines a variety of enzymes whose activity correlates with AD development.
  - This article will be used. It provides an amazing overview of a more complicated, more recent hypothesis for neurodegeneration.
  - Maruszak A, Zekanowski C. "Mitochondrial Dysfunction and Alzheimer's Disease." *Progress in Neuro-Psychopharmacology and Biological Psychiatry* 32, no. 5 (2011): 320–330.

Glucocorticoid Cascade Hypothesis

- "Depression and Alzheimer's Disease: Is Stress the Initiating Factor in a Common Neuropathological Cascade?"
  - This article suggests that there exists an important link between depressive disorders earlier in life and the development of AD later in life.
  - I had never heard of this linkage before, so this article spurred my desire to look further into this connection. I will use this and other articles on this topic.
  - Aznar S, Knudsen GM. "Depression and Alzheimer's Disease: Is Stress the Initiating Factor in a Common Neuropathological Cascade?" *Journal of Alzheimer's Disease* 23, no. 2 (2011): 177–193.

- "The Neuroendocrinology of Stress and Aging: The Glucocorticoid Cascade Hypothesis."
  - Overview of the glucocorticoid hypothesis. Good.
  - Sapolsky RM, Krey LC, McEwen BS. "The Neuroendocrinology of Stress and Aging: The Glucocorticoid Cascade Hypothesis." *Science of Aging Knowledge Environment* 38 (2002).

Risk Factors

Genetic Factors

- "'Fatso Gene' Linked to Alzheimer's."
  - This short report provided great, basic information on a possible correlation between a gene deemed the "fatso gene," linked to the condition of obesity, and susceptibility to develop Alzheimer's. This could be a major problem because scientists estimate that one in three Americans have this gene.
  - This is a slightly different article, and even though it provides little information, the small fact that it does provide is incredibly interesting. It will provoke more research on the topic.
  - James, Frank. "'Fatso Gene' Linked to Alzheimer's." National Public Radio, April 29, 2010, http://www.npr.org/blogs/thetwo-way/2010/04/fatso_gene_linked_to_alzheimer.html?ps=rs

- "Increased Risk of Dementia in Mothers of Alzheimer's Disease Cases: Evidence for Maternal Inheritance."
  - This study found increased risk of developing AD when an individual had an affected mother.
  - Findings were simple but proved important. I will use this source unless I find one with more conclusive, comprehensive results.

○ Edland SD, Silverman JM, Peskind ER, Tsuang D, Wijsman E, Morns JC. "Increased Risk of Dementia in Mothers of Alzheimer's Disease Cases: Evidence for Maternal Inheritance." *Neurology* 4 (1996): 254–256.

Early Life Risk Factors

See "Depression and Alzheimer's Disease: Is Stress the Initiating Factor in a Common Neuropathological Cascade?" above

- "Do Early Life Insults Contribute to the Late-Life Development of Parkinson and Alzheimer Diseases?"
  ○ Events that occurred early in life (as early as in the womb) appear to play a role in the development of AD. Also, AD and cardiovascular disease have similar risk factors. Obesity is a big risk factor. And AD may be linked to a unique form of insulin resistance in the brain.
  ○ This article discusses many of the lesser-known risk factors contributing to the development of AD. It is really fascinating.
  ○ Miller DB, O'Callaghan. "Do Early Life Insults Contribute to the Late-Life Development of Parkinson and Alzheimer Diseases?" *Metabolism* 57, suppl. 2 (2008): S44–S49.

- "Developmental and Vascular Risk Factors for Alzheimer's Disease."
  ○ The authors found that the constellation of risk factors depended on ApoE-ε4 status. In those with one or more ε4 alleles, developmental risk factors played an important role, whereas in those without ε4 alleles, vascular risk factors were more predictive of disease. More generally, they found that both vascular and developmental risk factors play a role in the development of AD.
  ○ I will use this article because it investigates other risk factors that people would not know about. People would not think of early life experiences as risk factors in the development of a disease like AD, but this study shows that there is probably a connection.
  ○ Borenstein AR, Wu Y, Mortimer JA, Schellenberg GD, McCormick WC, Bowen JD, McCurry A, Larson EB. "Developmental and Vascular Risk Factors for Alzheimer's Disease." *Neurobiology of Aging* 26, no. 3 (2005): 325–334.

Treatment Options

Preventative Measures/Lifestyle Changes

- "Evidence Lacking to Support Alzheimer's Prevention."
  - Keep your cardiovascular health up, no smoking, eat a bunch of omega-3 fatty acids, be physically active, and so on.
  - I may or may not use this source. I will look for a more comprehensive source that discusses these points along with others.
  - Raymond, R. "Evidence Lacking to Support Alzheimer's Prevention." National Public Radio, April 28, 2010, http://www.npr.org/blogs/health/2010/04/alzheimers_prevention_exercise.html?ps=rs.

- "How Exercise Might Help Keep Alzheimer's at Bay."
  - A panel convened by the NIH warned the public that it's not clear whether physical activity and associated exercises can prevent AD or other forms of mental decline. Mental exercise is another possible strategy that seems like a good idea to many AD researchers. After all, it appears to increase connections in the brain and perhaps makes the brain more resilient. Members of the panel found some benefit from omega-3 fatty acids like those in fish. But it found no convincing studies in people that antioxidants like vitamin E could make a difference.
  - The points presented here supplement basic findings. Overall, a strong report.
  - Hamilton, J. "How Exercise Might Help Keep Alzheimer's at Bay." National Public Radio, April 29, 2010, http://www.npr.org/templates/story/story.php?storyId=126370279.

- "Preventing Alzheimer's: Exercise Still Best Bet."
  - Physical activity appears to inhibit brain changes in mice. Of 6,000 women aged 65+, those who were physically active showed less decline in mental abilities. Mice bred to have AD-like brain plaques and allowed to exercise had fewer plaques than those that were sedentary.
  - This article does highlight the advantages of exercise within a rodent model and a general trend in humans.
  - Lunde, A. "Preventing Alzheimer's: Exercise Still Best Bet." *Mayo Clinic*, March 25, 2008, http://www.mayoclinic.com/health/alzheimers/MY00002.

Over-the-Counter Remedies

- "Vitamin B May Help Prevent Mental Decline in Elderly."
  - A University of Oxford study found that taking vitamin B tablets every day can reduce the rate of brain atrophy in older people with mild cognitive impairment by as much as half. Some B vitamins—folic acid, vitamin B_6, and vitamin B_{12}—control levels of the amino acid homocysteine in the blood, and high levels of homocysteine are associated with an increased risk of AD.
  - I will use this review in my report. It provides a simple understanding of a specific intervention the public knows.
  - Russell, P. "Vitamin B May Help Prevent Mental Decline in Elderly." WebMD.com, September 9, 2010, http://www.webmd.com/brain/news/20100909/vitamin-b-may-help-preventmental-decline-elderly.

- "Alternative Treatments."
  - This site provides an overview of some of the more well-known alternative treatments for AD as well as possible concerns that one should have before undertaking any treatment. Mainly, this site discusses the theories behind using caprylic acid and coconut oil, coenzyme Q_{10}, coral calcium, *Ginkgo biloba*, huperzine A, omega-3 fatty acids, phosphatidylserine, and Tramiprosate. It does a great job of highlighting whether or not researchers have found each treatment to be effective or not. Most have not been found to have much effect.
  - I will use this source in the report because it is a great overview.
  - "Alternative Treatments." Alzheimer's Association, 2012, http://www.alz.org/alzheimers_disease_alternative_treatments.asp.

- "Can Vitamins Prevent Alzheimer's?"
  - Some research has suggested that vitamin E might help, but a large-scale study published in the *New England Journal of Medicine* in June 2005 showed otherwise. It was the largest such study of vitamin E ever undertaken.
  - Factors that increase the risk of heart disease, such as high blood pressure and hardening of the arteries due to high cholesterol, appear to be related to an increased risk of AD. In fact, autopsies show that one-third of those who had "vascular dementia" (cognitive changes brought on by atherosclerosis of the brain) also had evidence of AD. Therefore, prevention and treatment of these risk factors and conditions may be good preventive medicine. That means keeping blood pressure under control, keeping cholesterol levels low, not smoking, and maintaining a

healthy weight. If diabetes is present, it also means keeping blood sugar controlled as tightly as possible.

○ Butler, R. "Can Vitamins Prevent Alzheimer's?" *Everyday Health*, November 20, 2006, http://www.everydayhealth.com/specialists/senior _aging/butler/qa/can-vitamins-prevent-alzheimers/index.aspx.

Medical Intervention

- Cognitive Symptoms
    - ○ The Alzheimer's Association reports on various known, used treatment options. It lays out the pros and cons of the treatments and how they work.
    - ○ The US Food and Drug Administration has approved two classes of drugs to treat cognitive symptoms of AD: cholinesterase inhibitors and NMDA receptor antagonists. Vitamin E supplements are frequently prescribed and have become part of a standard treatment regimen for most people with AD.
    - ○ Morris, J. "Cognitive Symptoms." Alzheimer's Association, 2012, http://www.alz.org/professionals_and_researchers_cognitive_symptoms.asp.

- "About Alzheimer's Disease: Treatment."
    - ○ This review of the basic treatment options provides a good overview and links to outside sources which provide more in-depth knowledge about options. This approach provides a great starting point.
    - ○ "About Alzheimer's Disease: Treatment." National Institute on Aging, 2011, http://www.nia.nih.gov/alzheimers/topics/treatment.

- "Alzheimer's Disease Medications Fact Sheet."
    - ○ This site provides an overview of the medication available for use for different stages of AD. It does a great job of highlighting what the drugs do, who should use them, possible side effects, and much more. Great source.
    - ○ "Alzheimer's Disease Medications Fact Sheet." National Institute on Aging, December 13, 2011, http://www.nia.nih.gov/alzheimers/publication/alzheimers-disease-medications-fact-sheet.

Work Remaining

I have had a few issues getting in touch with many physicians. Due to timing, most have been in and out of the office or out of town. As of now, I have not conducted any interviews, but I assure you, I am still trying.

I will research further on the hypotheses as well. I found different theories toward the end of my research marathon, so I will continue to look into those. I believe they provide an important basis of understanding.

I will need to fill in gaps in research as I begin to write. That will require extra research as the paper progresses.

Changes in the Initial Proposal

I have not made many drastic changes in the paper topic. I changed how I will structure the paper. I made one big addition: the section on hypotheses. Once I started reading, I had to learn more. There exist many different mechanical hypotheses, and I felt it important to investigate and report on them. A general overview cannot present the true intricacies of the disease the way that it needs to be done. It will make the report a little more technical, but if I succeed, it will not seem more technical.

Updated Outline

    I. Summary/Abstract
   II. Introduction
        a. What is Alzheimer's and why is it important
        b. What the reader needs to learn from this report
  III. Rationale for report
  IV. Physiological hypotheses
        a. Amyloid cascade hypothesis
            i. Amyloid protein plaques
        b. Mitochondrial cascade hypothesis
            i. Mitochondrial degradation
        c. Glucocorticoid cascade hypothesis
   V. Risk factors
        a. Who is at risk
            i. Genetic factors
                1. Maternal links
                2. Mitochondrial links
            ii. Early-life risk factors
  VI. Treatment options
        a. Over-the-counter remedies
        b. Preventative measures/lifestyle changes
            i. Exercise
            ii. Avoidance of head trauma
        c. Medical intervention
            i. Methods
            ii. Effectiveness
 VII. Summary and Conclusion

Updated Bibliography

"About Alzheimer's Disease: Treatment." National Institute on Aging, 2011, http://www.nia.nih.gov/alzheimers/topics/treatment.

"Alternative Treatments." Alzheimer's Association, 2012, http://www.alz.org/alzheimers_disease_alternative_treatments.asp.

"Alzheimer's Disease Medications Fact Sheet." National Institute on Aging, December 13, 2011, http://www.nia.nih.gov/alzheimers/publication/alzheimers-disease-medications-fact-sheet.

Aznar S, Knudsen GM. "Depression and Alzheimer's Disease: Is Stress the Initiating Factor in a Common Neuropathological Cascade?" *Journal of Alzheimer's Disease* 23, no.2 (2011): 177–193.

Bloss CS, Delis DC, Salmon DP, Bondi MW. "Decreased Cognition in Children with Risk Factors for Alzheimer's Disease." *Biological Psychiatry* 64, no. 10 (2008): 904–906.

Borenstein AR, Wu Y, Mortimer JA, Schellenberg GD, McCormick WC, Bowen JD, McCurry A, Larson EB. "Developmental and Vascular Risk Factors for Alzheimer's Disease." *Neurobiology of Aging* 26, no. 3 (2005): 325–334.

Butler R. "Can Vitamins Prevent Alzheimer's?" Everyday Health, November 20, 2006, http://www.everydayhealth.com/specialists/senior_aging/butler/qa/can-vitamins-prevent-alzheimers/index.aspx.

Edland SD, Silverman JM, Peskind ER, Tsuang D, Wijsman E, Morns JC. "Increased Risk of Dementia in Mothers of Alzheimer's Disease Cases: Evidence for Maternal Inheritance." *Neurology* 4 (1996): 254–256.

Hamilton J. "How Exercise Might Help Keep Alzheimer's at Bay." National Public Radio, April 29, 2010, http://www.npr.org/templates/story/story.php?storyId=126370279.

Hardy JA, Higgins GA. "Alzheimer's Disease: The Amyloid Cascade Hypothesis." *Science* 256, no. 5054 (1992): 184–185.

James F. "'Fatso Gene' Linked to Alzheimer's." National Public Radio, April 19, 2010, http://www.npr.org/blogs/thetwo-way/2010/04/fatso_gene_linked_to_alzheimer.html?ps=rs.

Jin M, Shepardson N, Yang T, et al. "Soluble Amyloid Beta-Protein Dimers Isolated from Alzheimer Cortex Directly Induce Tau Hyperphosphorylation and Neuritic Degeneration." *Proceedings of the National Academy of Sciences USA* 108 (2011): 5819–5824.

Lunde A. "Preventing Alzheimer's: Exercise Still Best Bet." *Mayo Clinic*. 25 March 2008. http://www.mayoclinic.com/health/alzheimers/MY00002.

Maruszak A, Żekanowski C. "Mitochondrial Dysfunction and Alzheimer's Disease." *Progress in Neuro-Psychopharmacology and Biological Psychiatry* 35, no. 2 (2011): 320–330.

Miller DB, O'Callaghan JP. "Do Early-Life Insults Contribute to the Late-Life Development of Parkinson and Alzheimer Diseases?" *Metabolism* 57, suppl. 2 (2008): S44–S49.

Morris J. "Cognitive Symptoms." Alzheimer's Association, 2012, http://www.alz .org/professionals_and_researchers_cognitive_symptoms.asp.

Raymond R. "Evidence Lacking to Support Alzheimer's Prevention." National Public Radio, April 28, 2010, http://www.npr.org/blogs/health/2010/04/ alzheimers_prevention_exercise.html?ps=rs.

Russell P. "Vitamin B May Help Prevent Mental Decline in Elderly." WebMD .com, September 9, 2010, http://www.webmd.com/brain/news/20100909/ vitamin-b-may-help-preventmental-decline-elderly.

Sapolsky RM, Krey LC, McEwen BS. "The Neuroendocrinology of Stress and Aging: The Glucocorticoid Cascade Hypothesis." *Science of Aging Knowledge Environment* 38 (2002).

US Department of Health and Health Services. *2010 Alzheimer's Disease Progress Report: A Deeper Understanding.* National Institutes of Health, 2010, http://www.nia.nih.gov/alzheimers/publication/2010-alzheimers -disease-progress-report-deeper-understanding.

APPENDIX C: Medication Chart [28]

Medications to Treat Alzheimer's Disease

This brief summary does not include all information important for patient use and should not be used as a substitute for professional medical advice. Consult the prescribing doctor and read the package insert before using these or any other medications or supplements. Drugs are listed in order of FDA approval, starting with the most recent.

DRUG NAME	DRUG TYPE AND USE	HOW IT WORKS	COMMON SIDE EFFECTS	MANUFACTURER'S RECOMMENDED DOSAGE	FOR MORE INFORMATION
Namenda® (memantine)	N-methyl D-aspartate (NMDA) antagonist prescribed to treat symptoms of moderate to severe Alzheimer's	Blocks the toxic effects associated with excess glutamate and regulates glutamate activation	Dizziness, headache, constipation, confusion	• Tablet: Initial dose of 5 mg once a day • May increase dose to 10 mg/day (5 mg twice a day), 15 mg/day (5 mg and 10 mg as separate doses), and 20 mg/day (10 mg twice a day) at minimum 1-week intervals if well tolerated • Oral solution: same dosage as above • Extended-release tablet: Initial dose of 7 mg once a day, may increase dose to 14 mg/day, 21 mg/day, and 28 mg/day at minimum 1-week intervals if well tolerated	For current information about this drug's safety and use, visit *www.namenda.com*. Click on "Prescribing Information" to see the drug label.
Razadyne® (galantamine)	Cholinesterase inhibitor prescribed to treat symptoms of mild to moderate Alzheimer's	Prevents the breakdown of acetylcholine and stimulates nicotinic receptors to release more acetylcholine in the brain	Nausea, vomiting, diarrhea, weight loss, loss of appetite	• Tablet: Initial dose of 8 mg/day (4 mg twice a day) • May increase dose to 16 mg/day (8 mg twice a day) and 24 mg/day (12 mg twice a day) at minimum 4-week intervals if well tolerated • Oral solution: same dosage as above • Extended-release capsule: same dosage as above but taken once a day	For current information about this drug's safety and use, visit *www.razadyneer.com*. Click on "Important Safety Information" to see links to prescribing information.
Exelon® (rivastigmine)	Cholinesterase inhibitor prescribed to treat symptoms of mild to moderate Alzheimer's	Prevents the breakdown of acetylcholine and butyrylcholine (a brain chemical similar to acetylcholine) in the brain	Nausea, vomiting, diarrhea, weight loss, loss of appetite, muscle weakness	• Capsule: Initial dose of 3 mg/day (1.5 mg twice a day) • May increase dose to 6 mg/day (3 mg twice a day), 9 mg (4.5 mg twice a day), and 12 mg/day (6 mg twice a day) at minimum 2-week intervals if well tolerated • Patch: Initial dose of 4.6 mg once a day, may increase to 9.5 mg once a day after minimum of 4 weeks if well tolerated • Oral solution: same dosage as capsule	For current information about this drug's safety and use, visit *www.fda.gov/cder*. Click on "Drugs@FDA," search for Exelon, and click on drug-name links to see "Label Information."
Aricept® (donepezil)	Cholinesterase inhibitor prescribed to treat symptoms of mild to moderate, and moderate to severe Alzheimer's	Prevents the breakdown of acetylcholine in the brain	Nausea, vomiting, diarrhea	• Tablet: Initial dose of 5 mg once a day • May increase dose to 10 mg/day after 4-6 weeks if well tolerated, then to 23 mg/day after at least 3 months • Orally disintegrating tablet: same dosage as above • 23-mg dose available as brand-name tablet only	For current information about this drug's safety and use, visit *www.fda.gov/cder*. Click on "Drugs@FDA," search for Aricept, and click on drug-name links to see "Label Information."

Index